AF880314

Yukinori Ono · Jun Kondoh
Editors

Recent Advances in Technology Research and Education

Selected Papers of the 20th International Conference on Global Research and Education Inter-Academia

Springer

Editors
Yukinori Ono
Research Institute of Electronics
Shizuoka University
Hamamatsu, Japan

Jun Kondoh
Graduate School of Science and Technology
Shizuoka University
Hamamatsu, Japan

ISSN 2367-3370 ISSN 2367-3389 (electronic)
Lecture Notes in Networks and Systems
ISBN 978-3-031-54449-1 ISBN 978-3-031-54450-7 (eBook)
https://doi.org/10.1007/978-3-031-54450-7

This Springer imprint is published by the registered company Springer Nature Switzerland AG
The registered company address is: Gewerbestrasse 11, 6330 Cham, Switzerland

Paper in this product is recyclable.

Preface

This volume contains selected papers as the outcome of the 20th International Conference on Global Research and Education, or "Inter-Academia" held from September 27 to 29, 2023, organized by Shizuoka University in Hamamatsu, Japan. It is the 5th volume in the series, following the editions from 2015 to 2022. The International Advisory Conference Committee was run by the members of the International Society of Inter-Academia community.

With the rapid advancement of research and educational technologies in recent years, the scope of the Inter-Academia has become increasingly relevant. Interdisciplinary research collaboration within the Inter-Academia Community has also become essential for the development of new ideas and concepts to support the progress of science and its social impact at international level.

In this milestone 20th edition of the Inter-Academia Conference, a rich diversity of topics provides the seeds for advanced research, combining "Materials Science and Manufacturing Technology", "Nanotechnology and Nanometrology", "Biotechnology and Plasma Physics", "Photonics and Optics", "Electrical and Electronic Engineering", "Signal/Image Processing and Informatics", "Robotics and Precision Engineering", and "Social Science and Education". The boundaries between the above fields and more can and should be crossed often by researchers through dynamic and complementary communication, implicitly leading to a smooth and impactful circulation of research findings across fields.

The editors are grateful to the authors for their excellent work. On behalf of the organizing committee, we also would like to express our appreciation to the reviewers for offering their time in reviewing the papers.

Yukinori Ono
Jun Kondoh

Lecture Notes in Networks and Systems 939

The series "Lecture Notes in Networks and Systems" publishes the latest developments in Networks and Systems—quickly, informally and with high quality. Original research reported in proceedings and post-proceedings represents the core of LNNS.

Volumes published in LNNS embrace all aspects and subfields of, as well as new challenges in, Networks and Systems.

The series contains proceedings and edited volumes in systems and networks, spanning the areas of Cyber-Physical Systems, Autonomous Systems, Sensor Networks, Control Systems, Energy Systems, Automotive Systems, Biological Systems, Vehicular Networking and Connected Vehicles, Aerospace Systems, Automation, Manufacturing, Smart Grids, Nonlinear Systems, Power Systems, Robotics, Social Systems, Economic Systems and other. Of particular value to both the contributors and the readership are the short publication timeframe and the worldwide distribution and exposure which enable both a wide and rapid dissemination of research output.

The series covers the theory, applications, and perspectives on the state of the art and future developments relevant to systems and networks, decision making, control, complex processes and related areas, as embedded in the fields of interdisciplinary and applied sciences, engineering, computer science, physics, economics, social, and life sciences, as well as the paradigms and methodologies behind them.

Indexed by SCOPUS, INSPEC, WTI Frankfurt eG, zbMATH, SCImago.

All books published in the series are submitted for consideration in Web of Science.

For proposals from Asia please contact Aninda Bose (aninda.bose@springer.com).

Contents

Photonics, Optoelectronics and High-Energy Physics

Computational Science

Materials Science and Technology

Properties of Vacuum-Evaporated Metal Halide Perovskite Absorbers on Planar and Nanotextured Silicon Substrates

Gagik Ayvazyan[1(✉)], Harutyun Dashtoyan[1], Ashot Khudaverdyan[1], and Lenrik Matevosyan[2]

[1] National Polytechnic University of Armenia, 105, Teryan, 0009 Yerevan, Armenia
agagarm@gmail.com

[2] Institute of Radiophysics and Electronics, 1, Alikhanian, 0203 Ashtarak, Armenia

Abstract. The optical and structural properties of metal halide perovskite layers as absorbers for top solar cells in tandem applications were studied. $CH_3NH_3PbI_{3-x}Cl_x$ perovskite layers were deposited on planar Si and glass substrates and nanotextured Si substrates by vacuum co-evaporation and sequential evaporation methods. Methylammonium chloride (CH_3NH_3Cl) and lead iodide (PbI_2) were used as the raw precursors. Nanotexturing of the Si substrates was carried out using reactive ion etching in black Si formation mode. The experimental results show that sequentially evaporated (after annealing at 100 °C for 30 min) and co-evaporated layers had practically identical structural-phase quality and transmission spectra. The obtained perovskite layers were dense, without pores or cracks, and were formed by spherical-like crystallites. In the case of a nanotextured substrate, perovskite crystallites filled the interneedle spaces of the black Si layer. It was found that the black Si interlayer leads to a significant increase in the root mean square roughness of the perovskite layers and reduces the reflection of tandem structures.

Keywords: Perovskite · Evaporation · Black Silicon · Tandem Solar Cell · Morphology · Reflectance

1 Introduction

At present, perovskite-based solar cells, which consist of transport hole and electron layers and a perovskite layer as an absorber between them, have begun to be intensively studied and used [1–3]. The characteristics of these solar cells have been greatly improved in recent years. In 2009, when perovskite-based solar cells were first reported by Japanese students, their power conversion efficiency (PCE) was 3.81% [4]. Currently, this value has increased to 26.0% [5]. This success can largely be ascribed to the excellent optoelectronic properties of organometal halide perovskites ($CH_3NH_3PbX_3$, where X could be either Cl, Br, I, or mixed halides) [1–3].

To date, the record PCE value of perovskite-based solar cells is close to the maximum theoretically possible limit. The limitation is mainly because these solar cells absorb

Y. Ono and J. Kondoh (Eds.): Inter-Academia 2023, LNNS 939, pp. 3–10, 2024.
https://doi.org/10.1007/978-3-031-54450-7_1

only certain wavelengths of the solar spectrum. One simple way to expand the spectral response range and improve the performance of perovskite-based solar cells is to combine them with silicon (Si) solar cells. In such tandem solar cells, the top cell comprising a high-bandgap perovskite generates a photocurrent from the short-wavelength part of the solar spectrum [6–8]. Longer-wavelength light is transmitted to an underlying bottom lower-bandgap Si solar cell. Currently, the record PCE value of tandem Si/perovskite solar cells is 33,7% [5].

The PCE of tandem solar cells can be significantly increased by reducing the reflection of solar radiation from the perovskite surface and the interface between the top and bottom solar cells. For this purpose, a micron-sized texture is formed on the surface of a Si substrate using anisotropic chemical etching [7, 8]. Tockhorn et al., in their recent studies, presented gentle sinusoidal nanotextures as promising candidates for monolithic tandem solar cells [9, 10]. It has been experimentally demonstrated that such nanotextures provide more effective antireflection behavior without compromising the optoelectronic quality of the absorber. However, these studies use a complex nanoimprint lithography process to form nanotextures.

In our recent papers [11, 12], we proposed using a black silicon (b-Si) interlayer as a nanotexture for tandem solar cells, which is formed in a self-organized process (without a mask and lithography). In all respects (cost, technology, and parameters), this material is superior to micron texture and has been successfully used in single-junction solar cells as an antireflection surface [13–15]. Currently, the b-Si layer has not yet been used in tandem Si/perovskite solar cells, although optical modeling shows that an intermediate b-Si layer with an intermediate refractive index and high radiation absorption for wavelengths of 300–800 nm significantly improves the optical transmission of light into the sub-cells [16].

This work is aimed at experimental research on the optical and structural properties of organometal halide perovskite layers on planar and nanotextured (with a b-Si layer) Si substrates for tandem solar cell applications. The main practical purpose is to obtain a perovskite layer with maximum crystal grain and surface coverage on a nanostructured Si substrate, as well as to improve the antireflective properties of tandem structures.

2 Experimental Details

The metal halide perovskite $CH_3NH_3PbI_{3-x}Cl_x$, which is a suitable candidate for the absorber role of the top solar cell in tandem applications, was used [1, 2]. This perovskite exhibits a high absorption coefficient with a sharp optical absorption edge and has a low production cost, good stability, and relatively low toxicity. One important advantage of $CH_3NH_3PbI_{3-x}Cl_x$ absorbers for solar cell applications is that they can be prepared using a variety of different techniques, particularly the evaporation method.

Experimental samples of tandem structures were manufactured on planar and nanostructured (with a b-Si layer) Si substrates. To perform transmission measurements to optimize the perovskite formation and annealing processes, we also used planar glass substrates. The b-Si layers were produced by the reactive ion etching (RIE) process using SF_6/O_2 plasma on the front surface of Si substrates. In this study, those values of the RIE process parameters that were previously chosen as the best for single-junction Si solar cells were used [14]. The process pressure was 55 mTorr, and the gas flow rates were 75 cm^3/min and 40 cm^3/min for SF_6 and O_2, respectively. The etching time was kept constant at 10 min. After plasma processing, the samples were washed with distilled water and dried in air. The average period, height, and diameter of the b-Si needles were 150, 640, and 100 nm, respectively.

A 30 nm thick TiO_2 film deposited by the low-temperature atomic layer deposition (ALD) method was used as an electron transport layer between the substrate and the perovskite layer. Titanium tetrachloride ($TiCl_4$) and DI H_2O precursor systems with pure N_2 as the carrier gas were used, similar to those reported in Ref. [14]. The samples were subsequently annealed at 300 °C in N_2 for 15 min.

The deposition of perovskite layers was carried out using the vacuum evaporation technique via two methods: sequentially (layer-by-layer) and jointly (co-evaporated). Methylammonium chloride (CH_3NH_3Cl, MACI) and lead iodide (PbI_2) were used as the raw precursors. To improve the stability of the layers, 2.3% cesium iodide (CsI) was added to the PbI_2 precursor. The temperatures of the crucibles of these precursors were maintained at 200 and 350 °C. At a precursor composition of 4:1 (MACl:PbI_2) and a process pressure of 1.3×10^{-5} mbar, the evaporation rate of MACl was maintained at a level of 1.2 Å/s, while that of PbI_2 was kept at 0.4 Å/s. The duration of the evaporation process was varied to obtain layers of different thicknesses. After sequential evaporation, the obtained layers were annealed at 100 °C for 10, 20, and 30 min. The thicknesses of the perovskite layers were 210, 330, 500, and 720 nm. Figure 1 shows schematic illustrations of the experimental samples.

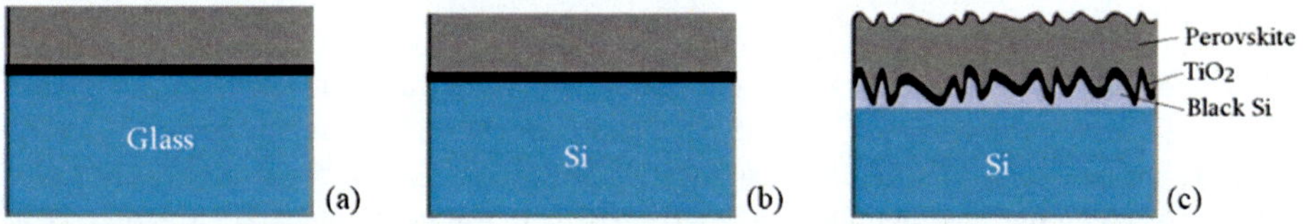

Fig. 1. Schematic illustrations of the experimental samples on planar glass (a), planar Si (b), and nanostructured Si (c) substrates.

A JEOL JSM-6700F Scanning Electron Microscope (SEM) and Solver Pro Atomic Force Microscope (AFM) were used for microstructural and surface investigations of the experimental samples. The open-source software Gwyddion was used to calculate the surface root mean square (RMS) roughness. The optical properties were measured by a UV-3101 PC spectrometer with an integrating sphere. The crystal structure of the perovskite layers was assessed by powder X-ray diffraction (XRD) using an Empyrean diffractometer equipped with a CuKα source with a wavelength of 1.54178 Å.

3 Results and Discussion

Figure 2 shows the transmission spectra of the initial (curve 1) and after annealing (curves 2–4) sequentially evaporated, as well as co-evaporated (curve 5) perovskite layers on planar glass substrates. Photographic images of the same layers are also presented.

The research showed that after annealing, the optical transmission spectrum of successively evaporated initial layers changes significantly, especially in the short-wavelength region of the spectrum. After long-term annealing (30 min), the transmittance spectra of sequentially evaporated and co-evaporated layers almost coincide. They acquire a dark brown color, typical of $CH_3NH_3PbI_{3-x}Cl_x$ perovskite layers. We also note that, as further studies showed, sequentially evaporated (after long-term annealing) and co-evaporated layers had practically the same structural, optical, and other properties. Therefore, the study results are presented below only for co-evaporated perovskite layers.

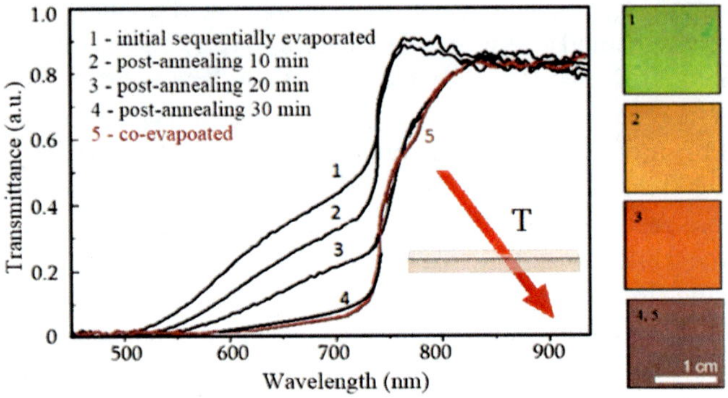

Fig. 2. Transmission spectra of the sequentially and co-evaporated perovskite layers on planar glass substrates.

The in-plane XRD patterns of the perovskite layers on the planar and nanotextured Si substrates are illustrated in Fig. 3. The perovskite layers on both substrates yield similar XRD patterns. Consequently, the structural and phase quality of these perovskite layers is the same. The perovskite layers have strong diffraction peaks at 14.2, 28.5, and 31.9°. These peaks are assigned to the (110), (220), and (310) crystal planes, respectively. The absence of clear peaks in the XRD spectra belonging to MACI and PbI_2 indicates that the perovskite is structurally well formed. The XRD patterns indicate the presence of a tetragonal crystal structure for the obtained layers, which is in good agreement with previous reports [1, 17, 18].

Figure 4 shows typical SEM images and grain size distribution of the perovskite layers on planar and nanotextured Si substrates. The obtained perovskite layers, regardless of the original substrate, are dense, without pores or cracks, and are formed by spherical-like crystallites. There are 2–3 grains located across the entire thickness of the perovskite layer from bottom to top. The characteristic grain size of perovskite on planar substrates was 87 nm. In the case of the nanotextured substrates, the average grain size was slightly larger (98 nm). It should be noted that the average grain size was also calculated from the corresponding XRD patterns using Scherrer's equation, but the results obtained were 10–15 nm smaller than the SEM results. A close-packed structure with large grains favors charge transfer, as more photogenerated charges without significant recombination at grain boundaries can successfully reach the electrodes. In the case of a nanotextured substrate, perovskite crystallites filled the interneedle spaces of b-Si.

The b-Si nanoscaffold, due to its high wettability [19], promotes the vertical growth of perovskite. Moreover, the b-Si texture with a certain degree of smoothing transferred to the upper surface of the perovskite. It should be noted that during the post-annealing of the sequentially evaporated perovskite layers, the structural properties of the initial b-Si layers remain almost unchanged due to the relatively low temperature of such treatment [20].

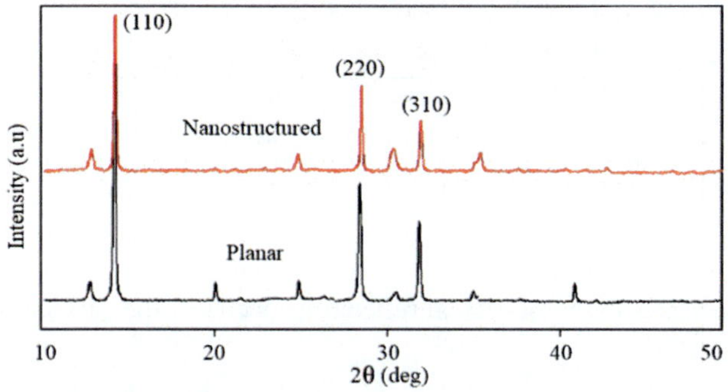

Fig. 3. Typical in-plane XRD patterns of the perovskite layers on the planar and nanotextured Si substrates.

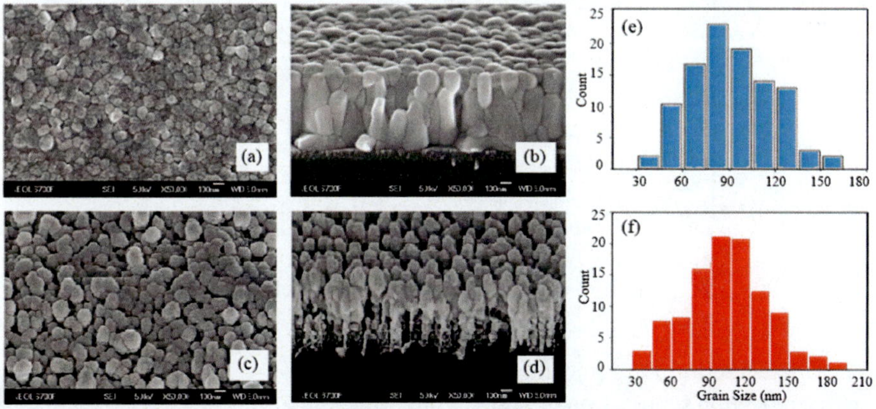

Fig. 4. Typical top view (a, c) and cross-sectional (b, d) SEM images and grain size distribution (e, f) of the perovskite layers on planar (a, b, e) and nanotextured (c, d, f) Si substrates.

The RMS roughness of the experimental samples supports the SEM observations (see Fig. 5). These results indicate that the obtained perovskite layers on planar substrates were very smooth. As expected, the presence of a b-Si layer leads to a significant increase in the RMS roughness of the perovskite layers. However, as the thickness of perovskite layers on nanotextured substrates increases, their top surface becomes noticeably smoother. In the case of planar substrates, the opposite structural evolution variations are observed, i.e., thicker layers have greater RMS roughness. This is because the thick perovskite layers tend to have larger grains, resulting in a rougher surface.

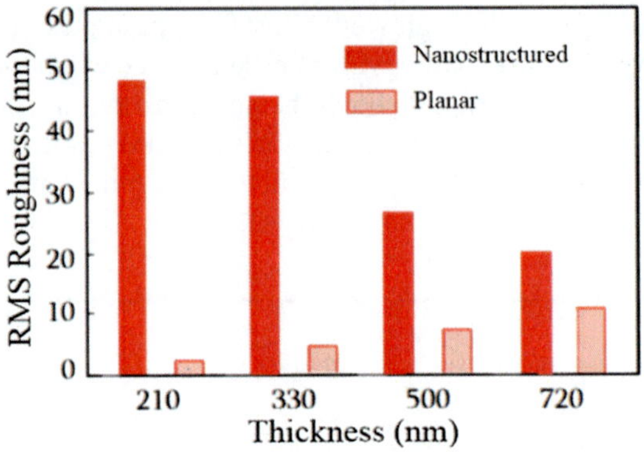

Fig. 5. RMS roughness of perovskite layers with various thicknesses on nanotextured and planar Si substrates.

Figure 6 illustrates the total optical reflection spectra of the surfaces of the obtained samples.

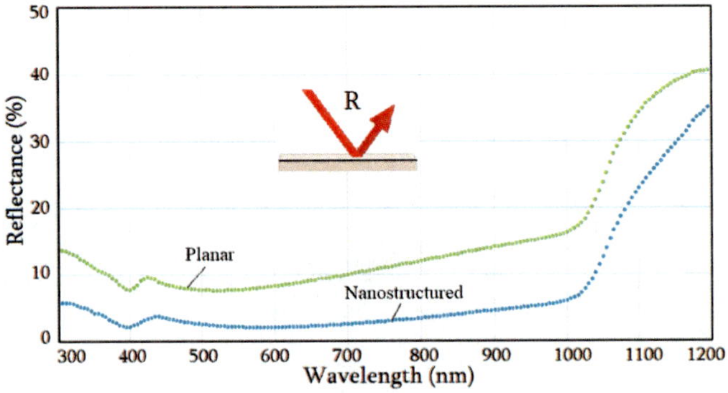

Fig. 6. Reflection spectra of the surfaces of the obtained samples.

The b-Si interlayer significantly reduces the reflection of tandem structures. The observed reflection suppression of samples on nanotextured substrates is associated with two mechanisms [16]. First, the rough surfaces of the perovskite and between the sub-cells will obliquely couple light into the semiconductor material and, as such, increase the optical path. Second, the b-Si layer is an "optically matching" medium, the refractive index of which is between perovskite and Si. It is well known that a gradual change in the refractive index in multilayer structures leads to a decrease in reflection.

4 Conclusion

In this work, metal halide perovskite absorber layers for tandem solar cell applications were deposited by vacuum thermal sequential evaporation and co-evaporation methods on planar and Si substrates with a b-Si layer. The structural and optical properties of the layers were characterized by SEM, X-ray diffraction, and reflection spectra measurements. The experimental results show that the b-Si interlayer in tandem structures does not prevent the formation of a perovskite layer with a high-quality structure and phase composition. In addition, optical reflection losses are significantly reduced. This antireflective behavior is explained by the high roughness of the sub-cell surfaces and the gradual change in the refractive index between perovskite and Si.

Acknowledgments. This work was supported by the Higher Education and Science Committee of the Republic of Armenia under project 21AG-2B011.

References

1. Hsu, H.P., Li, L.C., Shellaiah, M., Sun, K.W.: Structural, photophysical, and electronic properties of $CH_3NH_3PbCl_3$ single crystals. Sci. Rep. **9**, 13311 (2019)
2. Li, D., Shi, J., Xu, Y., Luo, Y., Wu, H., Meng, Q.: Inorganic–organic halide perovskites for new photovoltaic technology. Nat. Sci. Rev. **5**, 559–576 (2018)
3. Al-Asbahi, B.A., et al.: Effect of deposition method on the structural and optical properties of $CH_3NH_3PbI_3$ perovskite thin films. Opt. Mat. **103**, 109836 (2020)
4. Kojima, A., Teshima, K., Shirai, Y., Miyasaka, T.: Organometal halide perovskites as visible-light sensitizers for photovoltaic cells. J. Am. Chem. Soc. **131**, 6050–6051 (2009)
5. Green, M.A., et al.: Solar cell efficiency tables (version 62). Progr. Photovolt.: Res. Appl. **31**(7), 651–663 (2023)
6. Hu, Y., Song, L., Chen, Y., Huan, W.: Two-terminal perovskites tandem solar cells: Recent advances and perspectives. Sol. RRL. **3**, 1900080 (2019)
7. Hamada, K., et al.: Vacuum deposition of $CsPbI_3$ layers on textured Si for Perovskite/Si tandem solar cells. Jpn. J. Appl. Phys. **58**, SBBF06 (2019)
8. Cojocaru, L., et al.: Detailed investigation of evaporated perovskite absorbers with high crystal quality on different substrates. ACS Appl. Mater. Interfaces **10**, 26293–26302 (2018)
9. Tockhorn, P., et al.: Improved quantum efficiency by advanced light management in nanotextured solution-processed perovskite solar cells. ACS Photonics **7**, 2589–2600 (2020)
10. Tockhorn, P., et al.: Nano-optical designs for high-efficiency monolithic perovskite–silicon tandem solar cells. Nat. Nanotechnol. **17**, 1214–1221 (2022)
11. Ayvazyan, G.Y., Kovalenko, D.L., Lebedev, M.S., Matevosyan, L.A., Semchenko, A.V.: Investigation of the structural and optical properties of silicon-perovskite structures with a black silicon layer. J. Contemp. Phys. (Armenian Acad. Sci.) **57**, 274–279 (2022)
12. Ayvazyan, G., Hakhoyan, L., Dashtoyan, H., Matevosyan, L.: Preparation and investigation of vacuum-deposited $CH_3NH_3PbI_{3-x}Cl_x$ perovskite films on black silicon. J. Contemp. Phys. (Armenian Acad. Sci.) **58**, 85–91 (2023)
13. Liu, X., Coxon, P., Peters, M., Hoex, B., Cole, J., Frayc, D.: Black silicon: fabrication methods, properties and solar energy applications. Energy Env. Sci. **7**(10), 3223–3263 (2014)
14. Ayvazyan, G.Y., et al.: Anti-reflection properties of black silicon coated with thin films of metal oxides by atomic layer deposition. J. Contemp. Phys. (Armenian Acad. Sci.) **56**(3), 240–246 (2021)
15. Otto, M., et al.: Black silicon photovoltaics. Adv. Opt. Mater. **3**, 147–164 (2015)
16. Ayvazyan, G., Gasparyan, F., Gasparian, V.: Optical simulation and experimental investigation of the crystalline silicon/black silicon/perovskite tandem structures. Opt. Mat. **140**, 113879 (2023)

17. Salhi, B., Wudil, Y.S., Hossain, M.K., Al-Ahmed, A., Al-Sulaiman, F.A.: Review of recent developments and persistent challenges in stability of perovskite solar cells. Renew. Sustain. Energy Rev. **90**, 210–222 (2018)
18. Guesnay, Q., Sahli, F., Ballif, C., Jeangros, Q.: Vapor deposition of metal halide perovskite thin films: Process control strategies to shape layer properties. APL Mater. **9**, 100703 (2021)
19. Peschenyuk, Y.A., Semenov, G.Y., Gatapova, E.Ya.: The final stage of droplet evaporation on black silicon by Schlieren technique with a graded filter. Exp. Fluids. **64**(1) (2023)
20. Ayvazyan, G., Hakhoyan, L., Ayvazyan, K., Aghabekyan, A.: External gettering of metallic impurities by black silicon layer. Phys. Status Solidi A **220**(5), 2200793 (2023)

Spin Relaxation in GaAs/AlGaAs Coupled Quantum Wells

Keigo Shimada[1,2], Takuya Matsumoto[1,2], Masao Ichida[3], Hideki Gotoh[4], and Tetsu Ito[1,2](✉)

[1] Graduate School of Integrated Science and Technology, Shizuoka University, Hamamatsu, Japan
ito.tetsu@shizuoka.ac.jp
[2] Research Institute of Electronics, Shizuoka University, Hamamatsu, Japan
[3] Department of Physics, Konan University, Kobe, Japan
[4] Research Institute for Nanodevices, Hiroshima University, Higashi-Hiroshima, Japan

Abstract. Spin relaxation times of electrons spatially separated from holes were evaluated by polarization and time resolved pump and probe reflection measurements. A coupled quantum well (CQW) with a structure in which two quantum wells with a width of 10 nm and 15 nm are separated by 11 nm barrier layers were used to separate electrons and holes spatially. We found that the spin relaxation time was longer in CQW compared to that of isolated well. This was attributed to the fact that the CQW had a thin (11 nm) barrier layer between the 10 nm and 15 nm wells in width, so that electrons excited in the 10 nm well tunnel into the 15 nm well, but holes do not tunnel because of their large effective mass. In other words, the electron-hole exchange interaction is weakened because the electrons and holes can be spatially separated. Suppression of the electron spin relaxation by separating electrons from holes spatially is clearly demonstrated. These results are important to control the electron spin state in semiconductor nano structure.

Keywords: spin relaxation · coupled quantum well · electron-hole exchange interaction

1 Introduction

Spin, one of the internal degrees of freedom of quanta such as electrons and holes, has a two-level system with up and down states and their superposition states. Therefore, spin is expected to be applied to quantum bit for quantum information technology [1, 2]. Electron and hole spin relaxation in a semiconductor nano structure, such as quantum wells and quantum dots has been reported by means of time resolved photoluminescence (PL) and non-linear reflectivity measurements. A lot of studies for electron spin have been performed [3–7], and they are mainly related to the electron spin relaxation phenomena, for example band mixing of conduction and valence band [the Eliott-Yafet, (EY) process] [8], electron spin splitting in non-centrosymmetric media [the Dyakonov-Perel', (DP) process] [9], and electron-hole exchange interaction [the Bir-Aronov-Pikus,

Y. Ono and J. Kondoh (Eds.): Inter-Academia 2023, LNNS 939, pp. 11–19, 2024.
https://doi.org/10.1007/978-3-031-54450-7_2

(BAP) process] [10]. Contrast to the electron, the hole spin relaxation time is difficult to measure and evaluate because it is shorter than electron because of strong band mixing effect in valence band due to complex band structure [11–13]. Short relaxation time of hole spin may affect to the electron spin through the electron-hole exchange interaction [10].

In this study, we focus on the electron-hole exchange interaction, which is one of the causes of spin relaxation. Expecting to suppress this electron-hole exchange interaction, we evaluated the spin relaxation time in a coupled quantum well (CQW) structure sample, in which electrons and holes can be spatially separated. The electron spin relaxation time was observed to be longer in CQW compared to that of isolated wells which indicate that the electron-hole exchange interaction is weakened in CQW by spatial separation of electrons and holes.

2 Experimental Methods

In the measurements, we used a coupled quantum well (CQW) sample in which two well layers (GaAs) with widths of 10 nm and 15 nm are separated by a barrier layer ($Al_{0.1}Ga_{0.9}As$) with a width of 11 nm and a single quantum well (SQW) sample in which two well layers (GaAs) with the same width as the CQW are separated by a barrier layer with a width of 50 nm. Schematic images of the structures are shown in the inset of Fig. 1. The CQW and SQW samples were illuminated by a Ti-doped sapphire laser with an excitation power density of 40 W/cm^2 at 4.7 K. The duration was 2 ps and the repetition rate was 80 MHz. PL spectra of CQW (a) and SQW (b) with an excitation photon energy of 1.5895 eV are shown in Fig. 1. Figure 2 shows temperature and emitted photon energy dependence of the PL intensity. The peak photon energy of the PL decreases with increasing temperature under Varshni's law [14]. The peak around 1.530 and 1.543 eV correspond to the heavy-hole (HH) emission from 15 and 10 nm wells, respectively. Each peak shifted 2.57 meV to lower energy by changing the sample temperature from 5 to 50 K.

The PL intensity from the 15 nm well for CQW and SQW were monitored in photoluminescence excitation (PLE) measurements to identify the HH resonance level of each well. The PLE spectra are shown in Fig. 3. Detected photon energies were 1.5284 and 1.5291 eV for CQW and SQW, respectively and polarization was anti-parallel component to excitation light (σ^-). Using peak levels obtained from Fig. 3 as a reference, we performed the polarization- and time-resolved measurements by pump and probe (PP) method. The laser pulse was divided into pump and probe pulses by beam splitter. Polarization of the pump pulse was converted to circularly polarization by quarter wave plate to generate spin polarization in the quantum wells. Linearly polarized probe pulse was focused to the area where the pump pulse was illuminated with a delay time relative to the pump pulse. Polarization of the probe pulse reflected form sample was examined as a function of delay time between pump and probe pulses. The electron- or hole-spin state can be detected from polarization of the probe pulse. We evaluated the polarization of σ^+ and σ^- obtained by PP measurements for the CQW and SQW samples, where σ^+ and σ^- are parallel and anti-parallel component to pump pulse, respectively.

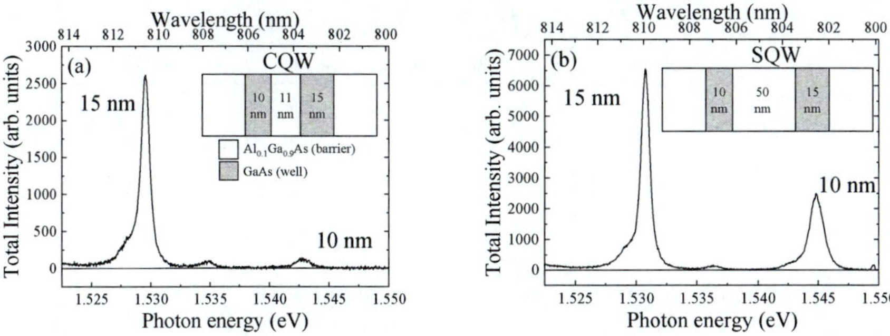

Fig. 1. PL spectra of (a) CQW and (b) SQW with an excitation photon energy of 1.5895 eV at 4.7 K. Excitation power density is 40 W/cm^2. Each sample structure images are shown in the inset.

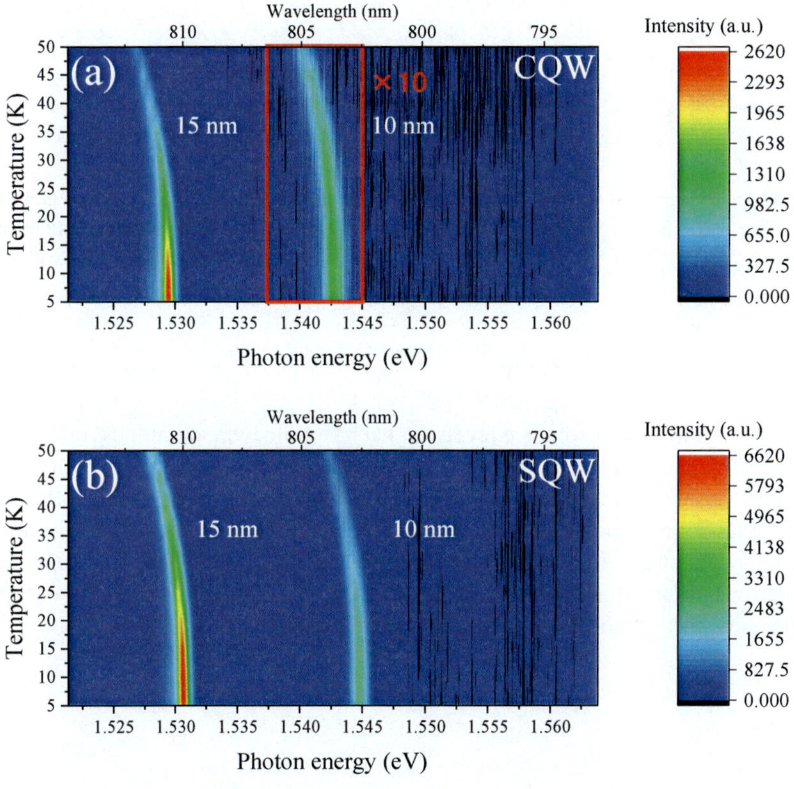

Fig. 2. Temperature and emitted photon energy dependence of the PL intensity of (a) CQW and (b) SQW with an excitation photon energy of 1.5895 eV. Excitation power density is 40 W/cm^2. Rectangular area in (a) is ten times magnified in intensity.

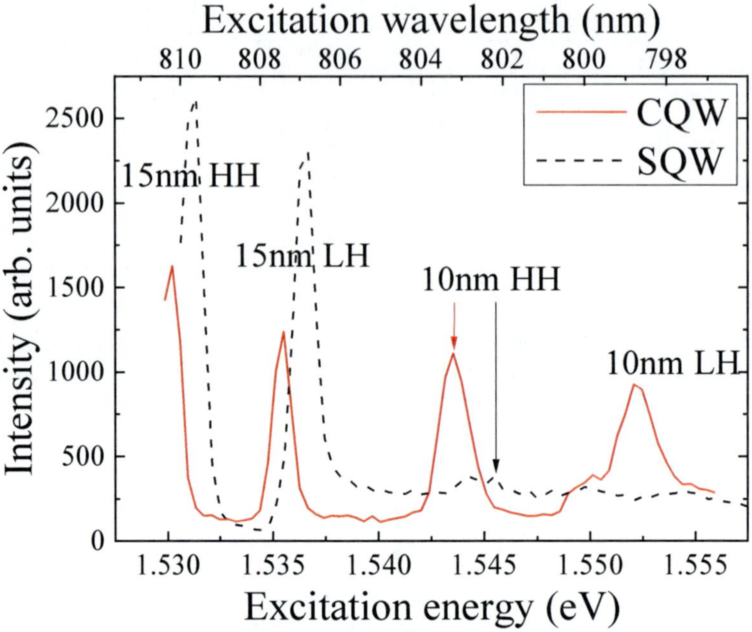

Fig. 3. PLE spectra of CQW (solid curve) and SQW (dashed curve). Detected photon energies were 1.5284 and 1.5291 eV for CQW and SQW, respectively and polarization was anti-parallel component to excitation light (σ^-). Arrows indicate HH resonance revel of 10 nm well for each sample.

3 Results and Discussion

The energies of the HH resonance levels for each sample were obtained from Fig. 3. The four peaks represent, from left to right, the 15 nm HH level, the 15 nm light-hole (LH) level, the 10 nm HH level, and the 10 nm LH level, respectively. Two peaks originate from 10 nm well are observed in CQW but not in SQW. This indicate that illumination at a resonance energy of 10 nm also emits photon in the 15 nm well. This also suggests that electrons tunneled from the 10 nm well to the 15 nm well in the CQW. In SQW the electrons couldn't tunnel because of thick barrier layer.

Figure 4 shows time traces of degree of polarization (DOP) in the CQW and SQW obtained by PP measurements at the HH resonance levels of the 10 nm well as shown arrows in Fig. 3. DOP was obtained by $|\sigma^+ - \sigma^-|/|\sigma^+ + \sigma^-|$. They were both fitted with a duble exponential function ($\text{DOP}(t) = A_1 exp(-2t/\tau_{short}) + A_2 exp(-2t/\tau_{long})$) as solid curves in Fig. 4. Two relaxation components are observed in time traces of DOP. They are generally considered to be hole spin relaxation for the short component and electron spin relaxation for the long component because the hole spin relaxation time is short due to complex structure in valence bands [11–13].

Spin relaxation mechanisms, EY [8] and DP process [9] show different temperature dependence for spin relaxation time. Spin up- and down- states are mixed by band mixing in the high energy region. So, the spin relaxation time decrease with increase the temperature because of thermal energy distribution change of electrons and holes. The band mixing is strong in valence band because of small energy difference between each state. This is the main reason for shorter spin relaxation time of hole than electron [11–13]. The strength of the mixing of electron is smaller than that of hole because conduction band is isolated from other band such as HH, LH band in energy domain. The electron spin relaxation mechanism caused by the band mixing between conduction and valence band is investigated as EY process [8]. In addition to the EY process, the electron spin in quantum well is affected by DP process. DP process is that the spin direction is changed by effective magnetic field caused by non-centrosymmetric crystal structure [4]. The electron spin flip by the effective magnetic field is reduced with increasing temperature because of motional narrowing [9]. This effect leads to the increasing of electron spin relaxation time.

To identify the origin of two decay components in Fig. 4 we examined temperature dependence of the relaxation time. Pump photon energy was determined from temperature dependence of the PL peak shown in Fig. 2. Band gap energy shifted 2.57 meV to lower energy by changing the sample temperature from 5 to 50 K. The polarization- and time-resolved measurements were performed considering this energy shift. Decay signals with two components like that in Fig. 4 were observed for each temperature and double exponential decay function fitting were performed for each signal. Figure 5 shows temperature dependence of long and short relaxation time for CQW and SQW. The short components of the relaxation time shown in Fig. 5(b) and (d) were decreased with increasing the temperature for 10 and 15 nm well in CQW and SQW. This behaver can be well described by strong band mixing effect in valence band. So, the origin of the short components is considered to be hole spin. On the other hand, the long components were constant or slightly increased with increasing the temperature. This trend can be described by the characteristics of the temperature dependence of the electron spin relaxation time which is both affected by EY and DP process [15, 16]. The effect of increasing the spin relaxation time with increasing the temperature by DP process is comparable or slightly stronger than that of decreasing spin relaxation time by EP process in this sample and this temperature region. The long and short components in Fig. 4 is corresponding to electron and hole spin relaxation, respectively.

Comparing the two relaxation components in CQW and SQW shown in Fig. 4, we can see that the relaxation time for the short component (holes) is almost same, but that for the long component (electrons) is longer in CQW than in SQW. The PL intensity of 10 nm well of CQW was smaller than that of 15 nm as shown in Fig. 1 and 2. This indicate that carriers created in 10 nm well transfer to the 15 nm well by tunneling effect. The electrons easily tunneled from the 10 nm well to the 15 nm well in the CQW, but the holes were difficult to tunnel to 15 nm well because of large effective mass. These electrons in 15 nm well are considered to be less susceptible to exchange interaction [10] with holes in 10 nm well. Suppression of the electron spin relaxation by separating electrons from holes spatially is observed in coupled quantum wells.

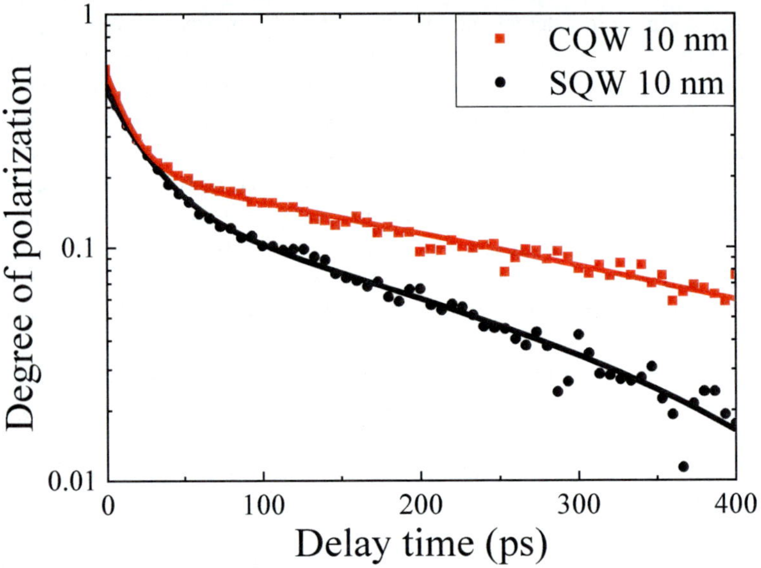

Fig. 4. Delay time variation of degree of polarization for 10 nm well in CQW (solid square) and SQW (solid circle) are shown on logarithmic axis. Curves indicate results of double exponential function ($\mathrm{DOP}(t) = A_1 exp(-2t/\tau_{short}) + A_2 exp(-2t/\tau_{long})$) fitting for each sample.

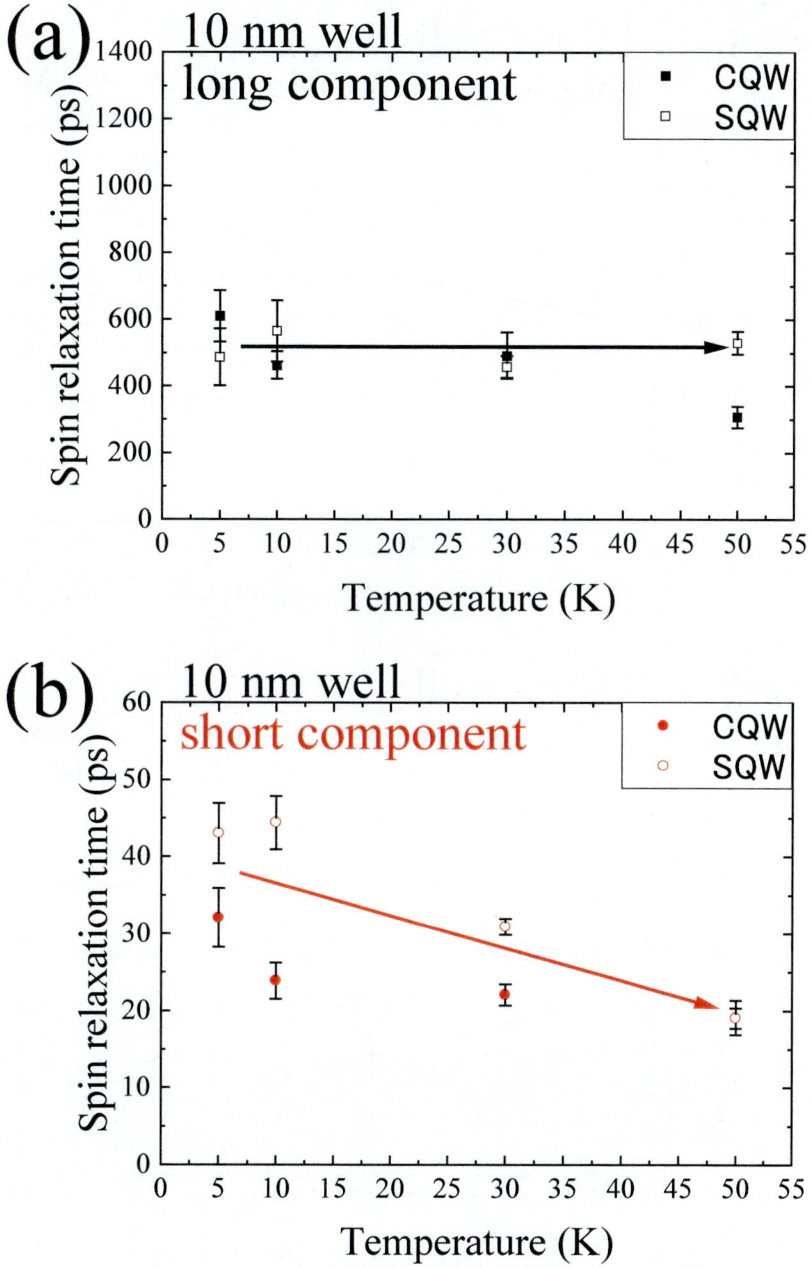

Fig. 5. Temperature dependence of spin relaxation time of CQW (solid symbol) and SQW (open symbol). Long and short component of 10 nm well HH resonance excitation are shown in (a) and (b), respectively. Those of 15 nm well are shown in (c) and (d). Arrows are guide for eyes.

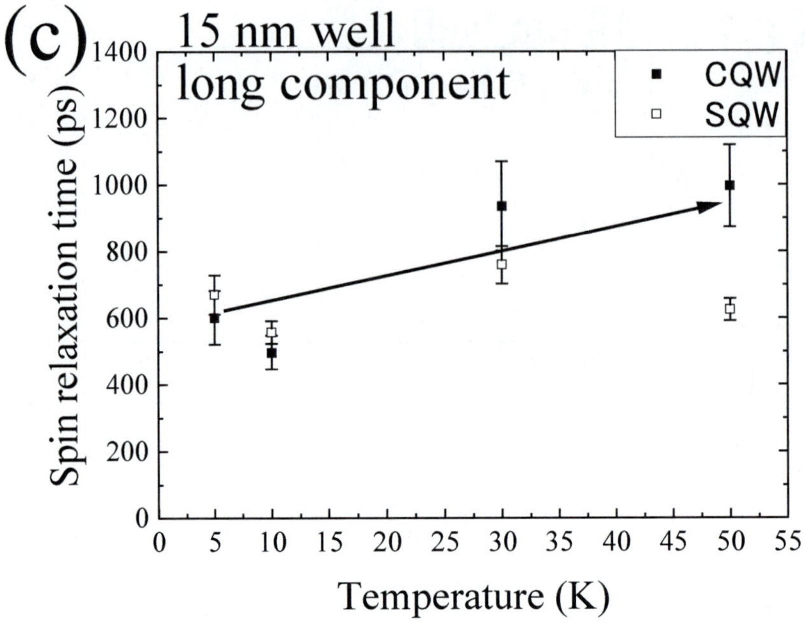

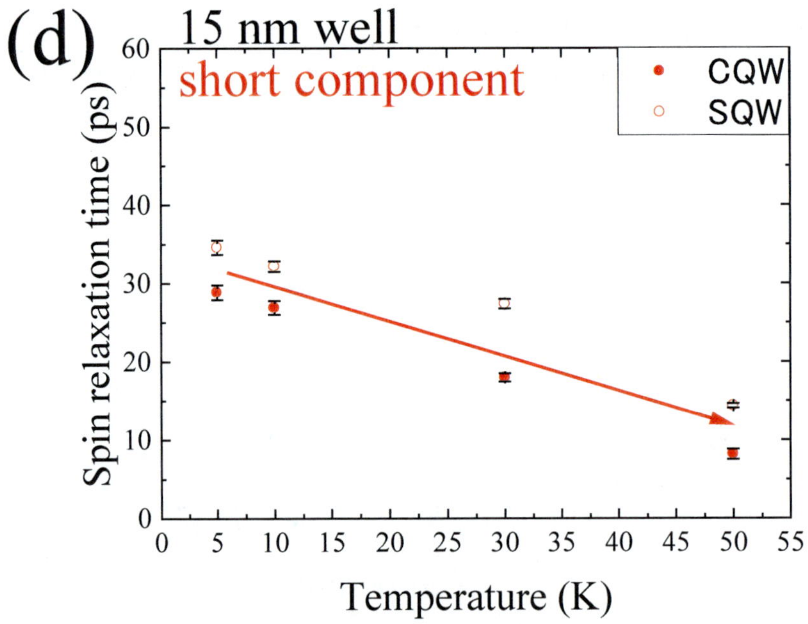

Fig. 5. *(continued)*

4 Conclusion

The long and short component in spin relaxation signal correspond to the electron and hole spin relaxation, respectively. PL and PLE measurements suggest that the coupled quantum well structure allowed spatial separation of the electron-hole distance, as electrons were thought to tunnel from the 10 nm well of the CQW to the 15 nm well. Separation of electrons and holes tended to increase the spin relaxation time of electrons. The electrons tunneled from the 10 nm well to the 15 nm well are less susceptible to exchange interactions with holes. The electron spin relaxation time can be controlled by spatially separating electrons and holes.

References

1. Loss, D., DiVincenzo, D.P.: Quantum computation with quantum dots. Phys. Rev. A **57**, 120 (1998)
2. Žutić, I., Fabian, J., Sarma, S.D.: Spintronics: fundamentals and applications. Rev. Mod. Phys. **76**, 323 (2004)
3. Kikkawa, J.M., Awschalom, D.D.: Resonant spin amplification in n-type GaAs. Phys. Rev. Lett. **80**, 4313–4316 (1998)
4. Ohno, Y., Terauchi, R., Adachi, T., Matsukura, F., Ohno, H.: Spin relaxation in GaAs(110) quantum wells. Phys. Rev. Lett. **83**, 4196–4199 (1999)
5. Oertel, S., Hübner, J., Oestreich, M.: High temperature electron spin relaxation in bulk GaAs. Appl. Phys. Lett. **93**, 132112 (2008)
6. Kosaka, H., et al.: Coherent transfer of light polarization to electron spins in a semiconductor. Phys. Rev. Lett. **100**, 096602 (2008)
7. Ito, T., Gotoh, H., Ichida, M., Ando, H.: Evaluation of hole-spin superposition in GaAs/AlGaAs quantum wells through time-resolved photoluminescence measurements. Appl. Phys. Lett. **104**, 252406 (2014)
8. Elliott, R.J.: Theory of the effect of spin-orbit coupling on magnetic resonance in some semiconductors. Phys. Rev. **96**, 266 (1954)
9. D'yakonov, M.I., Perel', V.I.: Spin orientation of electrons associated with the interband absorption of light in semiconductors. Sov. Phys. JETP **33**, 1053 (1971)
10. Bir, G.L., Aronov, A.G., Pikus, G.E.: Spin relaxation of electrons due to scattering by holes. Sov. Phys. JETP **42**, 705 (1976)
11. Damen, T., Vina, L., Cunningham, J.E., Shah, J., Sham, L.J.: Subpicosecond spin relaxation dynamics of excitons and free carriers in GaAs quantum wells. Phys. Rev. Lett. **67**, 3432–3435 (1991)
12. Syperek, M., et al.: Spin coherence of holes in GaAs/(Al, Ga)As quantum wells. Phys. Rev. lett. **99**, 187401 (2007)
13. Fang, S., Zhu, R., Lai, T.: Spin relaxation dynamics of holes in intrinsic GaAs quantum wells studied by transient circular dichromatic absorption spectroscopy at room temperature. Sci. Rep. **7**, 287 (2017)
14. Varshni, Y.P.: Temperature dependence of the energy gap in semiconductors. Physica **34**, 149 (1967)
15. Adachi, S., Miyashita, T., Takeyama, S., Takagi, Y., Tackeuchi, A.: Exciton spin dynamics in GaAs quantum wells. J. Luminescence **72** (1997)
16. Han, L., Zhu, Y., Zhang, X., Tan, P., Ni, H., Niu, Z.: Temperature and electron density dependence of spin relaxation in GaAs/AlGaAs quantum well. Nanoscale Res. Lett. **6**, 84 (2011)

Tritium Release Behavior of Biphasic Lithium Ceramic with Low Li_4SiO_4 Content

Qilai Zhou[1,2](✉), Yuguo Zhang[1], Asahi Sanfukuji[2], Yuzuka Hoshino[2], and Yasuhisa Oya[2]

[1] Wuhan University of Technology, Wuhan 430070, China
zhou.qilai@whut.edu.cn, zhou.qilai@shizuoka.ac.jp
[2] Shizuoka University, Shizuoka 422-8529, Japan

Abstract. Nuclear fusion is one of the promising options for the global energy supply. In a fusion reactor, tritium (T) is produced by the Li(n, α)T reaction. Lithium ceramics, e.g., Li_2TiO_3, Li_4SiO_4, are regarded as candidate materials for tritium breeding. In this study, the tritium release behavior of Li_2TiO_3-Li_4SiO_4 biphasic ceramic was investigated. Especially, the content of Li_4SiO_4 was controlled at 12.5% and 25%. The Li_2TiO_3-xLi_4SiO_4 ($x = 0.125, 0.25$) ceramic powders were prepared by the solution combustion synthesis. The neutron irradiation was conducted using Kyoto University Research Reactor (KUR) with a neutron flux of 5.5×10^{12} n cm^{-2} s^{-1}. The tritium thermal desorption spectroscopy (Tritium-TDS) device at Shizuoka University was upgraded to recover tritium under various conditions. It was found that the tritium desorption rate at higher temperatures was increased with increasing Li_4SiO_4 content and the amount of tritium was also increased. The tritium recovery process was significantly improved using the purge gas with water vapor. The tritium-TDS spectra were shifted to the lower temperature side with a higher desorption rate. Besides, the tritiated water (HTO) and tritium gas (HT) were separated successfully. The desorption rate of HTO was increased and the HT desorption was shifted to the high temperature for the ceramic irradiated under higher neutron fluence.

Keywords: Tritium · Nuclear fusion · Lithium ceramic · Neutron irradiation

1 Introduction

Tritium (T) and deuterium (D) are the fuels for the D-T fusion reactors. The deuterium can be extracted from the sea water. Tritium is formed at upper atmosphere by cosmic radiation from space and exists in our environment, but the amounts of tritium on earth are rare. Therefore, tritium has to be produced artificially. In the fusion reactors, the 14.1 meV neutron, which is produced by the fusion reaction, would be slowed down and reacts with lithium (Li) in the blanket to produce tritium based on the ^{6}Li(n, α)T nuclear reaction. Lithium ceramics are regarded as the candidate materials for tritium breeding. In recent years, ternary lithium ceramics have been intensively studied owing to their higher chemical inertness compared with Li_2O. Among them, Li_2TiO_3 draws attention

Y. Ono and J. Kondoh (Eds.): Inter-Academia 2023, LNNS 939, pp. 20–28, 2024.
https://doi.org/10.1007/978-3-031-54450-7_3

due to its stability. On the other hand, Li_4SiO_4 also gains acceptance due to the relatively higher lithium density. From the viewpoint of nuclear reaction, higher lithium density would result in higher tritium amounts. Hence, the Li_2TiO_3-Li_4SiO_4 mixture, biphasic system, and core-shell structure were proposed [1–4]. However, most of the available reports are focused on material design, ceramic preparation, and mechanical properties. The information on tritium release behavior is scarce because of the difficulties in neutron irradiation and tritium experiments. The reasons are as follows. On one hand, tritium (T or ^3H) is a radioactive isotope of hydrogen, which emits β-ray with an average energy of 5.7 keV during its decay. Hence, tritium should be handled in the radiation-controlled area. On the other hand, the neutron source is needed to generate tritium in the lithium ceramics. It is urgent to obtain tritium data to evaluate the tritium release property of the Li_2TiO_3-Li_4SiO_4 ceramics.

In our previous research, the Kyoto University Research Reactor (KUR) was used for neutron irradiation [5]. The tritium release behavior of single-phase ceramics and Li_2TiO_3-Li_4SiO_4 biphasic ceramics with different phase ratios was investigated. It was confirmed that the tritium release amount of Li_4SiO_4 was higher than that of Li_2TiO_3, but the major desorption peaks were located at higher temperatures. For the case of Li_2TiO_3-Li_4SiO_4 with a phase ratio of 1: 2, the tritium desorption rate at the higher temperature was decreased. With the further reduction of the phase ratio to 2: 1, most of the tritium can be desorbed as a major desorption peak at the low temperature side. This study focused on the discussion of tritium release behavior for Li_2TiO_3-xLi_4SiO_4 ($x = 0.125, 0.25$) biphasic ceramics with low Li_4SiO_4 contents. The separation of tritiated water and tritium gas was attempted. A new purge line was equipped into the tritium thermal desorption spectroscopy (Tritium-TDS) system to introduce water vapor. The effect of water vapor on the tritium release behavior was also investigated.

2 Experimental

2.1 Preparation of Ceramic

Li_2TiO_3-xLi_4SiO_4 ($x = 0.125, 0.25$) biphasic ceramic powders were prepared using the solution combustion synthesis (SCS) system at Wuhan University of Technology. The SCS is a wet-chemical route for the quick synthesis of oxide ceramics with high purity. The aqueous solutions of lithium nitrate, silicate nitrate, and titanium nitrate were prepared in advance and were mixed with the citric acid that acted as the fuel for the combustion reaction. Thereafter, the mixture solution was heated by the microwave with a output power of 1100 W, which led to the combustion reaction. Based on the reaction, the white ceramic powders were obtained. To control the Li_4SiO_4 content, the dosage ratios of Ti and Si were set at 8: 1 and 4: 1 respectively, to achieve the $x = 0.125$, 0.25 in Li_2TiO_3-xLi_4SiO_4. The flowchart of the SCS process is shown in Fig. 1. The preparation of silicate nitrate, titanium nitrate and detailed dosages can be found in the previous reports [6, 7].

2.2 Neutron Irradiation

The as-prepared samples were sealed into capsules under an argon atmosphere. The capsules were introduced into the core of KUR reactor. The KUR reactor is a light-water moderated tank-type reactor operated at the maximum rated thermal power of 5 MW. The temperature of the cooling water was about 328 K. In the experiment, the KUR was operated at 1 MW with a thermal neutron flux of 5.5×10^{12} n cm^{-2} s^{-1}. The irradiation time was changed to achieve different neutron fluence, which was listed in Table 1.

Table 1. Neutron irradiation parameters.

Neutron flux (n cm^{-2} s^{-1})	Irradiation time (h)	Fluence (n cm^{-2})
5.5×10^{12}	2	4.0×10^{16}
5.5×10^{12}	4	8.0×10^{16}

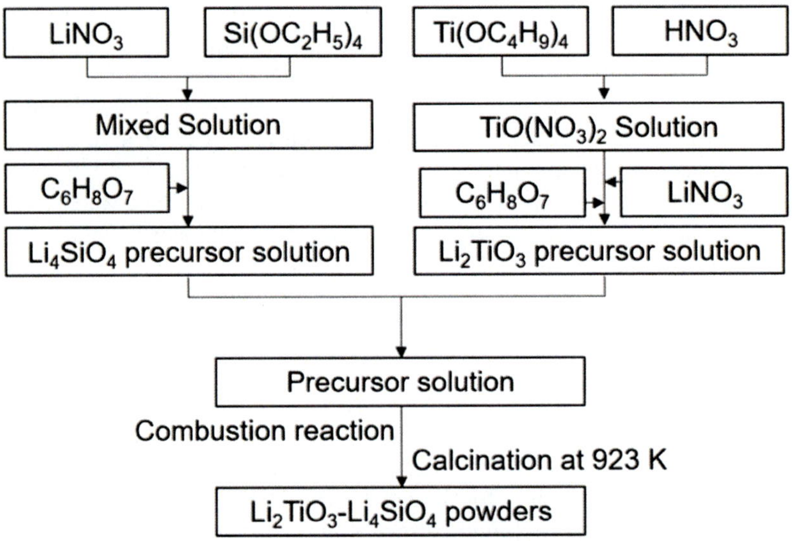

Fig. 1. The preparation flowchart of the Li_2TiO_3-Li_4SiO_4 biphasic ceramics by the solution combustion synthesis.

2.3 Tritium Release Experiment

Tritium was generated in the ceramics after neutron irradiation. These samples were transported to the radiation-controlled area at Shizuoka University. Tritium thermal desorption spectroscopy (Tritium-TDS) was used for the tritium release experiments. Figure 2 shows the schematic and photo of the tritium-TDS device in the radiation-controlled area. The neutron-irradiated sample was placed in the reactor quartz tube and was heated with a ramping rate of 30 K min^{-1} to 1113 K. Argon (Ar) was used as the purge and CH_4 was used as the counting gas. The system was equipped with two proportional counters for the monitoring of tritium release. The first water bubbler downstream trapped the HTO, and the HT was oxidized by CuO, which was trapped subsequently by the third bubbler. The tritium concentration in the water bubblers was measured by a liquid scintillation counter (LSC) to calculate the tritium release amount. The detailed information can be found in our previous report [8]. In the present work, a new water bubbler line was set upstream which can introduce water vapor in the Ar purge gas for the complete recovery of tritium released from the sample. Humid and dry purge gas can be switched. Besides, humid Ar+H_2 was purged after the sample was heated at 1113 K for 25 min for the recovery of the tritium absorbed on the surface of the quartz tube. In the experiment, some CuO was reduced into Cu by tritium and hydrogen. Therefore, O_2 gas was purged through CuO tubing to oxide Cu after the tritium release experiment.

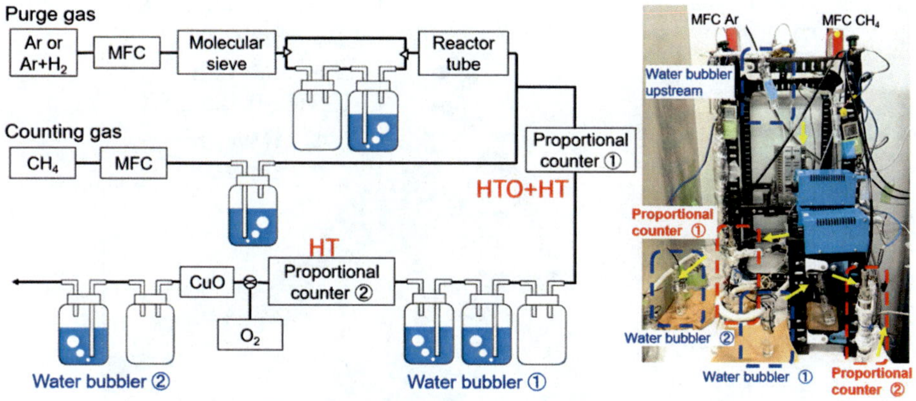

Fig. 2. The schematic and photo of tritium-TDS device.

3 Results

Figure 3 shows the tritium-TDS spectra of the Li_2TiO_3-xLi_4SiO_4 (x = 0.125, 0.25) biphasic ceramic in the case of different irradiation conditions. It can be seen that tritium started to desorb at a low temperature of 400 K for the Li_2TiO_3-0.25Li_4SiO_4 sample which has a higher Li_4SiO_4, indicating that some tritium was weakly adsorbed on the ceramic surface. In most cases, the tritium desorption rate increased significantly when the samples were heated above 500 K and the major desorption peaks were formed at around 610 K. It can be noticed that there was a shoulder peak at the high-temperature side. By comparing the TDS for Li_2TiO_3-0.125Li_4SiO_4 irradiated with different neutron fluence, it can be found that the intensity of the shoulder peak at 800 K was increased with the irradiation time, which means that this peak is associated to the desorption of tritium trapped by the irradiation defects. On the other hand, there was no clear change for the peak at 610 K. Besides, an additional peak only existed for the Li_2TiO_3-0.25Li_4SiO_4

sample, indicating that this peak can be ascribed to the desorption of tritium from Li_4SiO_4 phase. It was also found in our previous study [5] that there were two major desorption peaks for Li_4SiO_4 in the temperature range of 550–700 K.

Table 2 shows the tritium release amount of different samples measured by the liquid scintillation counters. It can be found that the tritium release amount of Li_2TiO_3-0.25Li_4SiO_4 was higher than that of Li_2TiO_3-0.125Li_4SiO_4 because of the higher lithium atom density in Li_4SiO_4. Also, the tritium release amount was linearly dependent on the irradiation time which showed that irradiation dose was not saturated. Besides, for the samples with the irradiation time of 2 h, the tritium retention in Li_2TiO_3-0.25Li_4SiO_4 was lower which can be ascribed to the tritium release during the neutron irradiation. Some tritium weakly trapped on the surface would be desorbed because of the heat generated by the (n, α) reaction.

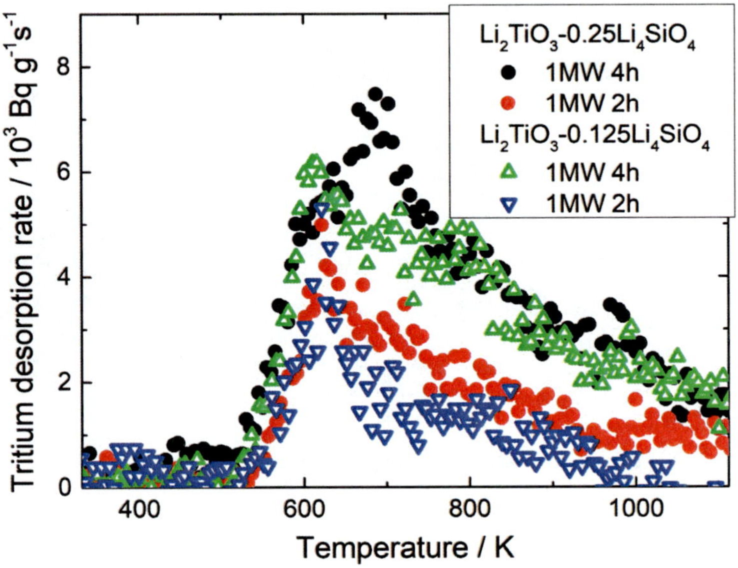

Fig. 3. Tritium-TDS spectra of the Li_2TiO_3-$x$$Li_4SiO_4$ ($x = 0.125, 0.25$) biphasic ceramic in the case of different irradiation conditions.

Table 2. Tritium release amounts measured by LSC.

Sample	Irradiation and purge condition	Amount (MBq g^{-1})
Li_2TiO_3-0.25Li_4SiO_4	1 MW 4 h, pure Ar	6.2
Li_2TiO_3-0.25Li_4SiO_4	1 MW 2 h, pure Ar	2.7
Li_2TiO_3-0.125Li_4SiO_4	1 MW 4 h, pure Ar	6.0
Li_2TiO_3-0.125Li_4SiO_4	1 MW 2 h, pure Ar	3.5

Figure 4 shows the effects of water vapor on the tritium-TDS spectra. The tritium recovery behavior for Li_2TiO_3-0.125Li_4SiO_4 is used as a typical example for the discussion. By passing through the water bubbler upstream, Ar became humid. The H_2O in the humid Ar gas can enhance the tritium desorption process, especially tritiated water, known as HTO, owing to the hydrogen isotope exchange. From the experimental result, it can be seen that the tritium desorption rate was increased significantly for the case of Ar+H_2O. The desorption was also shifted toward the lower temperature side. Besides, it can be also seen that the intensity of the desorption peak at 800 K, was not increased. Apart from the hydrogen isotope exchange effect, the H_2O would dissociate at higher temperatures and the hydrogen atom would diffuse into the bulk region and occupy the irradiation defects [9]. However, from the current experimental results, it was known that there was no clear change in the peak intensity corresponding to the strongly trapped tritium at irradiation defects, indicating that the diffusion of hydrogen into the bulk region was limited. Further, it can be seen that the tritium was almost recovered completely within 1000 K for the case of Ar+H_2O, while the reduction of the desorption rate at the higher temperature side was slow in the case of Ar. This phenomenon shows that some tritium was constantly trapped and de-trapping from the inner surface of the quartz tube. In the case of Ar+H_2O, the trapping sites in the quartz tube were occupied by hydrogen which facilitated the tritium recovery.

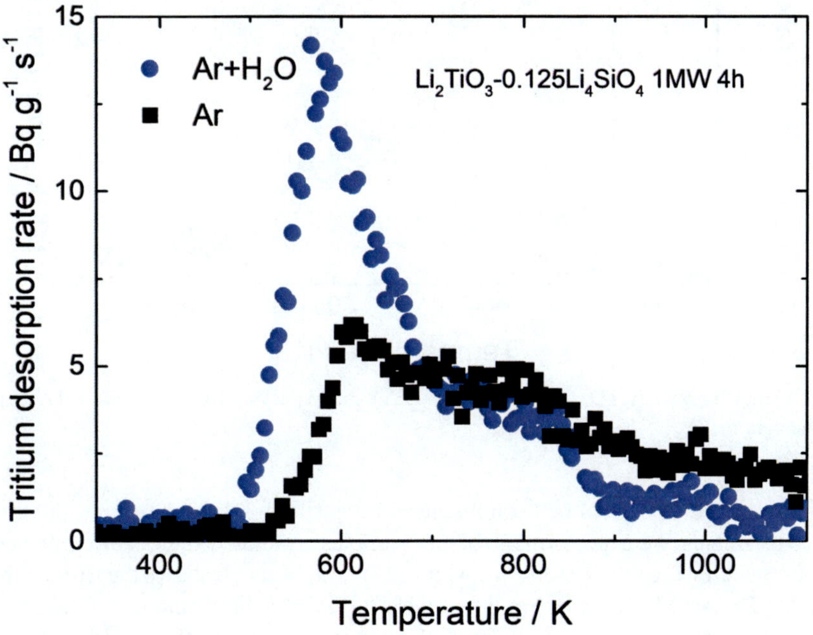

Fig. 4. Tritium-TDS spectra of the Li_2TiO_3-0.125Li_4SiO_4 biphasic ceramic using the dry Ar and humid Ar as the purge gas.

Figure 5 shows the separation result of the tritiated water and tritium gas, known as HTO and HT, respectively. Compared with HTO, HT was mainly desorbed at higher temperature side. Besides, different from the intense increase in the desorption rate of HTO, the HT desorption rate increased slowly and then dropped significantly at 900 K. There were two major desorption peaks in the HT-TDS spectrum. One small peak emerged at 630 K, which was slightly higher than the main desorption peak of HTO. Another large desorption peak was located at 820 K, which was near the temperature for the shoulder peak in HTO spectra. Tritium migrates as atoms or ions in the ceramic bulk, therefore tritium releases as HTO associated with the surface reactions with oxygen atoms. After the HTO desorption reached the maximum rate, the amount of trapping sites associated with the oxygen atoms was reduced, which facilitated the tritium release as HT gas. The HT desorption peak at higher temperatures can be also attributed to the tritium trapped in the bulk region because most of tritium in the surface region had been desorbed as HTO at lower temperatures.

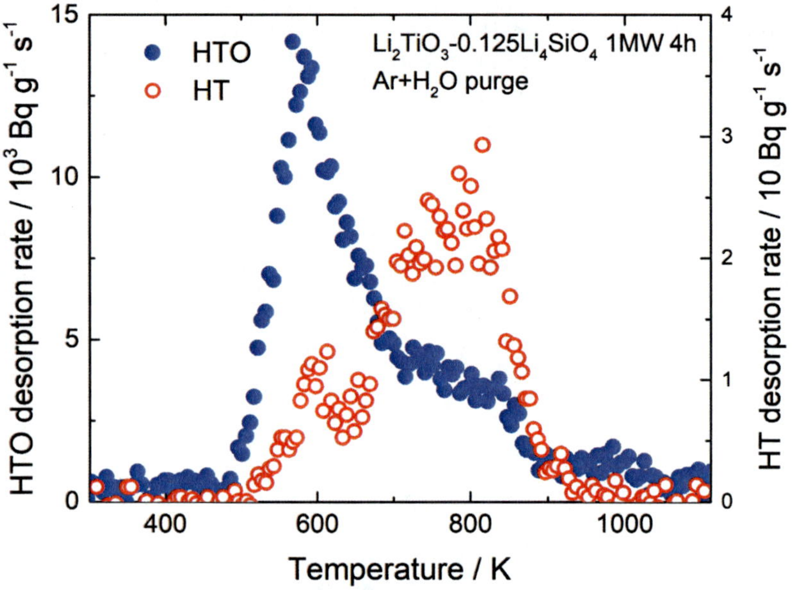

Fig. 5. Tritiated water (HTO) and tritium gas (HT) TDS spectra of the Li_2TiO_3-$0.125Li_4SiO_4$ biphasic ceramic.

To evaluate the effect of neutron fluence on the tritium release behavior, the tritium-TDS experiments were performed for the samples irradiated under different neutron fluences, which were noted as 1MW 2h and 1MW 4h. Figure 6 shows the tritium spectra of the Li_2TiO_3-$0.125Li_4SiO_4$ irradiated with different neutron fluence. Clear neutron fluence dependence can be found in the TDS spectra. The neutron fluence in the condition of 1MW 4h was twice higher than that of 1MW 2h. It can be seen that the intensity of the major desorption peak in HTO spectrum was twice higher. On the other hand, for the peak at 800 K, the desorption rate was also higher in the case of higher fluence. These results confirmed that higher neutron fluence not only increased the tritium generation amounts in the ceramic but also increased the density of trapping sites in the ceramic bulk. Besides, it can be seen that there was no clear peak shift for the peak position, indicating the binding energy for the tritium trapped at irradiation defects was not so high that most of tritium can be released at a temperature higher than 610 K. For the case of HT desorption, it can be seen that HT-TDS spectrum also shows a clear dependence on the irradiation condition. The desorption rates for both peaks were increased at higher neutron fluence. Most interestingly, it was found that the desorption peaks were shifted toward higher temperature side with increasing the neutron fluence, which was different from that of HTO results. This phenomenon can be explained as the amount of tritium trapped in the bulk region was increased in the case of higher neutron fluence with the stabilization of the irradiation defects related to oxygen atoms, e.g., O^--center. With the further increase of temperature, these defects were quickly recovered which led to the tritium release as HT gas.

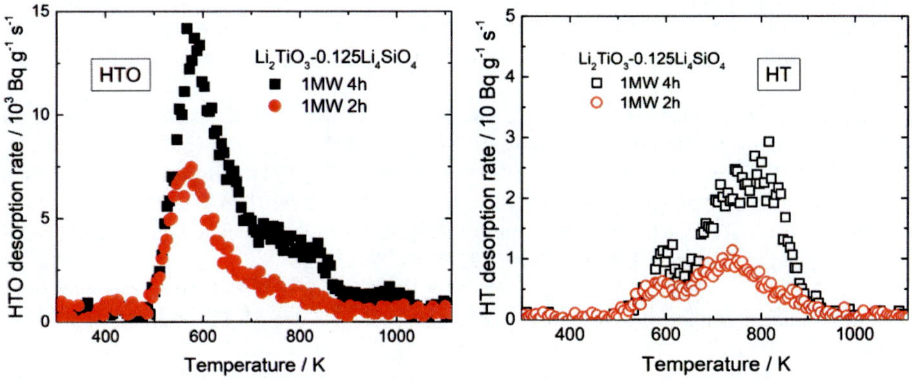

Fig. 6. Tritiated water (HTO) and tritium gas (HT) TDS spectra of Li_2TiO_3-$0.125Li_4SiO_4$ biphasic ceramic irradiated with different neutron fluence.

4 Summary

The tritium release experimental data of the Li_2TiO_3-Li_4SiO_4 biphasic ceramic with low Li_4SiO_4 content were obtained. There were two major desorption temperature regions for both samples. The desorption peak at around 610 K can be ascribed to the tritium diffusion process, and the peak at high temperature can be associated with the tritium de-trapping from irradiation defects. It was found that the Li_4SiO_4 content affects the distribution of tritium desorption peaks. There was another peak aside from the major desorption peaks in the case of higher Li_4SiO_4 content. The water vapor in Ar purge gas enhanced the tritium desorption in the near-surface region of the ceramic, while it had little effect on the de-trapping process. The neutron fluence affected the tritium desorption rate due to the change in the tritium generation amount, but it showed little effect on the desorption temperature of tritiated water. The desorption rate and temperature of HT gas are dependent on the neutron fluence. With increasing the neutron fluence, the HT desorption was shifted toward higher temperature side. In future work, the investigation on the tritium release kinetics and annihilation behavior of irradiation defects is scheduled to have an overall understanding of the tritium migration process in biphasic ceramics.

Acknowledgement. This work has been performed by using facilities of the Institute for Integrated Radiation and Nuclear Science, Kyoto University. This study is supported by the National Natural Science Foundation of China (No. 11905164) and the China Scholarship Council (award to Qilai Zhou for 1 year's study abroad at Shizuoka University as a Visiting Scholar, No. 202206955006).

References

1. Wang, Y., Zhou, Q., Xue, L., Li, H., Yan, Y.: Synthesis of the biphasic mixture of Li_2TiO_3-Li_4SiO_4 and its irradiation performance. J. Eur. Ceram. Soc. **36**(16), 4107–4113 (2016)
2. Zhou, J., Wang, Q., Li, S., Li, H., Zhou, Q.: Phase composition control of biphasic tritium breeding ceramics and correlation behavior of lithium and tritium. J. Nucl. Mater. **557**, 153294 (2021)
3. Hanaor, D.A.H., Kolb, M.H.H., Gan, Y., Kamlah, M., Knitter, R.: Solution based synthesis of mixed-phase materials in the Li_2TiO_3-Li_4SiO_4 system. J. Nucl. Mater. **456**, 151–161 (2015)
4. Qi, Q., et al.: Comparison of tritium release behavior in Li_2TiO_3 and promising core-shell Li_2TiO_3-Li_4SiO_4 biphasic ceramic pebbles. J. Nucl. Mater. **539**, 152330 (2020)
5. Zhou, Q., et al.: Release kinetics of tritium generation in neutron irradiated biphasic Li_2TiO_3-Li_4SiO_4 ceramic breeder. J. Nucl. Mater. **522**, 286–293 (2019)
6. Zhou, Q., Mou, Y., Ma, X., Xue, L., Li, H., Yan, Y.: Effect of fuel-to-oxidizer ratios on combustion mode and microstructure of Li_2TiO_3 nanoscale powders. J. Eur. Ceram. Soc. **34**(3), 801–807 (2014)
7. Zhou, Q., et al.: Solution combustion synthesis of lithium orthosilicate as the tritium breeder: effects of microwave power and fuel-to-oxidizer ratio on phase, microstructure and sintering. Ceram. Inter. **47**(15), 22006–22015 (2021)
8. Zhou, Q., et al.: Tritium and deuterium release behavior of Li_2TiO_3-$0.5Li_4SiO_4$-Pb ceramic. Ceram. Inter. **49**(16), 26778–26785 (2023)
9. Kobayashi, M., Toda, K., Oya, Y.: Dependency of irradiation damage density on tritium migration behaviors in Li_2TiO_3. J. Nucl. Mater. **447**(1–3), 1–8 (2014)

Synthesis and Characterization of Functionalized Zinc Phthalocyanine-Natural Polymer Systems

Tamara Potlog[1(✉)], Victor Suman[1,2], Ana Popusoi[1], Ion Lungu[1,2], and Hidenori Mimura[3]

[1] Laboratory of Organic/Inorganic Materials for Optoelectronics, Moldova State University, Chișinău, Republic of Moldova
tpotlog@gmail.com

[2] Doctoral School of Natural Sciences, Moldova State University, Chișinău, Republic of Moldova

[3] Research Institute of Electronics, Shizuoka University Hamamatsu, Hamamatsu, Japan

Abstract. New photosensitizers based on tetracarboxy- and octacarboxy- ZnPc derivatives conjugated to natural polymers for photodynamic therapy (PDT) are proposed. The tetracarboxy- and octacarboxy- ZnPc powders prepared by the direct condensation and hydrolysis were conjugated to chitosan (Ch). The $ZnPc(COOH)_4$ and $ZnPc(COOH)_8$ powders were solubilized in DMSO:H_2O mixture solvent, ratio 1:1 and then self-assembled with degraded chitosan solution. The resulting $ZnPc(COOH)_4$/Ch and $ZnPc(COOH)_8$/Ch solutions were used for the investigation of the some photophysics properties. The UV-Vis spectra of both solutions show Q broad bands red shifted in comparison with unsubstituted of ZnPc in DMSO solution. At the excitation of the $ZnPc(COOH)_4$/Ch with 643…702 nm a single band structure centered at 710 nm was revealed. For the ($ZnPc(COOH)_8$/Ch, at the excitation with 363 nm and 707 nm the band centered only at 480 nm was observed. The fluorescence decay of ($ZnPc(COOH)_8$/Ch shows 2.04 ns and 9.54 ns lifetimes of excited states, while for $ZnPc(COOH)_4$/Ch indicates 2.77 ns.

Keywords: tetracarboxy · octacarboxy · ZnPc · chitosan · photophysics properties

1 Introduction

The combination of valuable photophysical characteristics and high (thermo) chemical stability allows us to obtain various optoelectronic devices based on phthalocyanines such as: charge memory and the active layer of CD/DVD disks [1], gas sensors [2–4], light-emitting devices [5], optical limiters and photorefractive materials [6, 7]. A number of phthalocyanine derivatives are used catalysts for the purification of hydrocarbons from sulfur compounds and in the neutralization of toxic wastes [8, 9], and as sensitizers in preparations for photodynamic therapy (PDT) of oncological diseases [10–12]. Metal

Y. Ono and J. Kondoh (Eds.): Inter-Academia 2023, LNNS 939, pp. 29–35, 2024.
https://doi.org/10.1007/978-3-031-54450-7_4

Phthalocyanines (MPcs) and their derivatives have been considered excellent photosensitizers (PSs) (second generation) for PDT in numerous types of tumors. This type of molecule strongly absorbs in the red and near infrared regions of the visible spectrum, which corresponds to the tissue optical window. In addition, Pcs present high photo and chemical stability. Zinc phthalocyanine (ZnPc) is an attractive material to use in pharmaceutical medicine due to the photoactivity that exhibits under irradiation in the near infrared (NIR) region. The potential application of ZnPc lies in the inherent properties such as fluorescence, for optical imaging, and singlet oxygen generation, for their use in a broad variety of technologies such as PDT. ZnPc is insoluble in water. Chemical substitution at the periphery or axial positions of the macrocycle allows fine-tuning their solubility, aggregation, and optoelectronic properties. The main anionic functional group comprises carboxylate functionalities, directly attached to the macrocycle or linked through various spacers. Solubility of Pcs substituted by hydrophilic groups depends on the number of such groups in a macroheterocycle molecule. The hydrophilicity provided by these groups is strongly pH-dependent. Recently, the phthalocyanine-polymer composites have been attracted much attention, either main-chain or side-chain type offer a unique combination of good properties of polymers and phthalocyanines [13]. Therefore, improving the effectiveness and tolerability of the applied PS is a constant need. In the current study, we designed $ZnPc(COOH)_4$ -chitosan, $ZnPc(COOH)_8$-chitosan systems for the first time and subsequently studied its photophysical properties in $DMSO/H_2O$ solvent. In addition, the fluorescence quantum yield and lifetime of systems are presented for the comparison.

2 Materials and Methods

1, 2, 4-Benzene tricarboxylic anhydride (trimellitic anhydride) and pyromellitic dianhydride or (1, 2, 4, 5-tetracyanobenzene) and other chemicals were purchased from Aldrich and used as received for the synthesis. Mechanical and magnetic stirring methods are used for reaction processes. Tetracarboxy- and octacarboxy- zinc phthalocyanine were prepared according to the well reported procedure with slight modification.

FTIR spectra of the compounds were recorded in KBr pellets on Bruker ALPHA Platinum-ATR spectrometer in the wavelength range of 4000–400 cm^{-1}. Electronic spectra of the complexes are recorded in $DMSO/H_2O$ solution by using UV-Vis Perkin Elmer LS 55 spectrophotometer, quartz cells, 1 cm thick.

3 Synthesis of $ZnPc(COOH)_4$ and $ZnPc(COOH)_8$

The carboxylic groups in ZnPc molecules make the compound well soluble in water [14]. The zinc tetracarboxy-phthalocyanine ($ZnPc(COOH)_4$) was prepared by the direct condensation of trimellitic acid anhydride, urea and zinc chloride as the nucleation metal. Nitrobenzene was used as solvent. A minimum amount of DBU was used as the catalyst. The mass of Zn phthalocyanine containing COOH groups is 0.62 g (79%) (Fig. 1).

The zinc octacarboxy- phthalocyanine ($ZnPc(COOH)_8$) is synthesized by condensation of pyromellitic dianhydride or 1,2,4,5-tetracyanobenzene in presence of urea, the

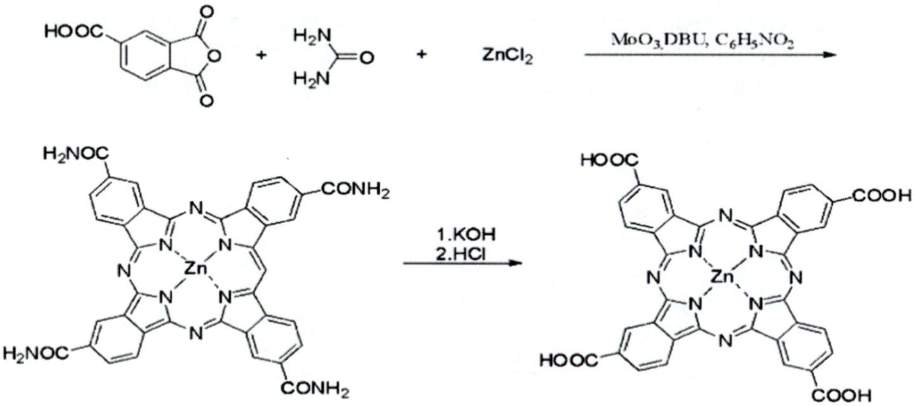

Fig. 1. Synthesis of ZnPc(COOH)$_4$.

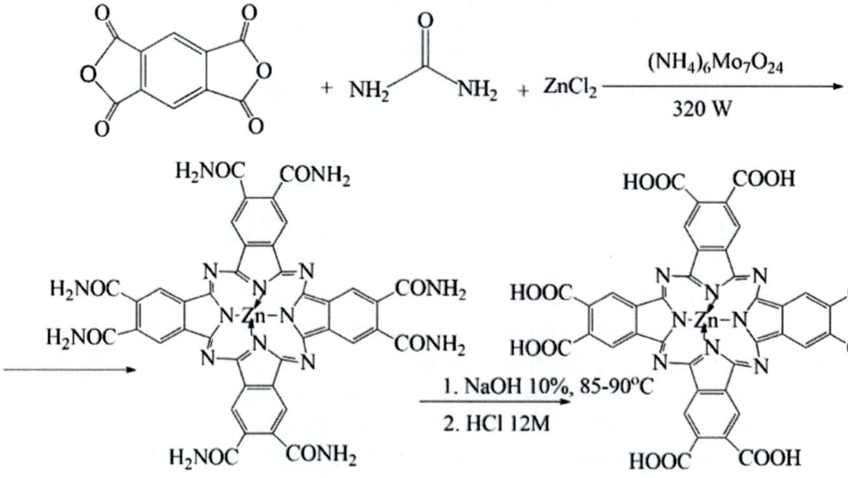

Fig. 2. Synthesis of ZnPc(COOH)$_8$.

corresponding metal salt, and DBU or aluminum molybdate, respectively. The resulting tetraimide phthalocyanine is subsequently hydrolyzed in acid or basic media (Fig. 2).

FTIR spectra of the ZnPc derivatives measured in the region 400 cm^{-1} to 4000 cm^{-1} are presented in the Fig. 3. The presence of peaks at 1615 – 1686 cm^{-1} and 1516 cm^{-1} are assigned to aromatic – C=N– and –C=C– in plane skeletal vibrations of the phthalocyanine core respectively. The FTIR spectrum of tetracarboxy-ZnPc synthesized from trimellitic acid anhydride exhibited a band at 1623 cm^{-1}, which corresponds to a characteristic C=O vibration. In addition, FTIR vibrations, ν, cm^{-1}: 3397, 1702, 1578, 1524, 1420, 1343, 1282, 1250, 1049, 928, 861, 832, 766, 740, 672 were observed. The FTIR spectrum of octacarboxy-ZnPc gave a characteristic imide group pattern in the region from 1600 – 1800 cm^{-1}. Also, are present C-H vibrations at 3300 cm^{-1}, 2900 cm^{-1}, C–O vibrations at 1750 cm^{-1} and 1720 cm^{-1}, and C–C vibration at 1480 cm^{-1}. The

FTIR spectrum of synthesized from pyromellitic dianhydride gave a characteristic pattern of imide groups in the region of 1600–1800 cm^{-1}. In the FTIR spectra of the both Zn subsituted phthalocyanine derivatives, very similar spectra were obtained except for small shifts.

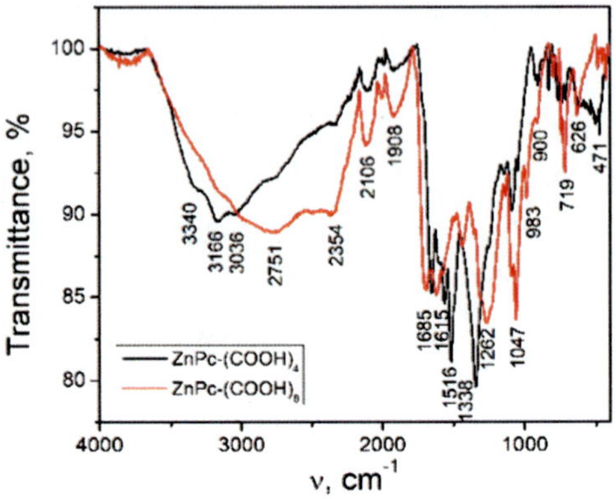

Fig. 3. FTIR Spectra of $ZnPc(COOH)_4$ and $ZnPc(COOH)_8$ derivatives.

4 Synthesis of Zinc ZnPc Derivatives/chitosan Systems

The above-mentioned ZnPc derivatives are able to immobilize on chitosan, forming systems with new useful properties. Chitosan is a biodegradable, biocompatible polysaccharide and, in comparison with many other polymers, has many free –OH and –NH^2 groups that can serve as anchors for conjugation of therapeutics and targeting ligands. Conjugation were formed between $ZnPc(COOH)_4$, $ZnPc(COOH)_8$ and chitosan (Ch). Chitosan was degraded in aqueous solution formed from 1% concentrated acetic acid and 3% H_2O_2 mixture. The FTIR spectrum of the chitosan showed vibrational bands at 3340 cm^{-1} (OH/NH), 2871 cm^{-1} (C–H), 1644 cm^{-1} (C–C), 1568 cm^{-1} (NH_2), 1373 cm^{-1} (C–N) and 1023 cm^{-1} (C–O) and $ZnPc(COOH)_4$–Ch shows bands at 3251 cm^{-1} (OH/NH), 2917 cm^{-1} (C–H), 1639 cm^{-1} (C–C), 1550 cm^{-1} (O = C–NH), 1370 cm^{-1} (C–N) and 1022 cm^{-1} (C–O). The observed shifts in the FTIR bands of both systems confirm the binding of chitosan to $ZnPc(COOH)_4$ and $ZnPc(COOH)_8$. The Fig. 4 compare the FTIR shifts in both chitosan systems based on ZnPc substituted with different COOH group.

5 Photophysical Properties

The UV-vis spectroscopy provides more information than other spectroscopic techniques, because phthalocyanines are rich in π-electron. The bands exzibited by ZnPc are the result of π–π* and deeper π–π* transitions [15]. These are Q-band occurring between 800 and 500 nm and B or Soret band seen between 450 and 285 nm and in some molecules, charge transfer transitions caused by metal–ligand or ligand–metal

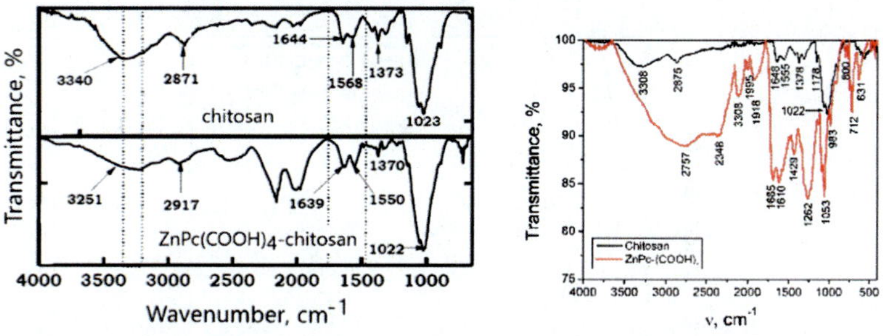

Fig. 4. FTIR Spectra of ZnPc(COOH)$_4$/Ch (left) and ZnPc(COOH)$_8$/Ch (right) systems.

transitions can also be observed. The absorption spectra were recorded in DMSO/H_2O at different concentrations 1.08 mol/m^3… 0.042 mol/m^3 (Fig. 5). The absorption maxima recorded in the Q band are due to the π–π* transitions that can occur within the molecule. The characteristic Q-bands of these ZnPc(COOH)$_4$/Ch system appeared at 640 nm and 695 nm, respectively and depend on the concentration of ZnPc(COOH)$_4$ in the system.

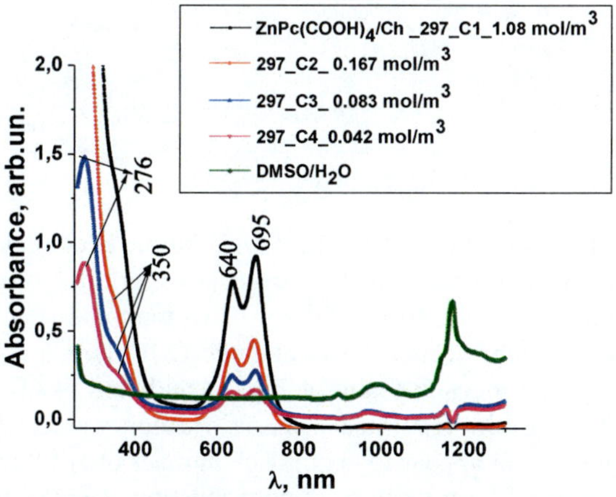

Fig. 5. Electronic spectra of ZnPc(COOH)$_4$/Ch system at different concentration values.

For the comparison the UV-Vis spectra of both ZnPc(COOH)$_4$/Ch and ZnPc(COOH)$_8$/Ch systems presented in Fig. 6. The UV-Vis spectra of both ZnPc(COOH)$_4$/Ch and ZnPc(COOH)$_8$/Ch systems show Q broad band red shifted in comparison with unsubstituted of ZnPc in DMSO solution [16]. The UV-Vis spectra of ZnPc(COOH)$_8$/Ch show more aggregation compared to ZnPc(COOH)$_4$/Ch (Fig. 6). The molar absorption coefficient value matches the values of 5.47 10^4 $M^{-1}cm^{-1}$ at maximum intensity 702 nm for ZnPc(COOH)$_4$/Ch and 8.52 10^4 $M^{-1}cm^{-1}$ for ZnPc(COOH)$_8$/Ch at 707 nm.

The fluorescence emission parameters for both systems are shown in Table 1. At the excitation of the ZnPc(COOH)$_4$/Ch with 643 nm a single band structure centered

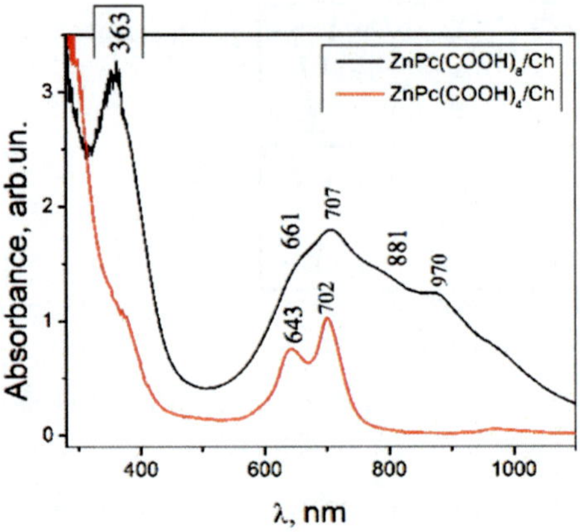

Fig. 6. Electronic spectra of ZnPc(COOH)$_4$/Ch and ZnPc(COOH)$_8$/Ch systems.

Table 1. Fluorescence quantum yields and lifetimes of the carboxy substituted ZnPc/chitosan systems.

Photosensitezers	λ_{abs}, nm	λ_{emis}, nm	Φ_F, %	τ_{1F}, ns	τ_{2F}, ns
ZnPc(COOH)$_4$/Ch	643 702	710 –	9.31	2.77	–
ZnPc(COOH)$_8$/Ch	363	480	0.17	2.04	9.54

at 710 nm was revealed. For the ZnPc(COOH)$_8$/Ch, at the excitation with 363 nm, the band centered at 480 nm was observed. The steady state and time-resolved fluorescence emission studies, indicate that the ZnPc(COOH)$_4$/Ch is monomeric with the fluorescence lifetime of 2.77 ns. The fluorescence decay of ZnPc(COOH)$_8$/Ch is best fitted by two-exponential curves with a times constant of 2.04 ns and 9.54 ns (Table 1). Our results indicate that ZnPc(COOH)$_4$ incorporated in chitosan is monomeric, while for conjugated ZnPc(COOH)$_8$, aggregation appear to occur. The number of substituent groups has an effect on the fluorescence lifetime values of both substituted ZnPc/chitosan systems.

6 Conclusion

The red-shift of the absorbance in substituted ZnPc(COOH)$_4$/Ch and ZnPc(COOH)$_8$/Ch systems makes them attractive as potential photosensitizers in PDT as the Q-band lies firmly in the range of the optical window of 650–800 nm, providing deeper penetration of light into biological tissues. The ZnPc(COOH)$_4$/Ch and ZnPc(COOH)$_8$/Ch solutions exhibit fluorescence at 710 nm and 480 nm, respectively. In the case of total fluorescence decay curves, the fitting was performed on the basis of the multiexponential decay law. The time-resolved fluorescence decay of ZnPc(COOH)$_4$/Ch is monoexponential, while for ZnPc(COOH)$_8$/Ch system is biexponential. This fact indicates that interactions occur between the carboxy substituted ZnPc and chitosan polymer. The presence of natural polymer in systems leads to an elongation of the excited state.

References

1. Belogorokhov, I., et al.: Photoluminescence semiconductor structures based on butyl-substituted phthalocyanines erbium FTP. Fizika Tekh. Poluprovodnikov **42**(3), 327–330 (2008)
2. Xie, D., Jiang, Y., Pan, W., Jiang, J., Wu, Z., Li, Y.: The characteristics and gas-sensing property of bys[phthalocyaninato] rare earth complexes-based charge flow transistor. Sens. Actuat. B **81**, 210–217 (2002)
3. Xie, D., Pan, W., Jiang, Y.D., Li, Y.: Erbium bis[phthalocyaninato] complex LB film gas sensor. Mater. Lett. **57**, 2395–2398 (2003)
4. Lam, M., et al.: Heterojunction OLEDs fabricated by Eu ternary complexes with conducting secondary ligands. Opt. Mater. **28**, 709–713 (2006)
5. Krasnovsky, A., Rogers, M., Galpern, M., Kinney, M., Lukyanets, E.: Quenching of luminescence of singlet molecular oxygen by phthalocyanines and naphthalocyanines. Bioorg. Chem. **16**(10), 1413–1418 (1990)
6. Dini, D., Hanack, M.: Physical Properties of Phthalocyanine-Based Materials. The Porphyrin Handbook, 17th edn. Academic Press, San Diego (2003)
7. Dini, D., Calvete, M., Hanack, M.: Nonlinear optical materials for the smart filtering of optical radiation. Chem. Rev. **116**(22), 13043–13233 (2016)
8. Liu, L., Hu, A.: Synthesis of soluble functional dye phthalocyanines and perylene tetracarboxylic derivatives by microwave irradiation and their photoelectric performances. J. Porphyrins Phthalocyanines **7**(08), 565–571 (2003)
9. Beletskaya, I., Tyurin, V., Tsivadze, A., Guilard, R., Stern, C.: Supramolecular chemistry of metalloporphyrins. Chem. Rev. **109**(5), 1659–1713 (2009)
10. Mroz, P., Tegos, G., Gali, H.: Photodynamic therapy with fullerenes. Photochem. Photobiol. Sci. **6**(11), 1139–1149 (2007)
11. Zhang, X., Xi, Q., Zhao, J.: Fluorescent and triplet state photoactive J-type phthalocyanine nano assemblies: controlled formation and photosensitizing properties. J. Mater. Chem. **20**(32), 6726–6733 (2010)
12. Bazyakina, N.: Metallocenes in the process of obtaining phthalocyanine catalytic systems. Abstract of thesis. Chem. Sci. Nizhny Novgorod (2006)
13. Akhmadullina, A., Akhmadullin, R.: On new developments and implementations in the field of desulfurization of hydrocarbon raw materials. Chem. Technol. Fuels Oils **44**(6), 371–378 (2008)
14. Borzęcka, W., Domiński, A., Kowalczuk, M.: Recent progress in phthalocyanine-polymeric nanoparticle delivery systems for cancer photodynamic therapy. Nanomaterials **11**(9), 2426 (2021)
15. Potlog, T., Popusoi, A., Lungu, I., Robu, S., Bulimestru, I.: Photophysics of tetracarboxy-zinc phthalocyanine photosensitizers. RSC Adv. **12**(49), 31778–31785 (2022)
16. Gurol, I., Durmus, M., Ahsena, V., Nyokong, T.: Synthesis, photophysical and photochemical properties of substituted zinc phthalocyanines. Dalton Trans. **34**, 3782–3791 (2007)

Study of the Phase Changes in ZnS-Based Phosphors Crystal Structure Impact on Their Radioluminescent Performances

Elena Vladimirovna Zelenina[1,2(✉)], Olga Olegovna Osmak[1], and Vadim Vladimirovich Bakhmetyev[1]

[1] Saint-Petersburg State Institute of Technology, St. Petersburg, Russia
elena.v.zelenina@gmail.com, zelenina@khlopin.ru
[2] Khlopin Radium Institute, St. Petersburg, Russia

Abstract. To improve the radioluminescence performances of ZnS-based phosphors various thermal and non-thermal methods have been tested to control the phase composition and structure. Particle surface and size and phase composition of phosphors are significantly affected either by treatment of the initial batches with high-energy electron flow and nitrogen plasma or by changes in atmosphere of synthesis. The correlations between phosphors phase content and their spectral radioluminescent characteristics have been established in ZnS:Cu, Br phosphors. The electron-beam treatment (EBT) of the initial charge and the ready phosphor forms the mixed sphalerite-wurtzite phosphor structure and increases the brightness of radioluminescence by 80%. Designed laboratory prototypes of solid-state radioluminescent light sources (SRLS) turned to be as effective as industrial gas-filled RLS, but considerably more compact.

Keywords: Radioluminescence · ZnS modification · Phosphor structure · Electron-beam treatment · Spectral content · Brightness · Sphalerite · Wurtzite

1 Introduction

In our work we're looking for improving the radioluminescence performances of ZnS-based phosphors to develop more safe and efficient radioluminescent light sources. For zinc sulfide radioluminescent phosphors the performance characteristics will be determined not only by the physicochemical but also by the structural properties of the phosphor. Thus, it is important to create luminescent materials with optimal structure and optical properties. To increase the brightness of radioluminescent phosphors, various methods of modifying the crystal structure of the phosphor by changing the synthesis conditions and the concentration of the activator can be effective. The authors of [1] developed a method for increasing the brightness of zinc sulfide electroluminescent phosphors by performing solid-phase synthesis in a quasi-closed space, creating increased pressure and more reducing medium in the reaction volume and prevent the reactant gases removal. In our previous works [2, 3] it was reported that electron-beam

Y. Ono and J. Kondoh (Eds.): Inter-Academia 2023, LNNS 939, pp. 36–41, 2024.
https://doi.org/10.1007/978-3-031-54450-7_5

and plasmochemical treatment allows to increase the effectiveness of ZnS electroluminescent phosphors. So, these thermal and non-thermal methods were also considered for the radioluminescent phosphors.

2 Materials and Methods

Several series of radioluminescent phosphors ZnS:Cu (0.005–0.1%wt), Br (0.8%wt) were prepared by solid-phase method. Powders were synthesized from the charge by annealing of the initial blend for 1.5 h in an alumina crucible at 950 °C under a layer of active carbon as a reducing medium. The quasi-closed space was realized when a crucible with a charge for synthesis is placed inside another crucible filled with active carbon. Electron beam processing of the charge mixtures and ready phosphors was carried out using a resonance-transforming electron accelerator RTE-1V (Efremov Institute of Electrophysical Apparatus, Saint-Petersburg, Russia) under the following conditions: electron energy 900 keV, absorbed dose about 600 kGy. Plasma modification of phosphors was carried out in nitrogen plasma using the vacuum installation Bulat NNV-6.6-I1 with arc discharge: 1500 V; 20 s. Radioluminescence of the synthesized phosphors were excited by the tritium β-source (Khlopin Radium Institute, Saint-Petersburg, Russia) comprising a tritium-containing titanium layer with incorporated activity <0.4 Ci. The radioluminescence spectra were measured using a spectrofluorimeter AvaSpec-3648. Obtained spectra were analyzed by Gaussian deconvolution. The radioluminescence brightness was measured using a radiometer IL-1700. XRD phase analysis was performed using a Rigaku SmartLab 3 X-ray diffractometer. A quantitative analysis of the phase composition was carried out using the Rietveld method.

3 Results and Discussion

3.1 Radiophosphor Treatment

The initial charges were treated before synthesis with high-energy electron flow and nitrogen plasma. The initial charges of phosphors synthesized in a quasi-closed space weren't treated before annealing.

Microphotographs (Fig. 1) of the synthesized phosphors show that all phosphors have good crystalline shape without distinct tendency to aggregation, except only the pressure-treated sample (Fig. 1b). Plasma and electron-beam treatment visually reduces the grain size of the phosphors from 5–7 to 1–3 μm, but the pressure treatment increases it up to 10 μm.

Increasing of dispersity of the phosphors may reveal more luminescence centers from the grain volume to the surface and may help the low-energetic tritium β-particles to reach the larger surface area of the phosphor grains.

Figure 2 represents the radioluminescence spectra of the brightest phosphors in the series synthesized without treatment (Cu 0.03%wt) and under pressure (Cu 0.01%wt), plasma (Cu 0.01%wt) and electron-beam (Cu 0.03%wt) treatment. Radioluminescence spectra on the Fig. 2 shows the prevailing blue spectral band on 430 nm for the non-treated phosphors. This band corresponds to the Zinc vacancies and $Cu_{(Zn)}$-$Cu_{(i)}$ associate luminescence centers [4]. All ways of treatment reduces the blue band intensity

Fig. 1. Microphotographs of radioluminescent phosphors synthesized: a – without treatment; b – with pressure treatment; c – with plasma treatment; d – with EB treatment.

and likely helps Cu to replace Zinc vacancies. Green spectral band corresponds to the $Cu_{(Zn)}$-$Br_{(S)}$ associate luminescence centers and activated Sulfur vacancies. Obviously, treatment restructurates the crystal lattice and helps the activator (Cu) and co-activator (Br) incorporation into the ZnS matrix.

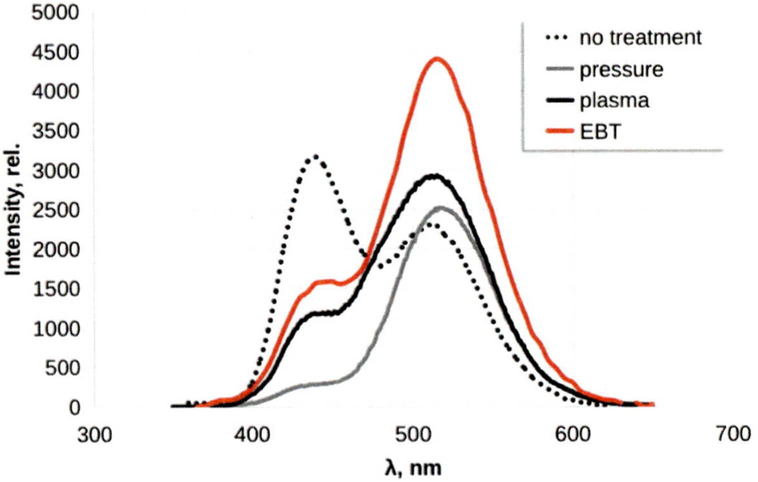

Fig. 2. Radioluminescence emission spectra of ZnS:Cu, Br phosphors synthesized under various modifying conditions (without treatment, under pressure treatment, under plasma treatment and under electron-beam treatment (EBT)).

According to XRD analysis, all synthesized phosphors has two-phase sphalerite-wurtzite structure. Earlier [5] it was found that radioluminescence brightness increases by 80% after EBT and correlates with the phase content. The analysis of phase and spectral content of radioluminescence for processed and non-processed ZnS:Cu,Br radioluminescent phosphors helped us to reveal the correlation between phosphors phase content and their spectral characteristics of tritium radioluminescence (Fig. 3). Besides the EBT, plasmochemical treatment and Increasing pressure during synthesis promotes the same correlation. Increasing of the wurtzite content in sphalerite crystal matrix of ZnS raises the intensity of radioluminescence. Brightness of radioluminescence (Fig. 4) demonstrates good agreement with the «green» 524 nm intensity data and confirms the EBT optimity for the radioluminescence performance enhancement.

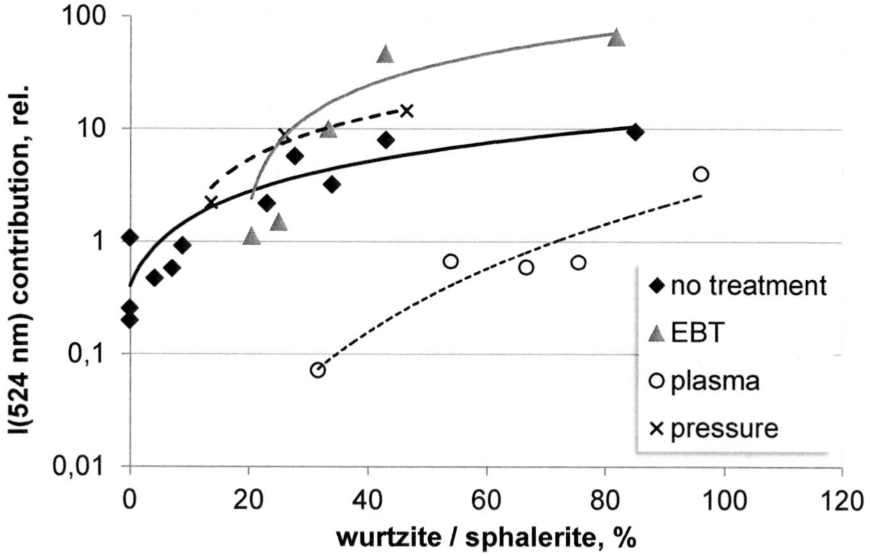

Fig. 3. Correlation of wurtzite-to-sphalerite phase ratio and relative intensity of the «green» (524 nm) band in radioluminescence emission spectra for the phosphors synthesized under various modifying conditions (without treatment, under pressure treatment, under plasma treatment and under electron-beam treatment (EBT)).

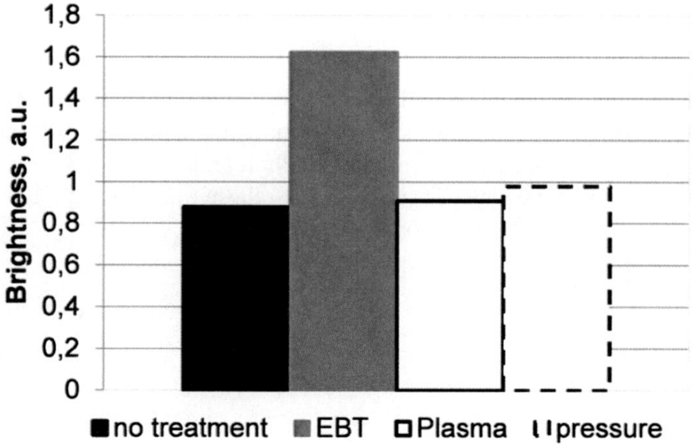

Fig. 4. Brightness of radioluminescence for the phosphors synthesized under various modifying conditions (without treatment, under pressure treatment, under plasma treatment and under electron-beam treatment (EBT)).

3.2 SRLS Manufacturing

Technique of coating the phosphor screens was described in our previous studies [6]. Using this technique we have designed the SRLS prototypes as a tritiated 0.5 μm titanium layer on the steel plate, saturated with tritium under vacuum and covered with a phosphor screen of optimal thickness. SRLS prototype design scheme is given in Fig. 5. Phosphor screen of optimal thickness (3.5–4.5 mg/cm^2) was deposited directly on the solid-state tritium source with area 1 cm^2 and activity of 0.4 Ci. Table 1 gives the comparison of our SRLS prototypes characteristics and commercial gas-filled RLS produced by PO "Mayak" [7].

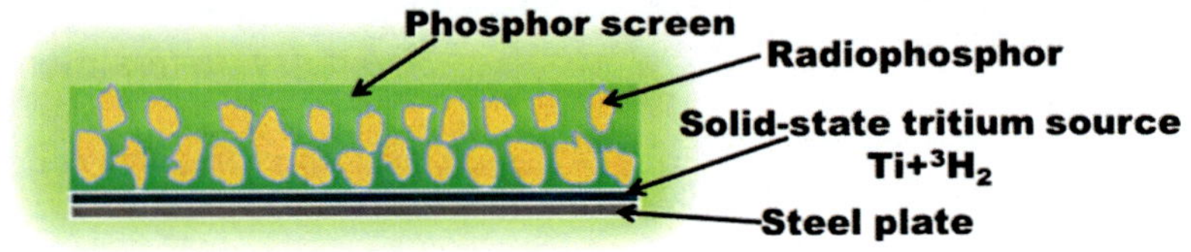

Fig. 5. SRLS prototype design scheme.

Table 1. Characteristics of SRLS prototypes and commercial radioluminescent devices

RLS	Enabled activity Amax, Ci	Source volume, cm^3	Brightness, Cd/cm^2	Effectiveness, Lm/Wt
Gas-filled RLS tube PO "Mayak"	2.2–19.8	10.9–43.7	0.2	10–40
SRLS prototype	0.40	0.2	0.15	37

It is obvious that our SRLS have less incorporated activity, greater reliability and safety, smaller dimensions. Gaseous radioactive phase is absent. The efficiency of SRLS is close to efficiency of industrial gas-filled RLS. Photographs of SRLS prototypes given in Fig. 6 helps to realize the way of using SRLS as alarm marks or light boundary lines by simply being sticked to a wall, floor or other surface.

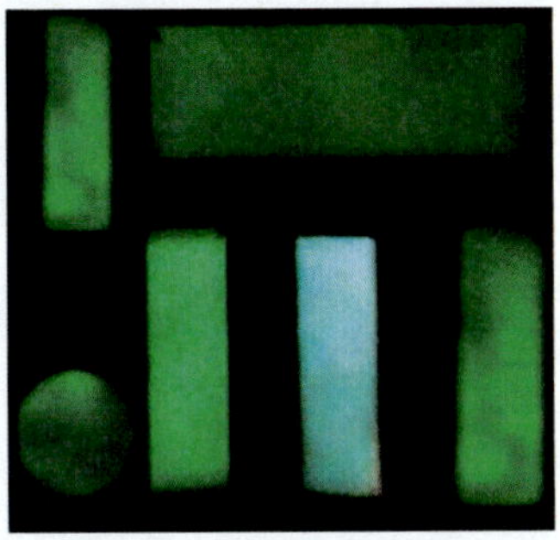

Fig. 6. Photograph of SRLS prototypes.

4 Conclusion

ZnS matrix non-thermal treatment before synthesis changes the dispersity and phase content of the phosphors, synthesized below the phase transition temperature. Formation of mixed sphalerite-wurtzite crystal structure of zinc sulfide increases the intensity of the "green" band in the spectra and affects the brightness improvement due to the activator redistribution in the ZnS matrix. Electron-beam treatment shows the best results in enhancement of spectral and brightness radioluminescent performances. The improved radioluminescent phosphors allows to create safe and compact SRLS with good efficiency.

References

1. Tishchenko, S.M., Ishchenko, D.A., Koval'kov, V.I., Golota, A.F., Ishchenko, V.M.: Influence of preparation conditions on the processes of heterophase structure formation in electrophosphors based on AIIBVI - $Cu_{2-X}S$. ChemChemTech **50**(7), 33–36 (2007)
2. Sychov, M., et al.: Modification of the EL properties of zinc sulfide powder phosphors by electron beam irradiation. In: Proceedings Conference of the Electronics Display Conference, Nagasaki, Japan, pp. 67–71 (2002)
3. Sychov, M., et al.: Plasmachemical modification effect on luminescence of AIIBVI phosphors. J. Lumin. **156**, 69–73 (2014)
4. Bakhmet'ev, V.V., et al.: Optimizing the electrooptic properties of phosphors for electroluminescent panels. J. Opt. Technol. **70**(7), 513–515 (2003)
5. Bakhmetyev, V.V., Zelenina, E.V., Shvindin, M.A. Synthesis of ZnS:Cu, Br radioluminescent phosphors using the electron-beam treatment and studying their characteristics. J. Phys.: Conf. Ser. **2056**(012047), 8 (2021)
6. Zelenina, E.V., Bakhmetyev, V.V., Pechertseva, E.A., Sychov, M.M.: A modified sol–gel technique of phosphor screens preparation for tritium solid-state radioluminescent light sources (SRLS). J. Sol-Gel Sci. Technol. **92**(2), 467–473 (2019)
7. Federal State Unitary Enterprise "Mayak Production Association", Manufacture of radioluminescent light elements. State Enterprise "Rosatom". https://www.po-mayak.ru/products_services/products/radioluminescent_light_elements/

Electronics and Device Physics

Analysis of the Resistance of Silicon Nanoscale Structures Highly Doped in Different Configurations

Shumpei Miura[1,2], Reon Asai[1,2], Yutaro Sasaki[1,2], and Daniel Moraru[1,2](✉)

[1] Research Institute of Electronics, Shizuoka University, Hamamatsu, Japan
moraru.daniel@shizuoka.ac.jp
[2] Graduate School of Integrated Science and Technology, Shizuoka University, Hamamatsu, Japan

Abstract. In this study, two types of silicon (Si) nanoscale-channel transistors were fabricated with high-concentration doping and their low-temperature current-voltage characteristics were evaluated in order to clarify the influence of channel design and doping conditions on the electrical conduction. The measured results of the resistance for nano-channels doped by phosphorus (P) donors were analyzed as a function of the actual channel width estimated from scanning electron microscope (SEM) images to reveal the impact of low dimensionality on the electrical properties of Si nanostructures. Although some weak signatures of Coulomb blockade transport are observed in a few devices doped with P-donors, these devices mostly behave as metallic resistors. On the other hand, devices with the nano-channels codoped with both P-donors and boron (B) acceptors exhibit Coulomb blockade features with high yield. These results can lead to further studies towards single-charge tunneling in silicon nano-transistors in which both donors and acceptors interplay in the formation of quantum dots (QDs).

Keywords: silicon-on-insulator (SOI) nano-transistors · low temperature characterization · coulomb blockade · codoping · silicon-on-insulator

1 Introduction

In recent years, as the Moore's Law is approaching its limits [1], single-electron tunneling (SET) using dopants as quantum dots (QDs) has attracted attention because this mechanism allows the development of atomic- and molecular-level electronics, i.e., *dopant-based electronics* [2, 3], while devices are still based on CMOS (complementary metal-oxide-semiconductor) technology. In addition, silicon (Si) transistors have become increasingly miniaturized so that their electrical characteristics can be significantly affected by random dopant fluctuations [4]. For obtaining low-resistance electrodes, high doping concentration is also necessary, but this would make the channels as well doped with a large number of dopants. Therefore, it becomes important to explore the possibility of SET operation in such devices, in which QDs would be formed due to the random distribution of dopants in nanoscale, while doping is raised at levels well above the metal-insulator transition (MIT), which is on the order of 4×10^{18} cm^{-3}.

Y. Ono and J. Kondoh (Eds.): Inter-Academia 2023, LNNS 939, pp. 45–50, 2024.
https://doi.org/10.1007/978-3-031-54450-7_6

In previous works, we have reported the observation of SET operation, even at room temperature, in a few nanoscale silicon-on-insulator field-effect transistors (SOI-FETs) in which the channel was doped at high concentration with phosphorus (P) donors [5, 6]. It was statistically demonstrated that high-concentration doping can allow the formation of QDs with deep energy levels due to the clustering of P-donors. However, such QDs can dominate electrical transport only when the dimensions of the channel are also reduced significantly, to avoid unwanted ohmic current paths. If such a QD controls the current in a nanoscale-channel SOI-FET, its larger barrier height can suppress thermally-activated transport, thus allowing the observation of SET mechanism, in some cases even at room temperature.

For extending this research and understanding the degree of controllability of such devices, we explore here two types of SOI transistors, doped at even higher doping concentrations: (i) transistors doped only with P-donors; (ii) transistors codoped with both P-donors and B-acceptors. The electrical resistance is measured mainly at low temperatures and plotted against the width of the channel, as evaluated from scanning electron microscope (SEM) measurements.

2 Experimental Details

In this study, as mentioned above, we fabricated two types of nanoscale-channel SOI transistors: one highly doped only with P-donors and another codoped with both P-donors and B-acceptors. Both doping concentrations are high and comparable: $N_D \approx 1.8 \times 10^{20}$ cm^{-3} and $N_A \approx 1.2 \times 10^{20}$ cm^{-3}, as estimated from four-point probe measurements on reference samples. The device structure of such SOI-FETs is shown schematically in Fig. 1, together with a basic electrical measurement setup. Figure 2 shows two examples of SEM images of the channel area for the two types of devices: (a) only-P-doped; (b) doped with P and B (codoped). The devices shown by the SEM images are designed with the same dimensions: channel length (L_{ch}) of 20 nm and channel width (W_{ch}) of 40 nm.

The devices were characterized by low-temperature ($T \approx 8.0$ K) current-voltage (I-V) measurements, mainly by monitoring drain current (I_D) as a function of drain voltage (V_D), in order to evaluate the electrical resistance of such channels. Typically, gate voltage (V_G) was not applied for such measurements (top gate was left floating).

The final channel width (different from the designed channel width) was estimated by SEM analysis at high resolution, with two examples shown in Fig. 2 for devices of the same design dimensions but different doping conditions: doped only with P-donors, in Fig. 2(a), and codoped with P-donors and B-acceptors, in Fig. 2(b). By plotting the measurement results for the low-temperature resistance against this actual channel width, the resistance can be correlated with the geometry of the nanoscale channels. From Fig. 2, it can be seen that the conditions of the two types of devices are similar, so that the effect of codoping can be, in principle, understood from the analysis of the electrical characteristics. Such analysis can reveal the importance of the doping configuration for controlling the transport in nanoscale SOI devices.

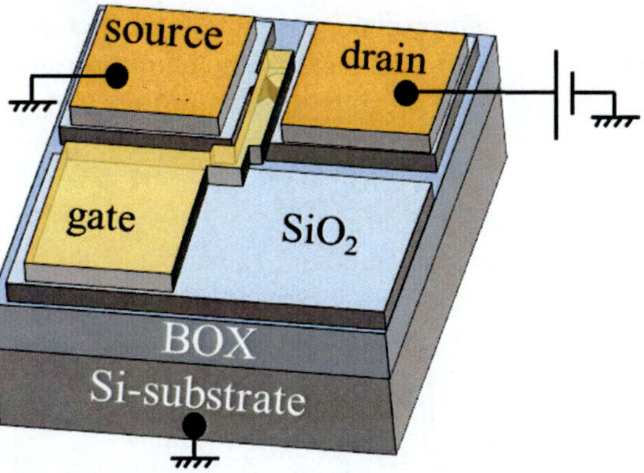

Fig. 1. Schematic device structure of an SOI-FET and its electrical measurement circuit. For electrical resistance measurements, source is grounded, while the top gate is left floating. Substrate Si is also grounded.

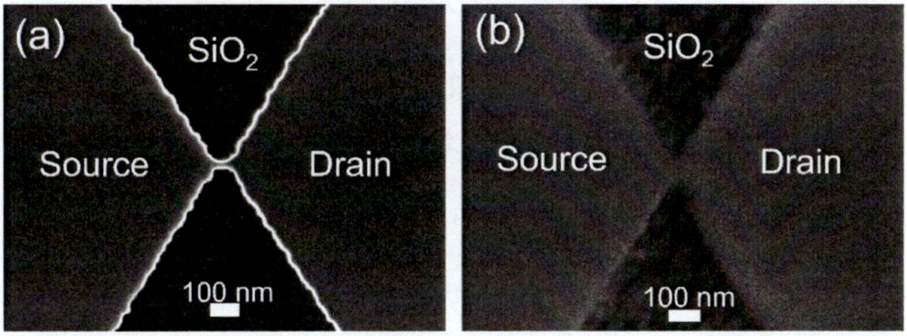

Fig. 2. Examples of SEM images of the channel area for devices with the same design parameters ($L_{ch} = 20$ nm, $W_{ch} = 40$ nm), but different doping configurations: (a) only-P-doped; (b) P and B doped (codoped).

3 Results and Discussion

The results of low-temperature current-voltage measurements (I_D-V_D characteristics) for 12 SOI-FETs doped only with P-donors are shown in Fig. 3(a). For this set, channel length is $L_{ch} = 20$ nm and channel width (W_{ch}) was varied, but similar sets were also taken for other channel lengths ($L_{ch} = 0$, 40, 1000 nm), raising the total number of measured devices of this type to 48. Since the curves are practically linear, it is understood that the nanoscale channels behave as metallic (ohmic) resistors. Thus, the electrical resistance (R) is easily estimated from the slope of the I_D-V_D characteristics. In order to have an accurate representation of the channel dimensionality, the resistance is plotted against the final (actual) channel width (assuming that channel length and thickness are the same for all devices designed in the same chip). As shown in Fig. 3(b), the electrical resistance R generally decreases as the channel width W_{ch} increases, and the devices generally exhibit clear ohmic properties. However, when $W_{ch} \leq 50$–60 nm, the current is significantly reduced, which corresponds to a noticeable increase of the resistance. It can be considered that this is due to the fact that the ohmic current path is restricted as W_{ch} becomes smaller. In addition, it can be seen from Fig. 3(b) (and from the inset at low W_{ch}) that the resistance is several times higher for long-channel transistors (L_{ch} = 1000 nm, as designed) than for short-channel transistors (L_{ch} = 0–40 nm, as designed). This confirms the controllability of the resistance by the geometric parameters.

Other previous reports in the literature illustrated the importance of low-dimensionality of nanowires doped with P-donors [7], with the doping done precisely using a scanning tunneling microscope (STM) atomic manipulation technique. A comparison with such works is beyond the scope of this paper and will be reported elsewhere. It should be noted, however, that the resistance increases significantly for the smallest nano-transistors.

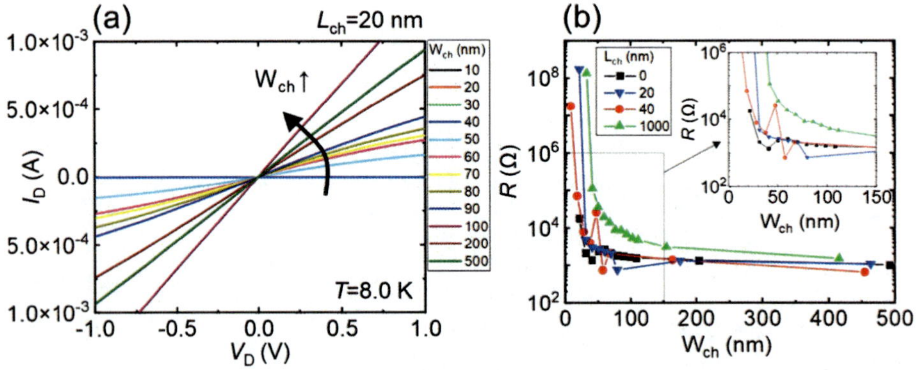

Fig. 3. (a) Low-temperature (T = 8.0 K) I_D-V_D characteristics for SOI-FETs doped only with P-donors at high concentration, as a function of W_{ch} for a fixed channel length, designed to be L_{ch} = 20 nm. (b) Analysis of electrical resistance R as a function of the final channel width (W_{ch}) estimated from SEM images. The inset shows the results of significant increase of resistance as width becomes lower than 50–60 nm.

For these devices, the I_D-V_D characteristics presented in Fig. 3(a) do not exhibit visible signatures of the Coulomb blockade effect, which would manifest as a low-current gap at low biases. However, in some devices of comparable parameters (albeit slightly lower doping concentration), it was shown that it is statistically possible that the smallest nano-channels (with all dimensions on the order of 10 nm or less) can contain one or just a few donor-induced QDs that dominate the electrical transport. This way, such QDs also possess sufficiently high tunnel barriers to preserve the SET operation even at practical temperatures, such as room temperature [6].

The results presented in this work indicate that such conditions are not easily achieved in highly-doped (only with P-donors) nanoscale transistors with even slightly larger dimensions, which generally exhibit ohmic conductance. One reason for such behavior may be the fact that dimensions are not sufficiently small to isolate the SET transport via one or a few QD(s) in this batch of devices.

In order to understand the possibility of using other approaches to further control of electrical properties of nanoscale highly-doped transistors, we also show here some results of the electrical characteristics of codoped nanoscale transistors, with comparable dimensionality. The analysis is limited to measuring the I_D-V_D characteristics for a set of 12 codoped SOI-FETs in a manner similar to the measurements for only-P-doped transistors. The results are shown in Fig. 4 as a function of the designed channel width (W_{ch}).

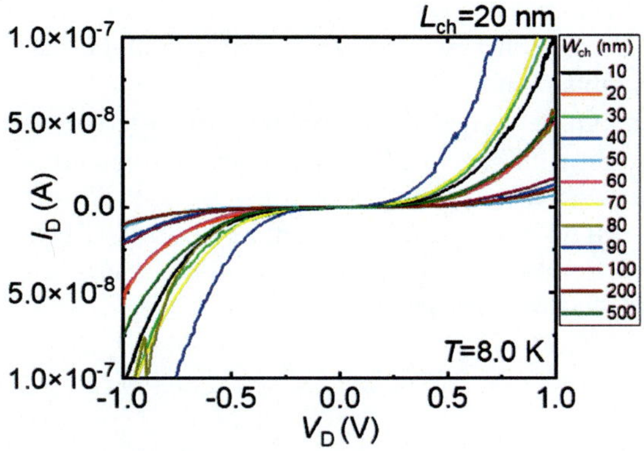

Fig. 4. Low-temperature ($T = 8.0$ K) I_D-V_D characteristics for SOI-FETs doped with both P-donors and B-acceptors (codoped) at high concentration. The characteristics are shown as a function of W_{ch} for a fixed channel length, designed to be $L_{ch} = 20$ nm.

The low-temperature ($T = 8.0$ K) I_D-V_D characteristics for these codoped SOI-FETs exhibit a clearly different behavior as compared to the SOI-FETs doped only with P-donors. The characteristics are not linear, but exhibit a low-current gap at low biases. This suggests that there is a mechanism that limits current flow at such low biases. One possibility is that Coulomb blockade works effectively in such codoped (highly compensated) nanoscale channels. Due to the fact that doping concentrations for P-donors and B-acceptors are comparable, the effective concentration ($|N_D$-$N_A|$) is actually significantly reduced if we assume that P-donors and B-acceptors compensate each other if they are in close proximity.

Resistance (R) is not evaluated from these non-linear curves and remains to be analyzed after the transport mechanism is clarified. In a previous work [8], it was shown that I_D-V_G characteristics for a large number of codoped SOI-FETs (with dimensions comparable to the devices shown in Fig. 4) exhibit clear current peaks, while the stability diagrams exhibit so-called Coulomb diamonds. These are clear signatures of Coulomb blockade working for single-electron tunneling (SET) operation in such devices. However, further study, including a statistical analysis of experimental and simulation data, is necessary to fully clarify the origin and demonstrate the SET transport mechanism.

Finally, it should be noted that, for each particular set of parameters (L_{ch}, W_{ch}, doping configuration), only one device was measured in this work. Although the reproducibility of the data for the same device was confirmed by multiple consecutive measurements, the repeatability on other devices with the same fabrication parameters has not been fully studied. Relatively large dispersion is expected due to the intrinsic randomness of the dopant distributions.

4 Conclusions

From the analysis of the current-voltage characteristics at low temperature for nanoscale SOI-FETs, differences are identified between devices doped only with P-donors and devices that are codoped (doped with both P-donors and B-acceptors). Codoping with comparable concentrations leads to a strong compensation effect, which results in a transition to a non-metallic (non-ohmic) behavior in transport. It is likely that QDs are formed due to the interaction between donors and acceptors, but further studies are needed for clarifying this point.

This study opens up new possibilities for investigating the effect of codoping in silicon nano-transistors, with multiple functionalities for dopant-based optoelectronics.

Acknowledgments. This work was partially supported by JSPS KAKENHI (no. 19K04529 and no. 22K04216). The authors thank K. Yamaguchi, T. Kaneko, C. Pandy, and T. T. Jupalli for their contributions to device fabrication and preliminary analysis.

References

1. Waldrop, M.M.: The chips are down for Moore's law. Nat. News **530**, 144–147 (2016)
2. Tabe, M., et al.: Single-electron transport through single dopants in a dopant-rich environment. Phys. Rev. Lett. **105**, 016803 (2010)
3. Moraru, D., et al.: Atom devices based on single dopants in silicon nanostructures. Nanoscale Res. Lett. **6**, 479–1–9 (2011)
4. Shinada, T., Okamoto, S., Kobayashi, T., Ohdomari, I.: Enhancing semiconductor device performance using ordered dopant arrays. Nature **437**, 1128–1131 (2005)
5. Samanta, A., et al.: Single-electron quantization at room temperature in a-few-donor quantum dot in silicon nano-transistors. Appl. Phys. Lett. **110**, 093107 (2017)
6. Jupalli, T.T., et al.: Room-temperature single-electron tunneling in highly-doped silicon-on-insulator nanoscale field-effect transistors. Appl. Phys. Express **15**, 065003 (2022)
7. Weber, B., et al.: Ohm's law survives to the atomic scale. Science **335**, 64–67 (2012). and references therein
8. Moraru, D., et al.: Single-charge tunneling in codoped silicon nanodevices. Nanomaterials **13**, 1911 (2023)

Study of *p-n* Junctions Created by Laser Implantation of In in Semi-Insulating *p*-CdTe:Cl

Dmytro Gnatyuk[1,2](✉), Serhii Levytskyi[1], and Alexander Stronski[1]

[1] V.E. Lashkaryov Institute of Semiconductor Physics, National Academy of Sciences of Ukraine, Nauky Ave, 41, Kyiv 03028, Ukraine

[2] Optical and Mechanical College, Taras Shevchenko National University of Kyiv, Volodymyrska Street, 60, Kyiv 01033, Ukraine

gnatyukdmytro@gmail.com

Abstract. The paper analyzes the measurements results of the electrical and photoelectric properties of the barrier structures created by laser implantation of In into the surface region of high-resistivity detector-grade *p*-CdTe:Cl crystals. The *I*-*V* characteristics and spectral photosensitivity measured in the photovoltaic mode of the formed structures are discussed. The conversion of the conductivity type of a thin surface layer in *p*-CdTe:Cl and formation of a shallow steep *p-n* junction at the *p*-CdTe:Cl/*n*-CdTe:In interface was approved. Etching of the CdTe:In surface region led to the disappearance of the photovoltaic effect in the barrier structures. The fabricated *p-n* junction diodes with high photosensitivity and low reverse dark current are promising as X/γ-ray sensors.

Keywords: CdTe semiconductor · (photo)electrical properties · spectral photosensitivity · high energy radiation detectors

1 Introduction

Semiconductors of the $A^{II}B^{VI}$ group, such as Cd(Zn)Te, are among the most common materials used for ionizing radiation detectors, photodetectors, infrared (IR) sensors, etc. [1–8]. The advantage of *X*- and γ-radiation detectors based on semi-insulating semiconductors Cd(Zn)Te is the high efficiency of radiation detection and the ability to operate at room temperature. The use of a potential barrier such as a Schottky diode or a *p-n* junction in these materials leads to a decrease in the dark current compared to homogeneous (barrier-free) detectors at the same bias voltages [2]. A decrease of the dark current in the detector usually ensures its noise reduction, which expands the implementation of the X/γ photon counting mode and spectroscopy. Considering the technological aspects of the detector material fabrication, specifically doping with Cl impurity followed by compensating thermal annealing [2], the formation of high-quality diffusion *p-n* junctions is complicated due to the limitation of the diffusion temperature and the self-compensation effect. Therefore, diode structures on such materials are created either with a Schottky barrier [3–5] or with a *p-n* junction, in particular formed by the method of laser doping of an In donor impurity into *p*-CdTe crystal [5–8].

Y. Ono and J. Kondoh (Eds.): Inter-Academia 2023, LNNS 939, pp. 51–56, 2024.
https://doi.org/10.1007/978-3-031-54450-7_7

It is known that laser irradiation of metal-semiconductor structures can change the Schottky contact properties [8, 9] and introduce metal atoms into the surface semiconductor region, providing doping of its thin layer [5–8]. There is a problem of distinguishing the effects of a laser-induced *p-n* junction or a laser-modified Schottky barrier. Therefore, the present study aims to find clear evidences of laser-induced doping with In and creation of CdTe-based diodes with a *p-n* junction, not only with a modified In/CdTe Schottky barrier.

2 Experimental Details

The electrical and photoelectric properties of *p-n* junctions at $T = 293$ K formed by laser implantation of In in high-resistance ($\rho \approx 2 \cdot 10^9\ \Omega \cdot$ cm) *p*-type CdTe single crystals doped with Cl (*p*-CdTe:Cl) are studied. Experimental samples with sizes of $6 \times 6 \times 0.5$ mm^3 were made of semi-insulating *p*-CdTe:Cl wafers with a diameter of 25 mm and a thickness of 1 mm, which were cut from the ingot using an abrasive string cutter. Later, the damaged layer on both sides of the plates was removed on a chemical-mechanical polishing machine using bromine-containing polishing etchants. Thin (0.8 μm) Au films (with an area of 5×5 mm^2) were sputtered on one side of the samples. Thermal annealing of the Au/*p*-CdTe:Cl structure in a pure hydrogen atmosphere at $T = 280°$ C for 8–10 min. Provided ohmic contact and increased the stoichiometry of the sample surface. Then thin (~0.5 μm) In films of similar size were sputtered onto the opposite surface. The entire area of the In film was irradiated with a nanosecond single pulse of a Nd laser with an energy density of 0.11 J/cm^2. Later, the In films were dissolved in hydrochloric acid solutions, and gold contact pads with a diameter of 0.8 mm were deposited onto the middle of the surface for electrical connection and measurements. Removal of the In film after laser irradiation ensured that the structure was a *p-n* junction diode rather than a Schottky one. The contact pads were shaded while measuring the spectral photosensitivity of multilayered structures Au/*p*-CdTe:Cl/*n*-CdTe:In/Au (barrier structures) in the photovoltaic mode.

3 Results and Discussion

The current-voltage (*I-V*) characteristics of the Au/*p*-CdTe:Cl/In structures in linear coordinates are shown in Fig. 1. A clear diode character can be seen. The negative polarity of the applied voltage to the In contact corresponds to the direct *I-V* characteristic. The resistivity of detector grade *p*-CdTe:Cl crystals is high (~1 GΩ). The forward current in the initial forward voltage range (0–8 V) was determined by a reverse biased low Schottky barrier at the Au/*p*-CdTe:Cl interface. Then ($V > 8$ V), the forward current obeyed the Ohm law. However, the lateral current component also affected.

Both bulk and surface components contributed to the total reverse dark current of the barrier structures. The measured value of the total reverse current was much lower than the current estimated on the base of the CdTe crystal resistivity because of the *p-n* junction barrier effect. Perhaps, the lateral surface component of the current dominated and determined the reverse dark current at low bias voltages (Fig. 1). Low reverse bulk current was determined by the generation-recombination mechanism in the space charge region of the barrier structures [3, 10].

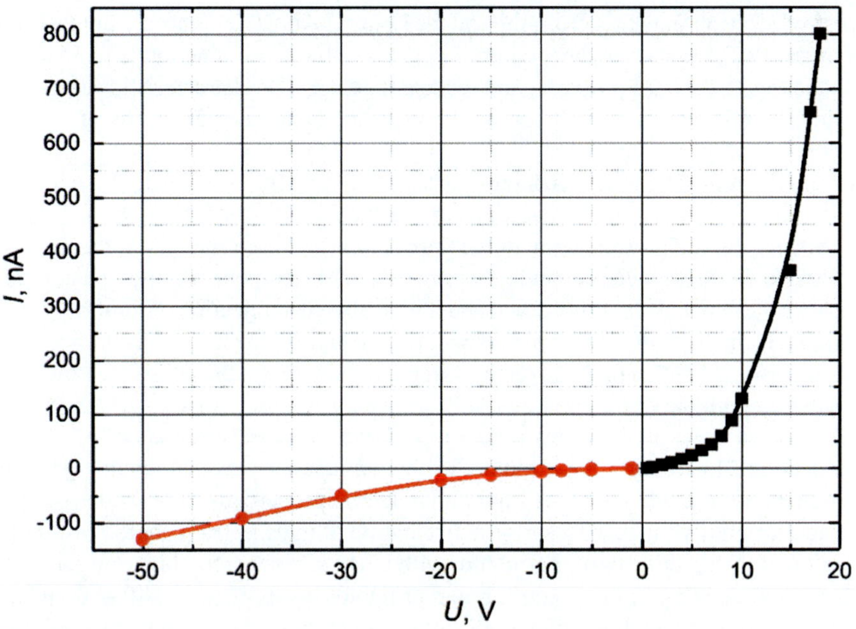

Fig. 1. *I-V* characteristic of the Au/*p*-CdTe:Cl/In structure after laser irradiation plotted in linear coordinates.

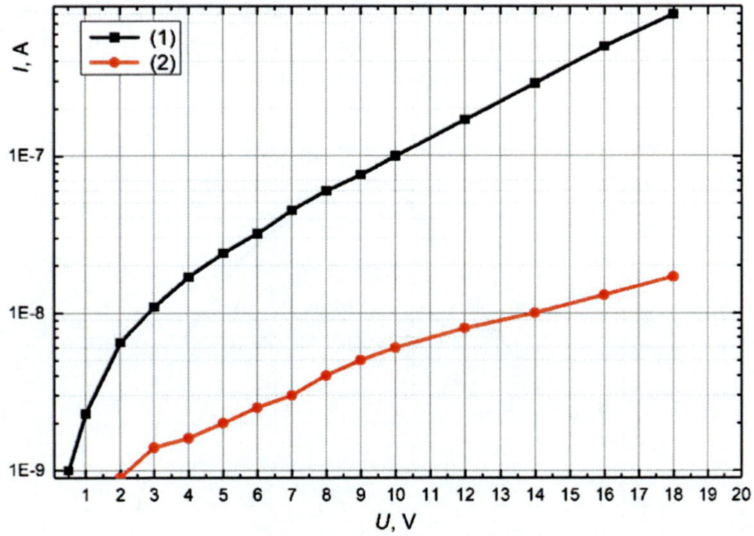

Fig. 2. *I-V* characteristic of the Au/*p*-CdTe:Cl/In structure after laser irradiation plotted in semi-logarithmic coordinates.

The reverse *I-V* curve does not have a region of dark current saturation (Fig. 1). At higher bias voltages, an additional increase in reverse current can be attributed to the process of electron injection from the forward-biased Au/CdTe Schottky contact to the bulk of the CdTe crystal and hence into the reverse-biased *p-n* junction [3].

In the Fig. 2, forward (curve 1) and reverse (curve 2) *I-V* characteristics are given in semi-logarithmic coordinates. It can be seen that the experimental results of the direct *I-V* (curve 1) at bias voltages $V \geq 6$ V are satisfactorily linearized, which indicates the exponential nature of the *I-V* behavior at significant injection of charge carriers into the *i*-region of the structures, according to the theory of *n-i-p* transitions [4]. At low bias voltages curve 2 corresponds to the carrier generation in the space charge region according to the Sah-Noyce-Shockley theory [11], then at higher voltages, an additional increase in the dark current is observed and it is attributed to injection of minority carriers (electrons) from the Au/CdTe ohmic contact (it is actually a low barrier Schottky contact which is forward-biased) to the reverse-biased *p-n* junction (at the *p*-CdTe:Cl/*n*-CdTe:In interface) through the CdTe crystal.

The typical spectral photosensitivity of the structure when illuminated from the side of the darkened In contact, measured in the photovoltaic mode, is shown in Fig. 3. High photosensitivity is evidence of the high photoelectric properties of the diode structures. The wide sensitivity range and non-selective nature of the spectrum are the result of the formation of an effective barrier structure with a low surface recombination rate [8].

It can be seen that in the spectral range of photosensitivity $\lambda = 0.40$–0.86 μm, the polarity of the photoresponse signal does not change its sign and a negative polarity is generated at the In contact, which indicates the presence of only one potential barrier in the studied structures. The maximum saturated photo-EMF (photo-electromotive force) generated in the structures illuminated by directed radiation from a quartz-halogen lamp with a power of 250 W was $U_X = (90 \pm 5$ mV) at $T = 293$ K. The spectra were normalized relative to the radiation source spectrum in order to avoid the effect of spectral inhomogeneity of quartz-halogen lamp radiation [12].

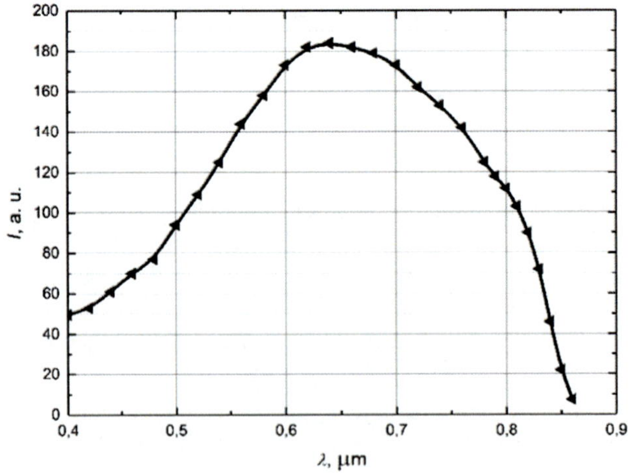

Fig. 3. Spectral photosensitivity (photo-EMF) of the Au/*p*-CdTe:Cl/In structure.

To determine the In-doped CdTe region thickness and the depth of the *p-n* junction formation, samples were subjected to polishing etching using a bromine-containing solution. The *I-V* characteristics of the laser-formed structures after chemical etching had a linear-like shape showing ohmic dependance, similar to the *I-V* curves of the initial Au/p-CdTe:Cl/In structures (before laser irradiation). Thus, etching of the CdTe:In surface region with a depth of ~0.12 μm led to the disappearance of the photovoltaic effect in the barrier structures and significantly changed their *I-V* characteristics reducing rectification. This evidenced laser irradiation of In-coated *p*-CdTe:Cl crystals resulted in *n*-type doping of the surface region and formation of a shallow steep *p-n* junction.

4 Conclusion

Using deposition of an In film on the *p*-CdTe:Cl crystal surface layer and removal of this film after laser irradiation of the In/*p*-CdTe:Cl structure allowed us to ensure that the obtained barrier structure is not a modified Schottky diode but a *p-n* junction diode created by laser-induced doping of a thin semiconductor layer by an In impurity. The obtained characteristics of the fabricated structures, such as low leakage current and wide spectral range indicate about high barrier characteristics. High photosensitivity of the fabricated barrier structures makes them promising for X/γ-ray detection. To decrease the leakage current in the *p-n* junction diodes, the ways of modification of the surface state of CdTe crystals before Au electrode deposition will be elaborated.

References

1. Zanio, K.: Cadmium telluride, Semiconductors and Semimetals, vol. 13. Academic Press, New York (1978)
2. Korbutyak, D.V., Melnichuk, S.V., Korbut, E.V., Borisyuk, M.M.: Cadmium telluride: impurity-defect states and detector properties. Ivan Fedorov, Kyiv (2000)
3. Kosyachenko, L.A., Sklyarchuk, V.M., Sklyarchuk, O.F., Gnatyuk, V.A.: Features of generation-recombination processes in CdTe-based Schottky diodes. Semicond. Sci. Technol. **22**(8), 911–918 (2007)
4. Sklyarchuk, V.M., Gnatyuk, V.A., Aoki, T.: Effect of CdTe crystal thickness on the efficiency of Cr/CdTe/Au Schottky-diode detectors. Nucl. Instrum. Methods Phys. Res. A. **953**, 163224 (2020)
5. Gnatyuk, V., Maslyanchuk, O., Solovan, M., Brus, V., Aoki, T.: CdTe X/γ-ray detectors with different contact materials. Sensors **21**(10), 3518 (2021)
6. Aoki, T., et al.: Development of energy discriminated CdTe imaging detector for hard X-ray. In: Proceedings of SPIE, vol. 5540, pp. 196–205. SPIE, San Diego, CA, USA (2004)
7. Gnatyuk, V.A., Aoki, T., Hatanaka, Y., Vlasenko, O.I.: Defect formation in CdTe during laser-induced doping and application to the manufacturing nuclear radiation detectors. Phys. Status Solidi C **3**, 1221–1224 (2006)
8. Gnatyuk, V.A., Aoki, T., Vlasenko, O.I., Levytskyi, S.N., Dauletmuratov, B.K., Lambropoulos, C.P.: Modification of the surface state and doping of CdTe and CdZnTe crystals by pulsed laser irradiation. Appl. Surf. Sci. **255**(24), 9813–9816 (2009)
9. Gnatyuk, V.A., Aoki, T., Hatanaka, Y., Vlasenko, O.I.: Metal-semiconductor interfaces in CdTe crystals and modification of their properties by laser pulses. Appl. Surf. Sci. **244**(1–4), 528–532 (2005)

10. Sze, S.M., Ng, K.K.: Physics of Semiconductor Devices, 3rd edn. John Wiley & Sons Inc., Hoboken, New Jersey (2007)
11. Sah, C.-T., Noyce, R.N., Shokley, W.: Carrier generation and recombination in *p-n* junctions and *p-n* junction characteristics. Proc. IRE **45**(9), 1228–1243 (1957)
12. Elvidge, C.D., Keith, D.M., Tuttle, B.T., Baugh, K.E.: Spectral identification of lighting type and character. Sensors **10**(4), 3961–3988 (2010)

Spectroscopic Studies of High Performance CdTe-Based Schottky Diode X/γ-Ray Sensors

Volodymyr Gnatyuk[1(✉)], Valeriy Sklyarchuk[2], Oleksandr Kulyk[3], and Toru Aoki[4]

[1] V.E. Lashkaryov Institute of Semiconductor Physics of the National Academy of Sciences of Ukraine, Prospekt Nauky 41, Kyiv 03028, Ukraine
gnatyuk@ua.fm

[2] Yuriy Fedkovych Chernivtsi National University, Kotsyubynsky Str. 2, Chernivtsi 58012, Ukraine

[3] V.N. Karazin Kharkiv National University, 4 Svobody Sq., Kharkiv 61022, Ukraine

[4] Research Institute of Electronics, Shizuoka University, 3-5-1 Johoku, Naka-Ku, Hamamatsu 432-8011, Japan

Abstract. The Ni/CdTe/Au Schottky diode X/γ-ray sensors, fabricated using the Ar-ion bombarding technique, exhibit low leakage current density (~20 nA/cm^2 at 1000 V) and high energy resolution (~1.0%@662 keV). We present and analyze the dependencies of the ^{137}Cs isotope energy spectra on the applied bias, the distance between the sensor and the radiation source, and the operation time. Despite the small (1 mm) CdTe crystal thickness, resulting in low radiation absorption and, therefore, a low probability of γ-photon capture and thus generation of electron-hole pairs, the sensors demonstrated high detection efficiency. This can be attributed to a high potential barrier at the Ni/CdTe Schottky contact, allowing the application of high bias voltages and ensuring a complete charge carrier collection.

Keywords: CdTe crystal · Schottky diode · *I-V* characteristic · charge transport · X/γ-ray sensor · isotope spectrum · energy resolution

1 Introduction

Demand for ionizing radiation detection instruments increases due to threats of radiation contamination arising from nuclear and radiological incidents. Optimal characteristics of semi-insulating CdTe crystals continue to make this semiconductor effective and attractive for compact solid-state room temperature high energy radiation detectors used in environment monitoring and other fields [1–3]. However, the resistivity of CdTe is not high enough that can result in a larger leakage current of the CdTe-based sensors and thus lead to deterioration in detection efficiency and spectral resolution. To overcome the problem of charge losses due to incomplete photogenerated carrier collection by applying a higher bias voltage, expanding thereby the depletion region in a diode to obtain high energy resolution, we have developed diode-type sensors based on either a

Y. Ono and J. Kondoh (Eds.): Inter-Academia 2023, LNNS 939, pp. 57–63, 2024.
https://doi.org/10.1007/978-3-031-54450-7_8

heterostructure [4], a *p-n* junction formed by laser-induced doping [5–7], or a Schottky barrier created using Ar plasma surface processing [7–10]. The spectroscopic properties of the Ni/CdTe/Au diode structures with a high-barrier Schottky and near-ohmic contacts at the Ni/CdTe and Au/CdTe interfaces, respectively, have been studied. The obtained dependencies of the ^{137}Cs isotope energy spectra, taken with the Ni/CdTe/Au Schottky diode sensors, on the applied bias, the distance to the radiation source, and the operation time have been analyzed.

2 Experimental Techniques and Procedures

Detector-grade Cl-compensated (111) oriented *p*-like CdTe semiconductor single crystal wafers produced by Acrorad Co. were used [3]. The shape and sizes (a), structural features (b), and characteristics of the CdTe samples (c) are shown in Fig. 1. The polar A- and B-faces, so-called Cd- and Te-terminated surfaces of a CdTe(111) crystal, differ in their physical and chemical properties, which are exhibited during various surface treatments (chemical etching, annealing, laser irradiation, plasma processing, etc.), and in measurements of optical and photoluminescence characteristics [11, 12].

The technique of Schottky barrier formation, based on Ar-ion bombardment of the CdTe(111)B surface, was developed and used to fabricate diode-type X/γ-ray sensors [8–10]. The result of different surface treatments (chemical polishing (a) and Ar-ion (b) etching) is shown in Fig. 2. The high-barrier Schottky contact was formed on the modified CdTe surface (Fig. 2(b)).

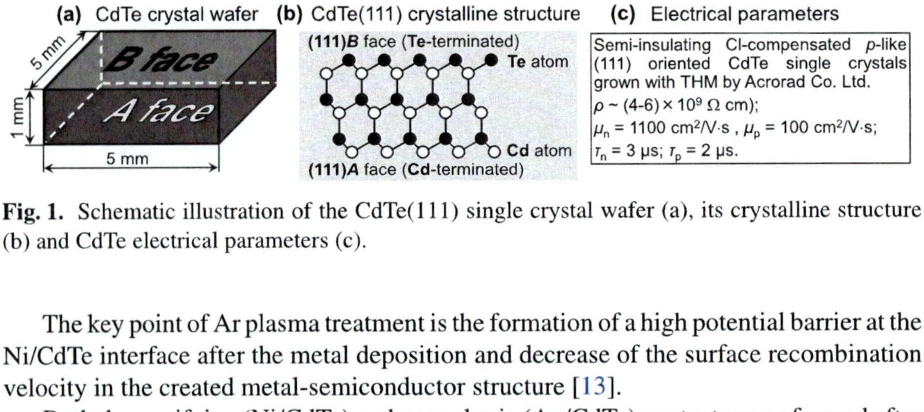

Fig. 1. Schematic illustration of the CdTe(111) single crystal wafer (a), its crystalline structure (b) and CdTe electrical parameters (c).

The key point of Ar plasma treatment is the formation of a high potential barrier at the Ni/CdTe interface after the metal deposition and decrease of the surface recombination velocity in the created metal-semiconductor structure [13].

Both the rectifying (Ni/CdTe) and near ohmic (Au/CdTe) contacts were formed after preliminary polishing etching of the parallelepiped-like CdTe wafers with sizes of 5 × 5 × 1 mm^3 in a Br-methanol solution, washing in methanol and following treatment with an Ar plasma at different regimes (ion energy, beam density, processing duration, etc.) for the B- and A-faces, respectively [8]. A Ni electrode was created on the CdTe(111)B surface by vacuum evaporation using a Mo mask with sizes of 3.16 × 3.16 mm^2. An Au electrode was formed on the entire CdTe(111)A surface by chemical deposition from a gold-chloride solution, i.e., the surface was entirely coated by an Au film (Fig. 3). Passivation of the lateral faces of the Ni/CdTe/Au diode structures was employed to

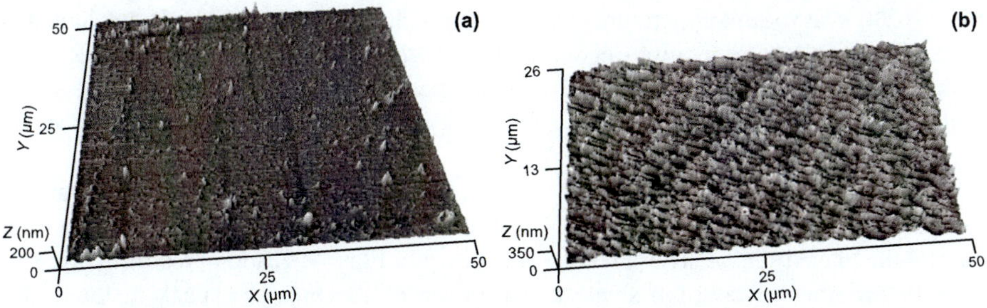

Fig. 2. AFM topomicrographs of the CdTe crystal with the mirror surface (a) after chemical polishing etching in a Br-methanol solution and with the modified surface (b), obtained by additional (after chemical etching) processing of the CdTe surface in an Ar-ion plasma under special regimes (at voltage of 500–550 V and current of 10–15 mA during 20–25 min).

eliminate leakage currents. The developed Ni/CdTe/Au Schottky diode sensors were studied by electrical and spectroscopic measurements carried out at room temperature.

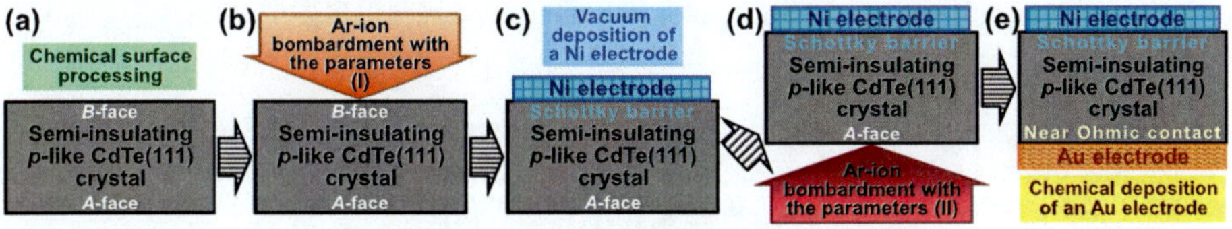

Fig. 3. Fabrication of Schottky-diode sensors: surface processing of the crystal (a); etching of the CdTe(111)B by Ar-ion bombardment with parameters (I) (b); thermal vacuum deposition of a Cr, Ti, or Ni electrode on the CdTe(111)B surface and formation of a Schottky barrier (c), etching of the CdTe(111)A by Ar-ion bombardment with parameters (II) (d); chemical deposition of an Au electrode on the CdTe(111)A surface and formation of a near ohmic contact (e).

The fabricated Ni/CdTe/Au structures were connected to the measuring electronic devices by clamping contacts using a conductive soft material. The X/γ-ray spectra were taken with a compact USB stick form factor spectrometer ANS-MNT004-GTK, produced by ANSeeN Inc. [14], using a collimated γ-ray source based on a ^{137}Cs isotope. The Ni/CdTe/Au sensors were exposed to ^{137}Cs γ-ray radiation from the Ni electrode side and the spectra were taken at applied bias voltage $V = 100$–1000 V with the acquisition time of 5 min.

3 Results and Discussion

The diode-type X/γ-ray-sensors operated in a reverse mode, i.e., the Ni contact was positively biased with respect to the Au one. The *I-V* characteristics of Ni/CdTe/Au structures with a Schottky barrier were similar to those of Cr/CdTe/Au diodes manufactured using the same Ar-ion bombardment technique [8, 10]. Having the rectifying contact area of about 10 mm^2, the Ni/CdTe/Au diodes demonstrated reverse dark currents $I \sim 2$–3 nA at $V = 1000$ V at room temperature. The dominant charge carrier transport mechanisms were determined as: generation-recombination in the space-charge region (SCR) at $V = 1$–100 V, charge transport in the conditions when the SCR width exceeded the semiconductor crystal thickness at elevated voltages, and finally, currents limited by space charge at even higher bias voltages [15].

The energy spectra of a ^{137}Cs isotope, taken with the Ni/CdTe/Au Schottky-diode, were measured at different bias voltages, distances to the radiation source, and operation times of the biased sensor. These results are shown in Figs. 4, 5, and 6, respectively. The intensity (number of counts), shape and position of the 662 keV peak in the ^{137}Cs isotope spectra depended on the applied bias voltage (Fig. 4). The broad shoulder and tail, extending toward the low-energy side from the peak, were due to the Compton scattering effect [3].

The number of counts and energy resolution increased with increasing applied bias voltage from 100 V to 1000 V (Fig. 4). As observed, $V = 1000$ V was the optimal operation voltage, providing the highest number of counts and excellent energy resolution, i.e., a low full width at half maximum (FWHM) value, demonstrating the best spectroscopic performance (FWHM ~1% for the 662 keV line). It should be noted that the peak position was not changing over a wide bias voltage range (200–1000 V), indicating a nearly complete collection of photogenerated charge carriers (Fig. 4).

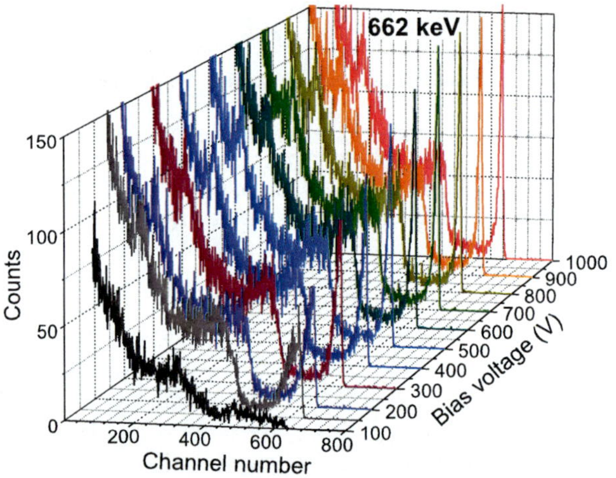

Fig. 4. The dependence of the ^{137}Cs isotope energy spectra, taken with the Ni/CdTe/Au Schottky diode sensor, on the applied bias voltage.

The dependence of the 662 keV peak height (number of counts) on the distance between the Ni/CdTe/Au sensor and the radiation source had a somewhat non-standard character, differing from the well-known inverse square law (Fig. 5). This could be due to the radiation source not being perfectly collimated or other reasons that require further investigation.

It is known that diode-type semiconductor X/γ-ray sensors suffer from the charge polarization effect, which leads to deterioration of detector parameters over operation time and limits their application at room temperature [1–3]. This phenomenon is specific to semiconductor barrier structures with blocking contacts, either Schottky or *p-n* junctions, which are usually used to suppress leakage current of sensors. A progressive decrease of the spectrum peak height, degradation of energy resolution, and the shifting of the peak toward lower energies over time after applying the bias can be explained by the non-uniform electric field due to the charge accumulation [16].

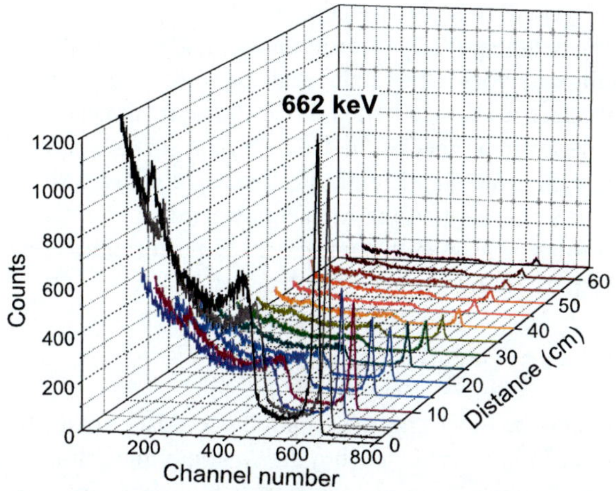

Fig. 5. The dependence of the ^{137}Cs isotope energy spectra, taken with the Ni/CdTe/Au Schottky diode sensor, on the distance to the radiation source.

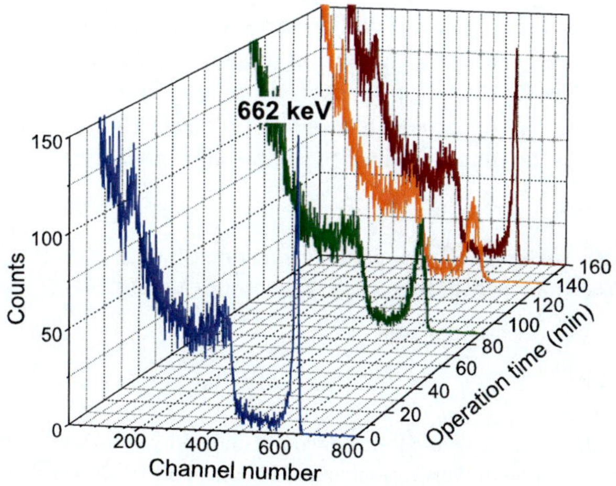

Fig. 6. Time evolution of the ^{137}Cs isotope energy spectrum, taken with the Ni/CdTe/Au Schottky-diode sensor just at applying bias $V = 1000$ V, after operation during 90 min, 140 min, and 160 min (before the last measurement, the bias was switched off for a few seconds).

Thus, the development of CdTe-based diode-type sensors operating at room temperature with negligible polarization and low leakage current remains a challenging task for practical applications [1–3]. As observed from the time transformation of the ^{137}Cs isotope energy spectrum, the Ni/CdTe/Au Schottky diode sensor exhibited quite high time tolerance. The spectrum did not deteriorate much during sensor operation for more than 2 h; moreover, the spectrum completely restored itself when the bias was turned off for a few seconds (Fig. 6).

It would be of interest to employ the elaborated plasma-assisted technology of Schottky diode formation in the development and fabrication of X/γ-ray pixel sensors for imaging applications. Such kind of sensors is generally formed on large-area CdTe wafers [1, 3]. It is well-known that the primary criterion influencing the quality and, thus, application fields of imaging detectors is the semiconductor homogeneity and equalization of the electrical and photoelectric characteristics across the entire area [1, 2]. The challenges of achieving homogeneous treatment of the CdTe crystal surface before Schottky contact formation, ensuring the time stability of spectroscopic characteristics in Ni/CdTe/Au diode X/γ-ray sensors, and addressing high flux tolerance and degradation resistance, continue to pose hurdles in the manufacturing of high-performance detectors for imaging applications and require future studies.

4 Conclusion

The obtained dependencies of the spectroscopic characteristics of the Ni/CdTe/Au Schottky diode X/γ-ray sensors, fabricated using the Ar-ion bombarding technique for high-barrier rectifying contact formation, showed quite a high detection efficiency and energy resolution (FWHM = 0.7–1.5%@662 keV). This was due to the sharp rectification of the Ni/CdTe Schottky contact that allowed us to apply high bias voltages (up to 1000 V) and thus enabled a complete collection of photogenerated charge carriers. The sensors exhibited sufficiently high electrical and spectroscopic characteristics, as well as time stability and degradation resistance under prolonged X/γ-ray irradiation. These qualities make the Ni/CdTe/Au diode-type sensors promising for spectroscopic and imaging applications, particularly for the localization and identification of radioactive sources. However, the issue of the time stability of spectroscopic characteristics still needs to be addressed.

Acknowledgement. The research was partially supported by the 2023 Cooperative Research Projects of the Research Center for Biomedical Engineering (grants No 2053 and 2078), Japan.

References

1. Triboulet R., Siffert, P.: CdTe and Related Compounds; Physics, Defects, Hetero- and Nano-Structures, Crystal Growth, Surfaces and Applications. Part II: Crystal Growth, Surfaces and Applications. Elsevier, Amsterdam and Oxford (2010)
2. Zappettini, A.: Cadmium telluride and cadmium zinc telluride. Chapter 8. In: Fornari, R. (ed.) Single Crystals of Electronic Materials, Growth and Properties, pp. 273–301. Woodhead Publishing, Cambridge (2019)
3. Takahashi, T., Watanabe, S., Ishikawa, S.: High-resolution CdTe detectors and application to gamma-ray imaging. Chapter 8. In: Iniewski, K. (ed.) Semiconductor Radiation Detection Systems, pp. 171–192. Taylor & Francis Group, New York (2010)

4. Brus, V.V., et al.: Graphene/semi-insulating single crystal CdTe Schottky-type heterojunction X- and γ-ray radiation detectors. Sci. Rep. **9**(1), 1065–1–8 (2019)
5. Aoki, T., et al.: Development of energy discriminated CdTe imaging detector for hard X-ray. In: Proceedings of SPIE, vol. 5540, pp. 196–205. SPIE, San Diego, CA, USA (2004)
6. Gnatyuk, V.: In/CdTe/Au p-n junction-diode X/γ-ray detectors formed by frontside laser irradiation doping. Nucl. Instrum. Meth. Phys. Res. A **1029**, 166397–1–8 (2022)
7. Gnatyuk, V., Maslyanchuk, O., Solovan, M., Brus, V., Aoki, T.: CdTe X/γ-ray detectors with different contact materials. Sensors **21**(10), 3518–1–20 (2021)
8. Sklyarchuk, V.M., Gnatyuk, V.A., Fang, X., Aoki, T.: Effect of the thickness of CdTe crystals on electrical and detection properties of Cr/CdTe/Au Schottky-diode detectors. In: Proceedings of SPIE, vol. 11114, pp. 111141S-1-7. SPIE, San Diego, CA, USA (2019)
9. Sklyarchuk, V. M., Gnatyuk, V. A., Aoki, T.: Effect of CdTe crystal thickness on the efficiency of Cr/CdTe/Au Schottky-diode detectors. Nucl. Instrum. Meth. Phys. Res. A **953**, 163224–1–5 (2020)
10. Sklyarchuk, V., Gnatyuk, V., Fochuk, P. Aoki, T.: Characterization of Schottky-diode X/γ-ray detectors based on CdTe crystals with different uncompensated impurity concentrations. Nucl. Instrum. Meth. Phys. Res. A **1027**, 166229–1–9 (2022)
11. Gnatyuk, D.V., Poperenko, L.V., Yurgelevych, I.V., Aoki, T.: Characterization of the surfaces of CdTe(111) single crystals after laser processing. In: Proceedings of SPIE, vol. 8142, pp. 81421N-1-7. SPIE, San Diego, CA, USA (2011)
12. Gnatyuk, D.V., Poperenko, L.V., Yurgelevych, I.V., Dacenko, O.I., Aoki, T.: Characterization of functional layers of CdTe crystals subjected to different surface treatments. IEEE Trans. Nucl. Sci. **62**(2), 428–432 (2015)
13. Sze, S.M., Ng, K.K.: Physics of Semiconductor Devices, 3rd edn. John Wiley & Sons Inc., Hoboken, New Jersey (2007)
14. ANSeeN Inc. http://anseen.com. Accessed 10 Nov 2023
15. Canali, C., Martini, M., Ottaviani, G., Zanio, K.R.: Transport properties in CdTe. Phys. Rev. B **4**(2), 422–431 (1971)
16. Cola, A., Farella, I. The polarization mechanism in CdTe Schottky detectors. Appl. Phys. Lett. **94**(10), 102113–1–3 (2009)

Check for updates

Comparative Evaluation of Electrical Properties of Carbon Nanotube Networks Deposited on CMOS-Compatible Platform

Hayato Kawanishi[1,2], Rohitkumar S. Singh[1], V. N. Ramakrishnan[3], Masaru Shimomura[1,2], and Daniel Moraru[1,2](✉)

[1] Research Institute of Electronics, Shizuoka University, Hamamatsu, Japan
moraru.daniel@shizuoka.ac.jp

[2] Graduate School of Integrated Science and Technology, Shizuoka University, Hamamatsu, Japan

[3] Vellore Institute of Technology (VIT), Vellore, India

Abstract. Carbon nanotubes (CNTs) can be used as channel material in a nanoscale or microscale gap between conducting electrodes. The contact potential of CNT-bundles deposited by inkjet-printing was measured by KFM (Kelvin probe force microscopy) using Pt-coated Si tips in ultra-high vacuum (~10^{-8} Pa). By comparing these results with electrical characteristics, SEM (scanning electron microscopy) and AFM (atomic force microscopy) measurements, we confirmed that the inkjet-printed CNT-bundles bridged the nanogap between different electrodes. Using this KFM technique, it can be shown how the potential of the CNT-bundles changes under the application of voltages. Combining these results, we investigate the electrical properties of CNT-networks deposited on CMOS-compatible platforms, aiming towards the development of hybrid nanoelectronics.

Keywords: carbon nanotubes (CNTs) · Kelvin Probe Force Microscopy (KFM) · Atomic Force Microscopy (AFM) · scanning electron microscope (SEM) · nanodevice · CMOS

1 Introduction

Miniaturization of silicon (Si) transistors has been a major driver for many advances in electronics, following the Moore's law [1] that indicates the trend of doubling the number of transistors per integrated circuit (IC) chip every approximately two years. However, Si-based transistors are reaching nanoscale limits since the size of the minimum feature (gate length) is well below 10 nm recently. In order to overcome the issues that may arise due to the need for lithography control at nanoscale size [2, 3], new materials intrinsically having dimensions in nanoscale are being actively investigated for transistor applications.

One of the most promising alternative materials are single-walled carbon nanotubes (SW-CNTs). SW-CNTs have excellent transport properties, high flexibility and are naturally formed in nanoscale [4]. It is known that, depending on their chirality and purity,

Y. Ono and J. Kondoh (Eds.): Inter-Academia 2023, LNNS 939, pp. 64–71, 2024.
https://doi.org/10.1007/978-3-031-54450-7_9

SW-CNTs exhibit metallic or semiconducting properties [5–7]. In our previous work, we have explored the possibility of fabricating hybrid devices in which the deposition of CNTs (CNT networks) was done by inkjet-printing between Al electrodes on a Si/SiO_2 surface, i.e., on a platform compatible with the complementary metal-oxide-semiconductor (CMOS) technology [8]. In this work, we extend this study by providing a comparative analysis to evaluate the electrical properties of CNTs (CNT-bundles) from different perspectives, especially by analyzing their electrostatic potential in the devices.

Kelvin probe force microscopy (KFM), which is based on measuring the electrostatic force existing between a sample and a probe (cantilever), was used previously to observe and analyze the properties of dopant-induced quantum dots (QDs) in nanoscale silicon-on-insulator (SOI) devices under operation [9–13]. Therefore, it can be considered a suitable method to characterize the electrical properties of other small devices, such as CNT-devices. In this research, we focus on measurements by KFM of the contact potential of CNT-bundles deposited by inkjet-printing, using for the KFM measurements a Pt-coated Si tip, while the sample is placed in ultra-high vacuum ($\sim 10^{-8}$ Pa). Also, we compare the results with SEM (scanning electron microscopy) and AFM (atomic force microscopy) measurements, while also confirming the current flow by current-voltage characterization.

2 Experimental Methods

Before the preparation of the CNTs, the CMOS-platform was prepared using cleanroom technology, namely oxidation, electron beam lithography (EBL), metal deposition and lift-off process. A 20-nm-thick SiO_2 layer was formed by thermal oxidation of the Si substrate, cleaned by a prior chemical treatment. This film serves as the gate oxide when the substrate-Si is used as a back gate. An Al film (approximately 70 nm thick) was deposited by an evaporation method onto a resist film that was priorly patterned using EBL processing, followed by lift-off. Depending on the EBL parameters, the gaps were opened between the Al electrodes with the width designed to be in the range of 500–3000 nm.

Dimethylformamide (DMF) was used as a solvent for highly-purified (99.9% purity) semiconductor-CNTs (NanoIntegris). The homogeneity and dispersion conditions of the SW-CNT solution were optimized for inkjet-printing, considering that the CNTs tend to strongly bundle. Thus, long-time, high-power sonication is needed (typically, 4 h immediately before inkjet-printing). It must be mentioned that CNT-bundles are expected to remain in the solution and, furthermore, re-agglomeration of CNTs may also occur, as the time between sonication and inkjet-printing becomes longer. When compared to regular deposition methods, such as spin coating, inkjet-printing allows the deposition of the CNT networks with relatively higher position control. Droplets of the CNT-solution (CNT-ink) were dispersed by the inkjet-printing system (SIJ-S050, SIJTechnology Inc., Tsukuba, Japan) with a sub-nL volume through a piezoelectric nozzle, using a sine wave at a frequency of 500 Hz and an amplitude of 300 V [8]. Based on the above conditions, it can be stated that inkjet-printing was used successfully to deposit the CNT-bundles in the gaps between the Al electrodes.

In Figs. 1(a)-(b), camera views are shown for an area of the sample containing several devices. However, the nozzle filled with the CNT-solution (CNT-ink) is aligned just above the gap for a selected device (marked by the green arrow) and brought in close proximity to the surface. Figure 1(a) shows the situation before the deposition. Figure 1(b) shows the droplet formed on the surface on an area of several tens of μm, but the droplet dries within a time of a few seconds. The schematic diagram of the device structure and the AFM/KFM measurement approach is shown in Fig. 1(c). During the KFM measurement, the cantilever follows the device structure and can detect the surface topography and corresponding electrostatic potential.

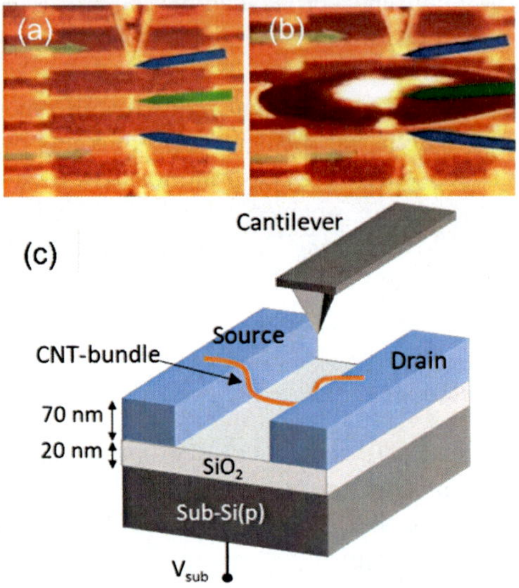

Fig. 1. (a)-(b) Camera views of several devices, with the gap between electrodes for one device aligned below a nozzle filled with CNT-ink, just before [(a)] and just after [(b)] inkjet-printing. The DMF/CNT droplet formed on the surface dries in a few seconds. (c) Schematic diagram of the device structure (note that the Si substrate is ~500 μm thick, i.e., not drawn to scale) prepared for AFM/KFM measurement by scanning a cantilever over the gap containing CNTs (or CNT-bundles) bridging the electrodes.

Measurements were taken at room temperature, while the substrate (labeled Sub-Si in Fig. 1(c)) was grounded or, alternatively, biased by a voltage (V_{sub}). Although the substrate is weakly *p*-type doped, it can still allow the control of the potential of the semiconducting-CNT structures deposited on top of the 20-nm-thick SiO_2 film. Since there is no control over the number and precise distribution of the CNTs or CNT-bundles by our method of deposition, a network of CNT-bundles is formed with random positions and layout patterns. A significant fraction of the measurement time is dedicated to locating CNT-bundles that bridge the two electrodes (potentially forming a conducting channel), rather than be connected to only one electrode or another.

The KFM measurements are supplemented by scanning electron microscope (SEM) measurements for direct observation of the distribution of the CNT-networks. Then, atomic force microscope (AFM) measurements, taken simultaneously to KFM measurements, allow us to identify the position of the measurement spot in a broader map of the devices (note that the AFM/KFM measurement area is only a few μm^2 while the SEM measurements can be taken on much larger areas). For each device, drain current vs drain voltage (I_D-V_D) measurements were also taken in a vacuum prober system to confirm the fact that current flows between the electrodes only after the CNT-networks were deposited by inkjet-printing.

3 Results and Discussion

SEM measurements were performed to observe the surface conditions of nanogaps containing CNT-bundles and to confirm that they form bridges between electrodes. An SEM image of a CNT-transistor with a few CNT-bridges deposited in a nanoscale gap (designed W = 500 nm) is shown in Fig. 2(a). CNT-bundles can be seen arranged as a random network not only in the gap between the electrodes, but also on top of the electrodes, as well. The tilted SEM image in Fig. 2(b) marks a few CNT-bundles as bridges.

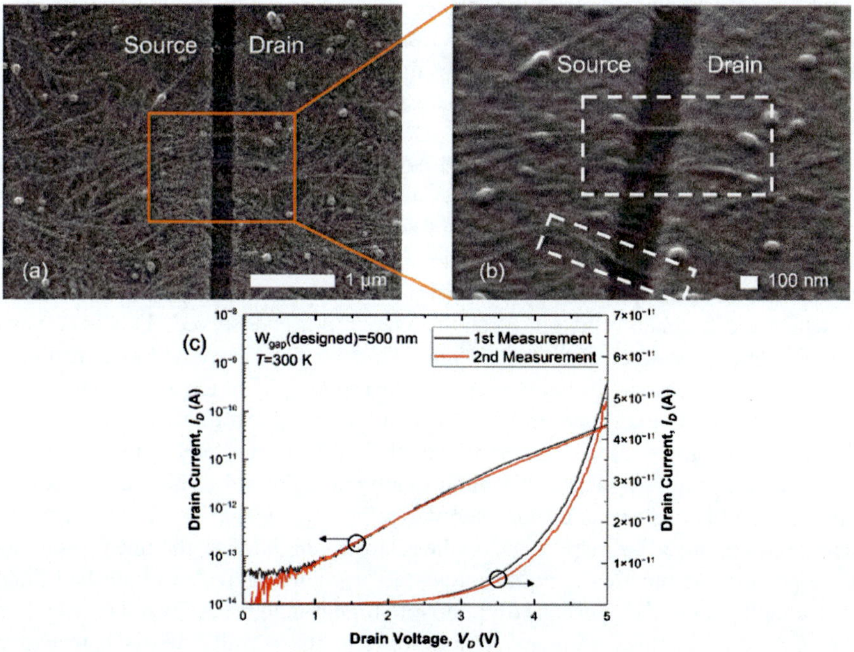

Fig. 2. (a) SEM image of a CNT-network deposited by inkjet-printing in a gap (designed width of 500 nm) between Al source and drain electrodes. The actual width is slightly smaller than the designed width of the gap due to specific issues related to the lift-off processes. (b) Zoom-in: close-up SEM image taken at a 30° tilt angle, showing a few CNT-bundles bridging source and drain. (c) Room-temperature (T = 300 K) I_D-V_D characteristics for a CNT-transistor, showing two consecutive measurements.

A low-density CNT-network is visibly formed by the inkjet-printing method presented above. Since the lateral size of each structure is, on average, about 20 nm, it is considered that the network is mainly containing CNT-bundles formed by a few to several tens of CNTs, since each CNT has a diameter of approximately 2 nm. This suggests that the sonication can be further continued and that time between loading the nozzle and inkjet-printing should be minimized in order to obtain a network formed dominantly of CNTs (not CNT-bundles) between the electrodes. Nevertheless, the CNT-bundles can conduct current between the two electrodes in a transistor configuration.

To confirm the current flow, as a first step, we measured the current-voltage characteristics of such devices, as shown in Fig. 2(c). The measurements were taken at $T = 300$ K (room temperature) by sweeping the drain voltage (V_D) and measuring the drain current (I_D), shown in the log scale on the left vertical axis and in the linear scale on the right vertical axis. For this device, the current gradually increases above the noise level (approximately $10^{-14} \sim 10^{-13}$ A) from a low voltage. In a previous work [8], measurements carried out on other devices with a lower number of CNT-bundles and at lower temperatures also reveal the fact that current switching frequently occurs in such devices, an indication of the possible interaction among the multiple CNT-bundles embedded in the network.

In the second step, we observe the surface topography and the electrostatic potential in the nanoscale gap and the adjacent electrodes (labeled source and drain), using the AFM and KFM techniques. Figure 3 shows the AFM images [Figs. 3(a)-(b)] and KFM images [Figs. 3(c)-(d)] for two different substrate voltages, $V_{sub} = 0$ V (upper maps) and $V_{sub} = 3$ V (lower maps).

In Fig. 3(a) and 3(b), the white dotted lines follow the topography of a CNT-bundle between two arrows, across the gap between the Al source and drain. From Fig. 3(a), the topography suggests that the CNT-bundles are attached to both electrodes, but also to the SiO_2 layer above the substrate, as schematically illustrated in Fig. 1(c). This can be confirmed for the upper area by the line profiles (black and red) shown in Fig. 3(e), from which a difference in height of about 65 ± 5 nm is observed. This is consistent with the height of the Al electrodes (~70 nm). The two AFM maps and line profiles also demonstrate good reproducibility in mapping the topography of the device. On the other hand, the potential maps shown in Figs. 3(c) and 3(d) suggest that the potential on the CNT-bundles is comparable to the potential on the Al electrodes (and significantly higher than for the SiO_2 regions where CNT-bundles are not deposited). This can be confirmed by the line profiles (green and blue) shown in Fig. 3(e) which are averaged from 10 adjacent lines in the KFM maps. Despite the relatively high noise, the line profiles show that the potential is practically equal across the segments indicated by dashed lines in Figs. 3(c) and 3(d). KFM noise is still relatively higher than the AFM noise, which may be due to the different measurement methodology, but the overall potential can clearly be controlled by V_{sub}. From this analysis, extended to multiple positions across the device, it was confirmed that CNT-bundles deposited by inkjet-printing can act as transistor channels even though they are not perfectly suspended bridges, but rather fallen bridges contacting the SiO_2 surface.

Figure 3 shows a CNT-bundle tilted at an angle relative to the direction normal to the gap, which may not be an optimal layout. As a proof-of-concept, it was previously shown on similar devices [8] that manipulation of such CNT-bundles is possible using an AFM tip moved over the area where CNT-bundles are found. Repositioning, rotating

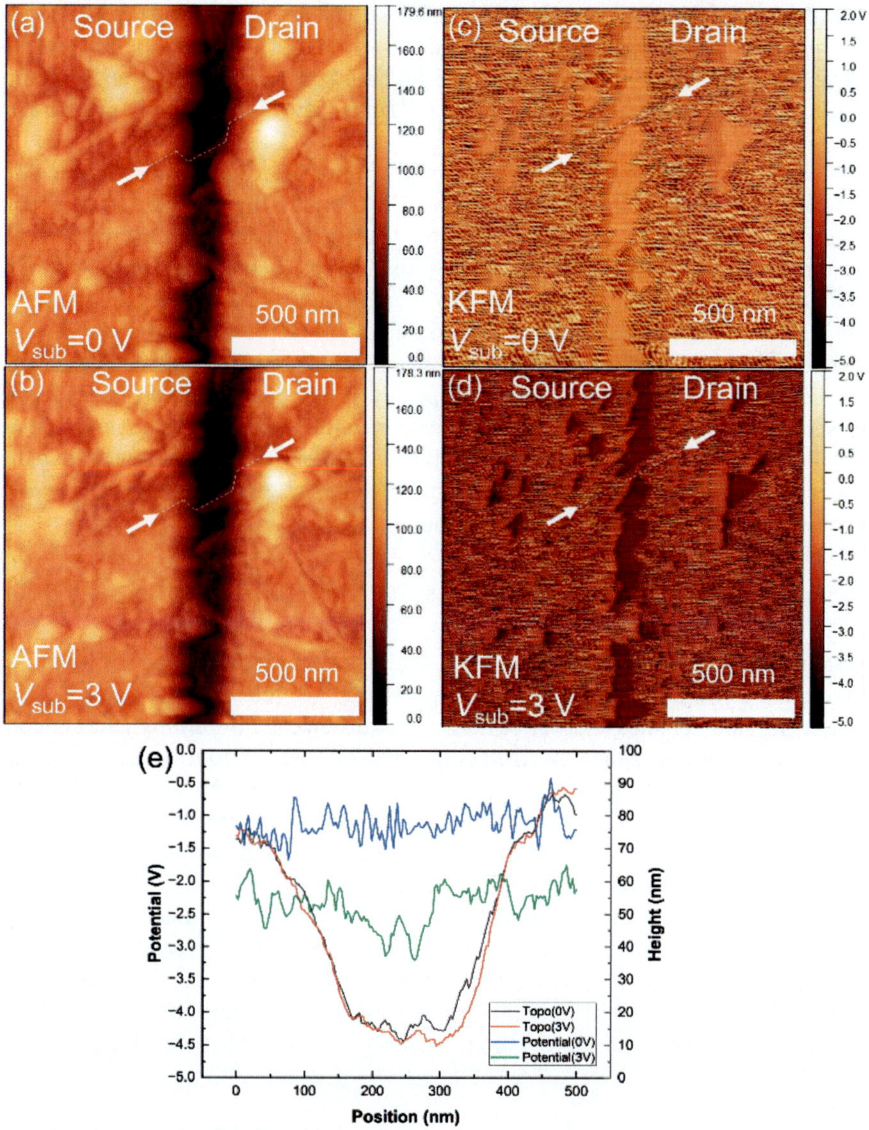

Fig. 3. (a)-(b) AFM images of a CNT-transistor where CNT-bundles bridge Al source and drain electrodes, as indicated by arrows and dashed lines. Different substrate voltages, V_{sub}, are applied (0 V in (a) and 3 V in (b)). (c)-(d) KFM images of same CNT-transistor taken simultaneously with (a) and, respectively, (b) at different V_{sub}. Source/drain electrodes and CNT-bundles marked by arrows and dashed lines are observed to have the same electrostatic potential. (c) Line profiles taken from the AFM and KFM images shown in (a)-(b) and (c)-(d).

and cutting of CNT-bundles are all feasible operations that provide further versatility to the design of the CNT-transistors in the future.

The analysis illustrated in Figs. 2 and 3 on an inkjet-printed CNT-transistor shows how a set of complementary measurements, such as SEM, current-voltage characterization, AFM, and KFM, can reveal the properties of CNT-bundles (and, eventually, of individual CNTs) when acting as channels between two electrodes with a nanoscale gap in between. In particular, it can be shown how CNT-bundles bridging the two electrodes with different configurations allow conducting paths to be formed in the transistor. Substrate voltage (V_{sub}) effect has also been investigated, showing that the electrostatic potential of the CNT-bundles is sensitive to V_{sub}. However, the detailed analysis of this effect is beyond the scope of the present work and will be presented elsewhere.

Repeatability is difficult to demonstrate in our devices because each device contains a unique CNT-bundle network, with its specific properties. It should be mentioned, however, that several devices have been measured with nominally the same (or similar) fabrication conditions. Although the details are different due to the reason explained above, all devices contain similar complex CNT-bundle networks and the types of the layouts of the CNT-bundles are basically the same.

4 Conclusions

CNT-bundles were deposited on a CMOS-compatible platform using inkjet-printing, with a relatively high degree of control at μm-scale. We relied on the current-voltage, i.e., I_D-V_D, characteristics to describe how much current can flow through the network of CNT-bundles, while we relied on KFM (supported by AFM and SEM) measurements to describe how the electrostatic potential is distributed locally within sections of the CNT network across the gap between Al source and drain. The combined methods provide a set of complementary information related to the layout and conducting properties of the CNT-bundles. As a result, it was confirmed that the CNT-bundles can function as channels of transistors formed largely by CMOS-compatible technology.

This analysis may enable the development of hybrid devices, leveraging the naturally small dimensions of organic structures, like the CNTs in this case, and the advanced CMOS technology based on Si. In the future, this research can be extended to CNT-networks deposited by a similar method in nano-gaps formed between highly-doped Si nano-leads.

Acknowledgments. This work was partially supported by JSPS KAKENHI (no. 19K04529 and no. 22K04216), and a Cooperative Research Project from the Research Institute of Electronics, Shizuoka University. The authors thank K. Takagi, T. Aoki, J. H. Moon, Y. Neo, F. Iwata, and H. Mimura for their support during the inkjet-printing and AFM manipulation, as well as for useful discussions.

References

1. Moore, G.: Cramming more components onto integrated circuits. Electron. Mag. **38**, 114–117 (1965)

2. Waldrop, M.M.: The chips are down for Moore's law. Nat. News **530**, 144–147 (2016)
3. Chau, R., Doyle, B., Datta, S., Kavalieros, J., Zhang, K.: Integrated nanoelectronics for the future. Nat. Mater. **6**(11), 810–812 (2007)
4. De Volder, M.F., Tawfick, S.H., Baughman, R.H., Hart, A.J.: Carbon nanotubes: present and future commercial applications. Science **339**, 535–539 (2013)
5. Lee, D., et al.: High-performance thin-film transistors produced from highly separated solution-processed carbon nanotubes. Appl. Phys. Lett. **104**, 143508 (2014)
6. Choi, S.J., Wang, C., Lo, C.C., Bennett, P., Javey, A., Bokor, J.: Comparative study of solution-processed carbon nanotube network transistors. Appl. Phys. Lett. **101**, 112104 (2012)
7. Choi, S.J., Bennett, P., Lee, D., Bokor, J.: Highly uniform carbon nanotube nanomesh network transistors. Nano Res. **8**, 1320–1326 (2015)
8. Singh, R.S., et al.: Precise deposition of carbon nanotube bundles by inkjet-printing on a CMOS-compatible platform. Materials **15**, 4935 (2022)
9. Tabe, M., et al.: Single-electron transport through single dopants in a dopant-rich environment. Phys. Rev. Lett. **105**, 016803 (2010)
10. Ligowski, M., Moraru, D., Anwar, M., Mizuno, T., Jablonski, R., Tabe, M.: Observation of individual dopants in a thin silicon layer by low temperature Kelvin Probe Force Microscope. Appl. Phys. Lett. **93**, 142101 (2008)
11. Anwar, M., et al.: Effect of electron injection into phosphorus donors in silicon-on-insulator channel observed by Kelvin probe force microscopy. Appl. Phys. Lett. **99**, 213101 (2011)
12. Tyszka, K., Moraru, D., Samanta, A., Mizuno, T., Jablonski, R., Tabe, M.: Comparative study of donor-induced quantum dots in Si nano-channels by single-electron transport characterization and Kelvin probe force microscopy. J. Appl. Phys. **117**, 244307 (2015)
13. Tyszka, K., Moraru, D., Samanta, A., Mizuno, T., Jablonski, R., Tabe, M.: Effect of selective doping on the spatial dispersion of donor-induced quantum dots in Si nanoscale transistors. Appl. Phys. Express **8**, 094202 (2015)

Dielectric SiO_2 Sol-Gel Coatings for Microelectronics

Dmitry Kovalenko(✉), Vasily Vaskevich, Vladimir Gaishun, and Sergei Khakhomov

Francisk Skorina Gomel State University, Sovetskaya 104, 246019 Gomel, Belarus
dkov@gsu.by

Abstract. Research has been conducted to investigate the selection of initial organic silicon compounds and the conditions for the formation of sol-gel coatings to achieve surface planarization. Experiments were conducted to determine the optimal modes of depositing the film-forming solution through spin-coating and the modes of heat treating the resulting coatings. The roughness and planarization of sol-gel coatings deposited on the integrated circuit surface with aluminum tracks were investigated using profilometry and scanning microscopy methods. The studies analyzed the thickness and homogeneity of the structure of the resulting materials using a scanning electron microscope. The capacitance-voltage and current-voltage dependences have been established. The dielectric constant, the value of the voltage shift of the flat zones, and the capacitance of the flat zones have been calculated.

Keywords: sol-gel · coating thickness · profilogram · planarization · capacitance-voltage characteristics · leakage current · dielectric constant

1 Introduction

The contemporary advancement of capacitor and transistor elements of integrated circuits requires the introduction of novel thin-film materials into the microelectronics industry. In order to increase the number of transistor structures without altering technological processes, it is necessary to develop accessible materials for multilevel systems. For such purposes, a coating is required to assure smoothing of the lower level of the completed integrated circuit, so that another integrated layer can be formed on the surface [1]. Currently, vacuum-formed thin layers (phosphosilicate glass, borophospho-silicate glass, etc.) are used to produce the required dielectric layer.

Sol-gel materials synthesised by a chemical method based on a set of various initial silicon compounds, such as methyltriethoxysilane, phenyltriethoxysilane, venyltriethoxysilane, etc., are of considerable interest due to their simplicity and low cost. The introduction of additives into the SiO2 structure enables the creation of a material with a set of predetermined properties: thermal, plastic, electrophysical, mechanical, etc. [2–4].

In recent times, significant advancements have been achieved in the field of organic-inorganic hybrids, with a specific focus on materials based on modified compounds of

Y. Ono and J. Kondoh (Eds.): Inter-Academia 2023, LNNS 939, pp. 72–80, 2024.
https://doi.org/10.1007/978-3-031-54450-7_10

silicon, titanium, phosphorus, and other substances [5, 6]. Within these coatings, fragments of organic compounds are incorporated into the structure of the initial matrix, exhibiting resistance to the thermal expansion of the composite surface of the microcircuit. Therefore, obtaining dielectric planarizing sol-gel coatings for microelectronic integrated circuits is a pressing matter.

2 Preparation of Materials

Currently, the main component for the production of organosilicon materials is tetroethylorthosilicate, which is confirmed by publications that describe a method for producing stable coatings based on the hydrolysis of TEOS [3]. However, such coatings are characterized by a lack of plasticity, which makes it possible to smooth out large differences in metallization up to 2 microns. To impart plasticity to the resulting coatings and modify their coefficient of thermal expansion, it is necessary to develop hybrid compositions in which organic compounds such as phenyltriethoxysilane, venylriethoxysilane, and methyltriethoxysilane will be used together with TEOS. It was experimentally found that the most promising for microelectronics is the use of methyltriethoxysilane (MTES). In this work, the optimal composition was experimentally selected based on organome-tallic silicon compounds, i.e., methyltriethoxysialane and tetraethyl orthosilicate produced by Sigma-Aldrich.

To establish the way the methyltriethoxysilane (MTES) concentration in a tetraethylor-thosilicate (TEOS)-based solution affects the planarizing and electrical properties, a set of film-forming solutions were prepared. These solutions consisted of different mass ratios of MTES:TEOS (2:1, 5:3, 4: 3, 1:1). A classic TEOS-based solution was prepared as well. Hydrolysis of the resulting solutions was carried out in an acidic medium; iso-propyl alcohol and isobutyl alcohol were used as solvents. In order to achieve complete maturation, the solutions were kept at room temperature (22 ± 2) °C for 2–3 days.

It is well known that solutions of film-forming substances used in the formation of homogeneous coatings must have a specific viscosity, with the optimal value ranging between 3 mm^2/s and 9 mm^2/s. At higher viscosity, the spreading of the solution is impeded, resulting in the creation of a film with nonuniform thickness. Conversely, at lower viscosity, a thin coating is produced that is inadequate for its intended purpose as a planarizing and interlayer dielectric. The research findings show that the resulting film-forming solutions have a viscosity in the range from 4.5 mm^2/s to 5.5 mm^2/s, making them suitable for the formation of uniform coatings on the surface of silicon wafers in integrated microelectronics. Experimentally determined modes for depositing sol-gel films via the centrifugation method were chosen to produce homogenous coatings of the required thickness. The process of selecting deposition modes was conducted via an Apogee Cee 200X thin film centrifuge. The resulting film-forming solutions were deposited on silicon wafers KDB-4 and KDB-12 with a diameter of 100 mm. After deposition, the wafers were placed in a muffle furnace and annealed in air. During heat treatment, there occurs evaporation of solvent and enhancement of polycondensation processes, formation of spatial structure of an organosilicon polymer. The only way to obtain a uniform coating on the substrate surface is to uniformly heat the component. The heat treatment mode was chosen to prevent the occurrence of cracks in the metallization areas and the transitions between non-metallized areas. The wafer in the oven was heated to 400 °C for 120 min (heating rate ≈ 3.3 °C per minute) and maintained at this temperature for 60 min. Then, the deposited substrate experiences a cooling process that occurs simultaneously with the oven for 60–80 min. After undergoing heat treatment and cooling, the wafers are removed from the furnace and placed in special plastic containers for further examination of their properties.

3 Results and Discussion

The thickness and uniformity of the resulting sol-gel coatings were studied by chipping a silicon wafer with the formed sol-gel coating using an S-4800 scanning electron microscope (Hitachi, Japan). The obtained SEM images clearly show the resulting layers of sol-gel coatings, which have a uniform structure throughout the entire thickness without visible inclusions and defects (Fig. 1). The thickness of the resulting coatings varies from 400 nm (for a coating based on TEOS) to 980 nm (for a coating with a MTES:TEOS ratio of 2:1).

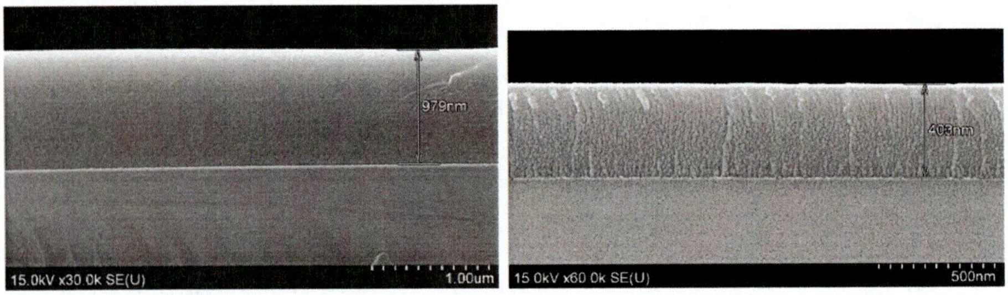

MTES:TEOS ratio of 2:1   Coating based on TEOS

Fig. 1. Images of longitudinal chips of the studied samples

The dielectric layer thicknesses that have been acquired will be necessary for the computation of the electrophysical values of the resulting sol-gel coatings.

Initial studies of the planarization of the resulting sol-gel coatings were carried out using a Surtronic 25 profilometer (Taylor Hobson, UK). This profilometer enables rapid evaluation of the planarization and surface roughness of the resulting samples (Fig. 2–3).

The results of studying the surface roughness of the resulting coatings using a profilometer show that the deposition of nanostructured sol-gel coatings can reduce the surface roughness of aluminum metallization from 46.5 nm to 9.3 nm.

A study of the metallization profile height in the wafer area with a metallization width of 15 μm at intervals of 15 μm indicates a smoothing of these areas with a height from 1 μm to a height of 0.4–0.5 μm.

The results of studying the profile height indicate a smoothing of the transition areas from the silicon base wafer to metallization from 1 μm (uncoated wafer) to 0.4–0.5 μm (sol-gel-coated wafer). Using a single-layer sol-gel coating improves surface

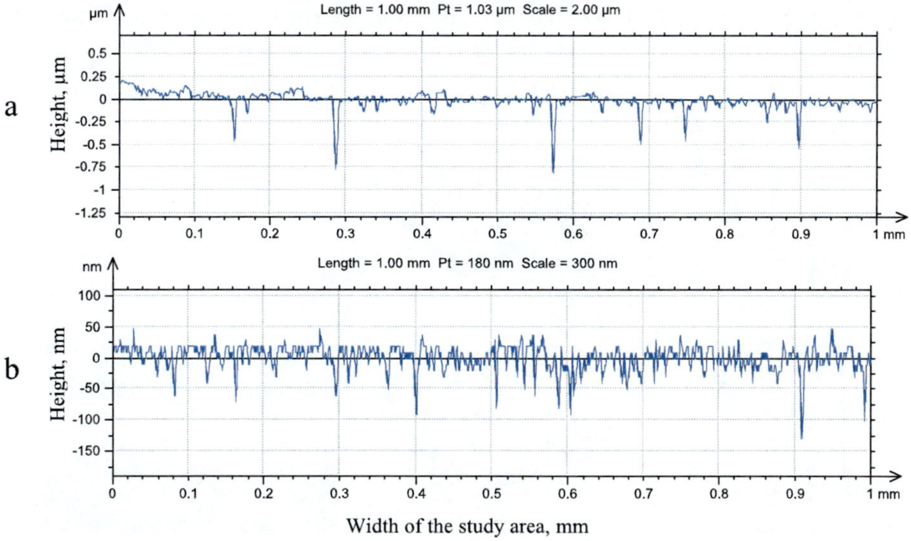

Fig. 2. The profilogram of the Sect. 1 mm with aluminum metallization: a – without cover, b – with sol-gel coating

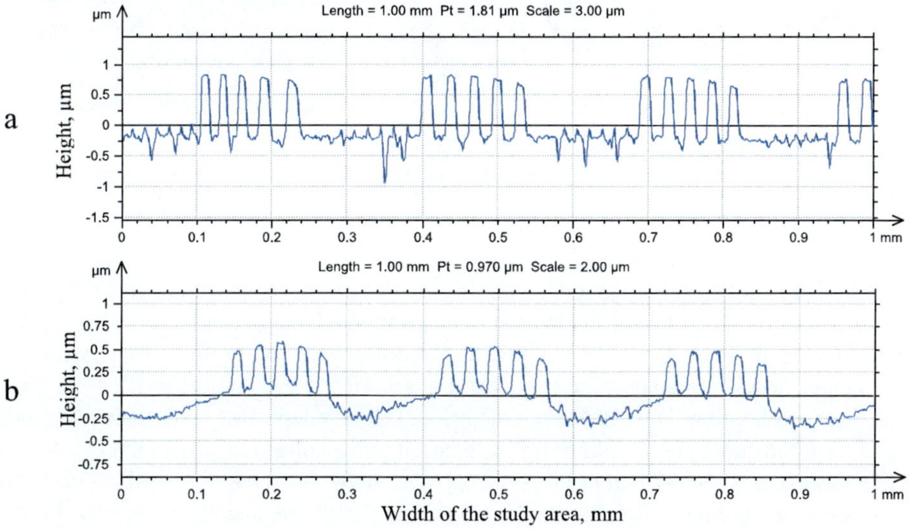

Fig. 3. The profilogram of the study area: a – without cover, b – with sol-gel coating

planarization by a factor of 2 due to the efficient solution filling of intermetallization gaps.

In order to conduct a comprehensive investigation on the planarization of integrated circuits using sol-gel coatings, surface analyses were performed via a scanning probe microscope SOLVER P47-PRO (NT-MDT LLC) and a scanning electron microscope S-4800.

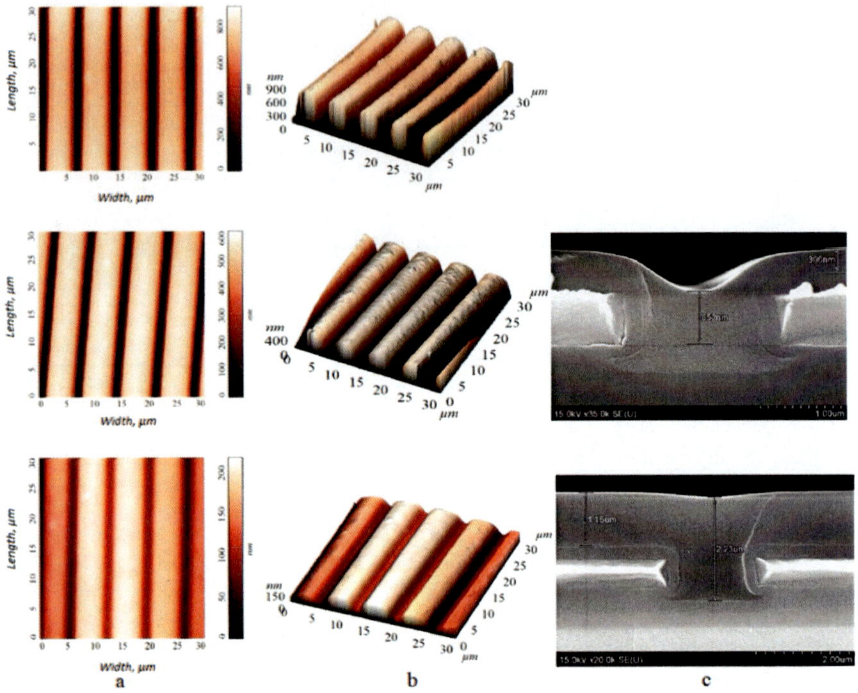

Fig. 4. Images of a test section of a wafer with a metallization width of 3 μm and a gap width of 3 μm

The AFM analysis of the specific area (Fig. 4, a-b) reveals smoothing of the structural area surface from 900 nm to 400 nm after depositing one layer of a sol-gel coating. After the formation of a two-layer sol-gel coating, smoothing occurs up to 150 nm, which is satisfactory and signifies that the film-forming solution has completely filled the gaps between metallization. The SEM images (Fig. 4, c) clearly depict the layers being formed, i.e., the silicon substrate, aluminum metallization and a sol-gel layer.

Irrespective of the number of layers employed, the sol-gel coatings exhibit a uniform structure over their entire thickness, devoid of any visible inclusions or defects. Moreover, they effectively fill the areas between metallization:

- thickness of a single-layer sol-gel coating: 306 nm in the metallization areas and 652 nm in the silicon substrate areas;
- thickness of a two-layer sol-gel coating: 1.16 μm in the metallization areas and 2.23 μm in the silicon substrate areas.

The investigational study of the electrical properties of capacitance-voltage (CV) and current-voltage characteristics was performed on a stand with a probe station using

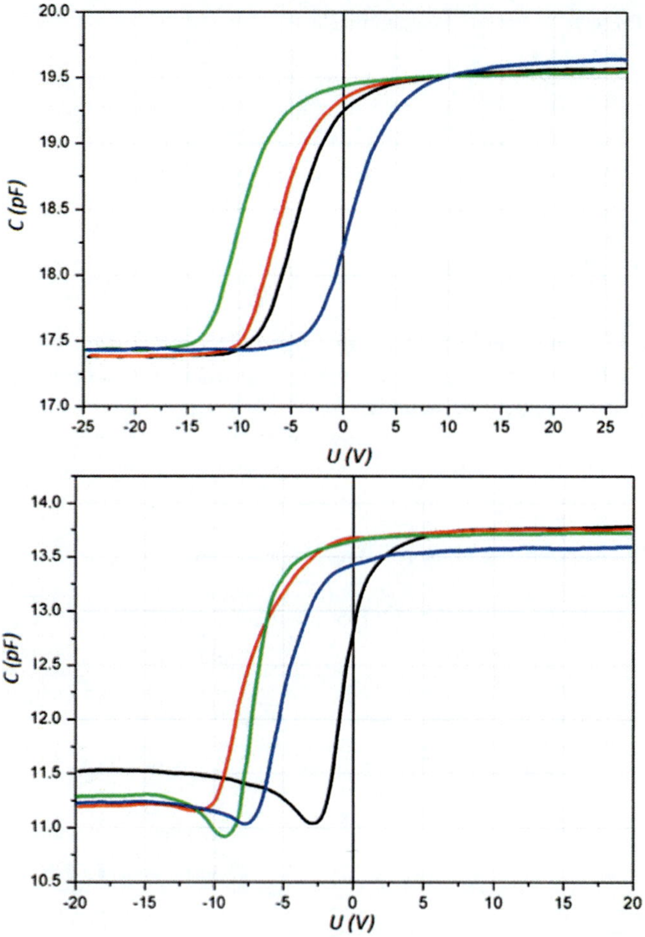

CV characteristics coatings with ratio MTES:TEOS – 2:1

CV characteristics coatings with ratio MTES:TEOS – 4:3

Fig. 5. Current-voltage characteristics of the obtained dielectric sol-gel coatings (4 measurements each).

a Keithley 4200A-SCS 3 parametric analyzer (Fig. 5). Additionally, to determine the capacitance of flat zones and the voltage of flat zones, we used formulas and methods that have been extensively presented and tested in the academic literature [6]. It is noteworthy that the outcomes of flat zone capacitance and flat zone stress estimation derived from the acquired graphs exhibit slight discrepancies when calculated using the arithmetic mean, as compared to the formula-based calculations. This discrepancy may be attributed to the use of an averaging approach based on a series of four measurements. The tabular values of certain quantities and constants were used to calculate the main characteristics of the resulting dielectric coatings (Table 1).

The dielectric constant of the SiO_2 layer was calculated using the following formula:

$$\varepsilon = \frac{C_{\max} d}{\varepsilon_0 S}$$

Table 1. Values of the quantities used.

Calculated values	Designation	Value
Vacuum permittivity	ε_0	8,854 pF/m
Metal-semiconductor contact area	S	$0{,}72 \cdot 10^{-9}$ m^2
Permittivity Si	ε_{Si}	11,7
Capacity SiO_2	C_{SiO2}	425,81 pF

where C_{max} is the dielectric capacitance, ε is the effective dielectric constant of the resulting coating, *d* is the thickness of the dielectric layer, S is the metal-semiconductor contact area, $\varepsilon 0$ is the dielectric constant of vacuum.

The main electrical characteristics of the resulting dielectric sol-gel coatings are presented in Table 2.

Table 2. CV characteristics of the resulting sol-gel coatings.

Calculated values	Designation	Result	
		Ratio 2:1	Ratio 4:3
Dielectric capacity (maximum value according to CV characteristics)	C_{max}	19,59 pF	13,56 pF
Dielectric capacity (minimum value according to CV characteristics)	C_{min}	17,39 pF	10,45 pF
Dielectric thickness	d	979 nm	729 nm
Dielectric constant of the coating	ε	2,51	1,85
Shear stress of flat zones (calculated)	V_{fb}	−6 V	−0,5 V
Shear stress of flat zones (from graph)	$V_{fb\text{-график}}$	−9,6 V	−6,0 V

The dielectric constant of the resulting coatings varies from 1.53 to 2.51, which is good for organic-inorganic vitreous dielectrics up to 1 μm thick. An increase in the MTES concentration leads to an increase in the dielectric constant of the resulting coatings, which may be due to the formation of organic-inorganic compounds that have a higher dielectric constant compared to the standard SiO2-based oxide. The results of the study of capacitance-voltage characteristics show that the magnitude of the voltage shift of flat zones is negative and at the maximum value is about minus 9.6 V, which corresponds to a change in the effective charge density at the interface of 7.2×10^{-8} m^2 and is a small and acceptable value for a given thickness of the dielectric layer. Using the measured characteristics (Table 2), theoretical calculations of the magnitude of the voltage shift of flat zones were made. For all samples, this value is less than that obtained in the experiment and does not exceed 7 V, which is the optimal value for a dielectric with a thickness of (0.7–1.0) microns.

To determine the leakage currents, the current-voltage characteristics of the resulting coatings were plotted (Fig. 6).

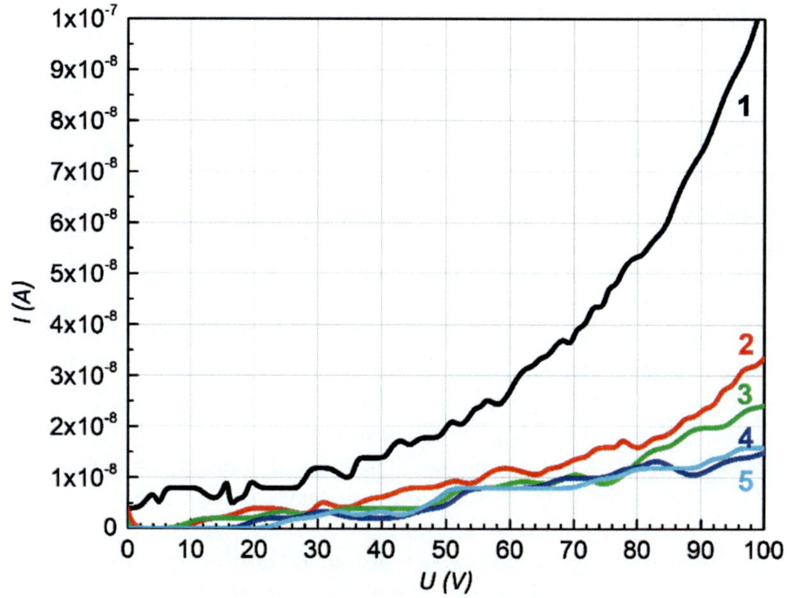

Fig. 6. Current-voltage characteristics of the obtained sol-gel coatings

The maximal values of currents for coatings derived from hybrid solutions based on TEOS + MTES are one order of magnitude less than for TEOS-based coatings. In this case, the increase in leakage current is clearly visible at U > 75 V (Fig. 6). The aver-age values of leakage currents for the coatings under study are presented in Table 3.

Table 3. The obtained data on the leakage current value.

Ratio (MTES:TEOS)	Curve number on the graph (Fig. 5)	Average leakage current at voltage + 5 V, nA	Average leakage current at voltage + 50 V, nA
2:1	4	0,9	2,0
5:3	5	0,9	2,1
4:3	3	0,9	1,9
1:1	2	1,0	2,7
without MTES	1	6,0	20

The results of studying the current-voltage characteristic reveals that the TEOS-based coating has is a significant leakage current, which is 20 nA at U = 50 V. For other samples, the leakage current is considerably lower, i.e., less than 1 nA for a voltage of plus 5 V and less than 3 nA for a voltage plus 50 V.

4 Conclusion

The results of studying the surface roughness of the resulting coatings using a profilometer show that the deposition of nanostructured sol-gel coatings can reduce the surface roughness of aluminum metallization from 46.5 nm to 9.3 nm. A study of the metallization profile height in the wafer area with a metallization width of 15 μm at intervals of 15 μm indicates a smoothing of these areas with a height from 1 μm to a height of 0.4–0.5 μm.

The results of studying of three-dimensional AFM images of the surface of structural elements show that the formation of a single-layer sol-gel coating makes it possible to reduce the difference in metallization heights from 900 nm to 500 nm, and the formation of a two-layer coating – 100 nm.

The dielectric constant of the resulting coatings varies from 1.53 to 2.51, which is a good result for organic-inorganic glassy dielectrics up to 1 μm thick.

For all samples, this value is less than that obtained in the experiment and does not exceed 7 V, which is the optimal value for a dielectric with a thickness of (0.7–1.0) microns.

The results of studying the current-voltage characteristic show that in the TEOS-based coating there is a large leakage current, which is 20 nA at U = 50 V. For other samples, the leakage current is significantly less, i.e., less than 1 nA for a voltage of plus 5 V and less than 3 nA for voltage plus 50 V.

Acknowledgements. Work is partially supported by the State Program of Scientific Research of the Republic of Belarus 2.16 "Nanostructure".

References

1. Redzheb, M., Armini, S.: Enabling bottom-up nanoelectronics fabrication by selective sol-gel dielectric-on-dielectric deposition. Mater. Sci. Eng. **263**, 114808 (2021)
2. Vaskevich, V., Kovalenko, D., Gaishun, V.: Synthesis and study of silicate sol–gel coatings for micro- and nanoelectronics. Nanosyst. Nanomater. Nanotechnol. **12**, 279–293 (2014)
3. Pilipenko, V.A., et al.: Planarization of an interlevel dielectric using liquid glass. Vestnik BSU **3**, 32–36 (2005)
4. Vaskevich, V., et al.: Dielectric sol-gel coatings based on silica dioxide for surface planarization of integral microcircuits. Probl. Phys. Math. Tech. **4**(45), 7–14 (2020)
5. Das, S.: Inorganic – organic hybrid nanoparticles from n-octyl triethoxy silane. J. Colloid Interface Sci. **252**, 82–88 (2002)
6. Munishamaiah, K.: Improved dielectric properties of microwave irradiated sol-gel derived SiO_2–TiO_2 thin film. Results Eng. **4**, 100033 (2019)

Fabrication and Characterization of Silicon Tunnel Diodes Doped by Short-Time Rapid Thermal Annealing

Yutaro Sasaki[1,2], Shun Masui[1,2], Shumpei Miura[1,2], and Daniel Moraru[1,2](✉)

[1] Research Institute of Electronics, Shizuoka University, Hamamatsu, Japan
moraru.daniel@shizuoka.ac.jp

[2] Graduate School of Integrated Science and Technology, Shizuoka University, Hamamatsu, Japan

Abstract. Band-to-band tunneling (BTBT) has attracted increasing attention as a mechanism for tunnel field-effect transistors (TFETs), promising higher switching speed than silicon (Si) metal-oxide-semiconductor field-effect transistors (MOSFETs). To investigate the BTBT mechanism in Si nanodevices, we fabricated and characterized nanoscale tunnel (Esaki) diodes based on silicon-on-insulator (SOI) substrates. Characteristics of Esaki diodes were observed in devices with large widths, but not in the target nanoscale devices. This can be related to the short-time rapid thermal annealing (RTA) process used in the fabrication of these diodes. This study presents the technological challenges and competing processes that can be tuned for fabrication of nanowire tunnel diodes in Si.

Keywords: tunnel (Esaki) diode · rapid thermal annealing (RTA) · band-to-band tunneling (BTBT) · doping · silicon-on-insulator (SOI)

1 Introduction

Conventional switching devices, in particular metal-oxide-semiconductor field-effect transistors (MOSFETs), have limited switching speed. This is due to the specific transport mechanism used in such MOSFETs, which involves thermally-activated carrier transport in the same band (conduction band or valence band, depending on the type of transistor). In recent decades, tunnel field-effect transistors (TFETs) have been proposed as an alternative because they use interband or band-to-band tunneling (BTBT) as their main mechanism [1–3]. Because of such design, thermally-activated carriers are eliminated from contributing to the current, a condition which makes TFETs more attractive due to their faster switching speed than MOSFETs in an ideal situation.

However, silicon (Si) TFETs suffer due to the indirect-bandgap nature of Si, which makes the BTBT rates relatively low. For such indirect semiconductors, BTBT requires phonon assistance, which subsequently limits the current level [4–7]. Previous studies by our group have suggested that dopants present at the *pn* junction (in the depletion layer) can affect the BTBT current in nanoscale (practically two-dimensional) Esaki diodes

Y. Ono and J. Kondoh (Eds.): Inter-Academia 2023, LNNS 939, pp. 81–88, 2024.
https://doi.org/10.1007/978-3-031-54450-7_11

[8, 9]. Therefore, it can be expected that dopant-induced energy states can localize the carrier wavefunctions and implicitly relax the BTBT conditions. However, an abrupt tunnel junction with nanoscale depletion layer must be achieved first.

It is known that rapid thermal annealing (RTA) can be used for a number of modifications in the semiconductor device fabrication and material optimization. For instance, RTA has been recently reported to be used to form homogeneous metal alloys in different phases [8]. RTA has also been reported as a technique to passivate interface traps and other defects, such as in compound-semiconductor diodes [9]. In Si/SiO_2 systems, low-temperature hydrogen annealing or rapid thermal oxidation are also used to passivate the interface traps [10]. However, for the purpose of maintaining an abrupt *pn* junction, RTA can be applied for the high-temperature drive-in of dopants into Si nanodevices.

In this study, we focus on the fabrication and characterization of such nanoscale tunnel diodes based on silicon-on-insulator (SOI) substrates using a short-time (few tens of seconds) rapid thermal annealing (RTA) process for the drive-in of dopants during the doping process. Tunnel-diode behavior is observed in large devices, but not in nano-patterned devices, which were the main purpose of the nano-fabrication. This suggests that the fabrication including short-time RTA drive-in process must be optimized.

2 Experimental Details

We fabricate lateral nanoscale tunnel (Esaki) diodes based on SOI substrates, as shown in Fig. 1. The left panel shows a schematic top view of the $p^+ - n^+$ diode (central area), while the right panel shows a 3D view of the device, also including the biasing circuit. All fabrication processes are carried out in a cleanroom environment, based on CMOS-compatible technologies, similarly to other previous reports [11, 12]. However, in order to fabricate diodes with steep *pn* junctions at high doping concentrations, the drive-in process of doping was performed at 1050 °C for approximately 20 s, carried out using a RTA unit (MILA-5050 mini-lamp annealer from Advance Riko, Japan). Temperature is gradually increased/decreased from/to 600 °C at a rate of usually 20 °C/s. The concentrations of doping obtained for the *n*-type and *p*-type areas were: $N_D \approx 2.2 \times 10^{20}$ cm^{-3} and $N_A \approx 2.0 \times 10^{20}$ cm^{-3}, which are, in principle, sufficiently high to ensure that both sides are degenerately doped. It should be mentioned, however, that a study of the effect of the RTA drive-in time is necessary to confirm the suitable balance between the competitive effects of side diffusion of dopants (detrimental for an abrupt *pn* junction) and diffusion of dopants in the entire depth (advantageous for a homogeneous *pn* junction).

Device parameters were changed systematically, with the width of the SOI nanostructure designed to be W = 100–7000 nm. The devices with designed width W = 7000 nm are patterned only before doping and nano-patterning is not carried out afterwards. For other devices, nano-patterning was carried out after the doping; this is an important factor in the operation of the devices, as will be discussed later.

Figure 2 shows two devices by the optical microscope and scanning electron microscope (SEM) images. The left figure in Fig. 2(a) shows a micrograph of a 7000-nm-wide device, and the SEM image of the device is shown on the right, focused on the central area. Similarly, Fig. 2(b) shows a micrograph and SEM image of a device designed

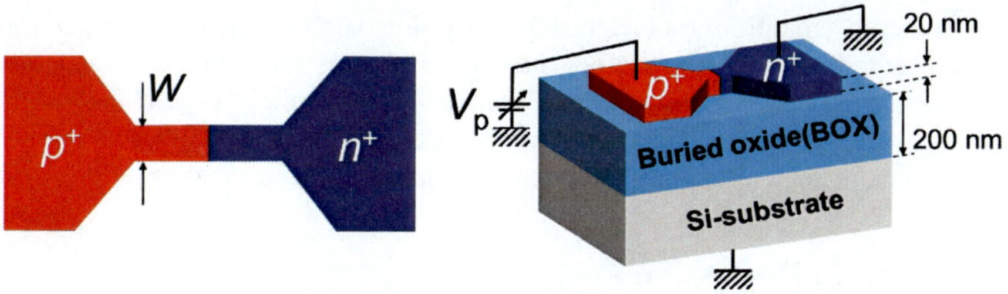

Fig. 1. Device structure for a SOI $p^+ - n^+$ diode (top view in the left panel and 3D view in the right panel), together with its measurement circuit. The approximate thicknesses of the layers are also indicated.

with a 100-nm-wide constriction. A codoped region (so-called overlap region), where P-donors and B-acceptors are both doped, was designed to be 0–500 nm, in order to increase the likelihood of the formation of the *pn* junction within the nanostructure. As references, *pin* diodes were also fabricated with the *i*-layer designed to be 100–500 nm. The analysis of these *pin* diodes is beyond the scope of this work and will be reported elsewhere.

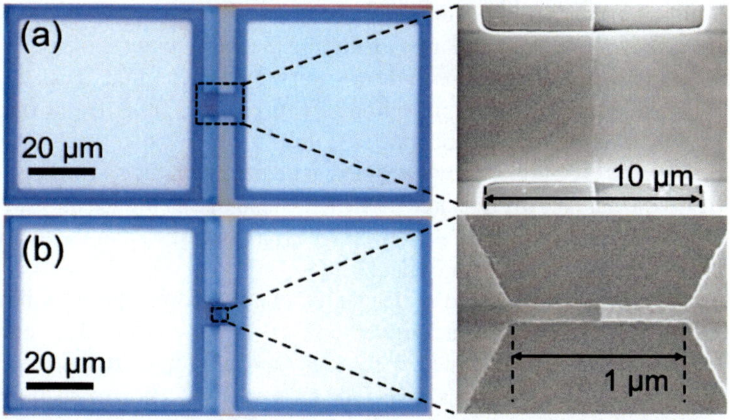

Fig. 2. (a) A device with a width of 7000 nm (without nano-patterning) shown by a microscope image (left) and by an SEM image (right). (b) A device with a width designed to be 100 nm shown by a microscope image (left) and by an SEM image (right).

From Fig. 2, the steps left in the thickness of the SOI layer can be used to identify the edges of the doping masks. In general, control of the doping masks can be achieved within a range of hundred-nm. The depth of doping inside the SOI layer is unknown, but it is expected that doping concentrations are significantly higher at the corners of the structures, where the doping sources surround the Si surface from multiple sides. For the nano-patterned device (as seen in the right panel of Fig. 2(b)), weak line edge roughness (LER) is observable, which may also influence the electrical characteristics. However, this observed LER is still small and it is only affected by the electron-beam

(EB) lithography conditions (not by the RTA conditions). In fact, the EB nano-patterning is subsequent to both doping processes (first with P-donors and then with B-acceptors).

To confirm the electrical properties of such devices, current-voltage ($I_{\rm p} - V_{\rm p}$) measurements were performed under low temperature conditions ($\approx$ 8.3 K) in a vacuum prober system, using a circuit as illustrated schematically in Fig. 1 (right).

3 Results and Discussion

Electrical characterization was performed by measuring $I_{\rm p} - V_{\rm p}$ characteristics while the *n*-lead and the substrate Si are grounded. The $I_{\rm p} - V_{\rm p}$ characteristics are measured in both reverse ($V_{\rm p} < 0$ V) and forward ($V_{\rm p} > 0$ V) regimes.

For devices with a width of 7000 nm, typical tunnel (Esaki) diode behavior is observed, i.e., the current increases rapidly in both the forward and reverse directions around $V_{\rm p} \approx 0$ V, and a negative differential conductance (NDC) peak is observed at $V_{\rm p}$ $\approx$ 100 mV, as shown for one example in Fig. 3.

This indicates that high-concentration doping was successfully achieved, even though only 20-s drive-in was used for both P-doping and B-doping. According to our rough estimations, considering the drive-in temperature and time, the diffusion length for P-donors and B-acceptors is about 100 $\pm$ 50 nm, larger for P-donors because they are exposed to twice the thermal budget (P-doping is done first, followed by B-doping).

The device shown in Fig. 3, although exhibiting the basic features of an Esaki diode, also exhibits many current inflections in the low forward-bias regime. It can be expected that phonon-assisted BTBT current inflections are mixed with other features, likely due to the 2D quantization of the thin codoped SOI film.

For a clearer analysis, Fig. 4 shows a different device, with the same designed width ($W = 7000$ nm) but a larger overlap area (500 nm instead of 200 nm for the device shown in Fig. 2). The first-derivative curve is also shown along with the $I_{\rm p} - V_{\rm p}$ characteristics in forward-bias regime in the zoom-in panel.

The NDC peak can be observed more clearly, but also some current inflections are more prominently noticeable. In particular, the first derivative has peaks around 20 and 60 mV of $V_{\rm p}$, which are consistent with the values expected for TA (transverse acoustic) and TO (transverse optical) phonons in Si. The clearer observation of such features for this device suggests that the indirect-bandgap nature of Si is preserved, even though the thickness of the SOI layer is below 20 nm. Therefore, the quantum-size effect is not sufficiently activated to modify the indirect-bandgap nature of Si for this device.

From the examples shown above, it can be understood that high doping concentrations are achieved in such lateral SOI diodes, at least for the case when nano-patterning is not performed after doping. As explained above, higher doping concentration is expected along the corners (edges) of the SOI layer.

For fabricating smaller devices – with the target being nanowire tunnel diodes – we use a subsequent electron-beam lithography (EBL) technique to pattern the nanostructures to 100–3000 nm. By this process, the sides of the central regions are etched away using a reactive ion etching (RIE) process, and current flows only through the core part of the region that had an initial width of 7000 nm.

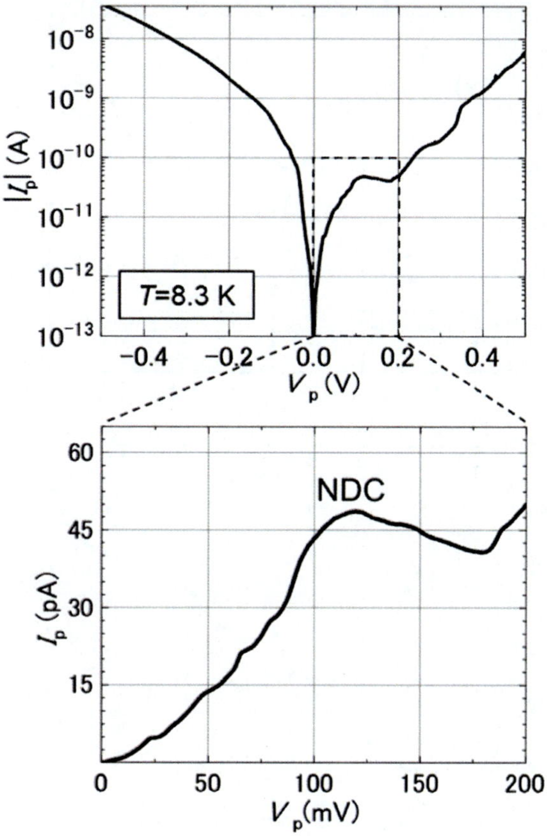

Fig. 3. Low-temperature ($T \approx 8.3$ K) $I_\mathrm{p} - V_\mathrm{p}$ characteristics of a device with a co-doped region designed to be 200 nm and a width designed to be 7000 nm, in reverse- and forward-bias regimes. The device exhibits the features of Esaki diodes.

Interestingly, the vast majority of the nanostructured devices do not exhibit the behavior of tunnel (Esaki) diodes. Figure 5 illustrates two examples of the characteristics that are obtained, with some not flowing any current in a wide range of bias, as shown in Fig. 5(a), and with others showing a current increase only after a no-current region at low biases, as shown in Fig. 5(b). We hypothesize that such a low yield is due to shallow doping as a result of the short-time drive-in processes. Current pathways are mainly formed at the edges of the large (non-patterned) area. These edges are removed by nano-patterning, which leaves behind only the lower-doped core parts available for transport. Another reason for such low yield can be related to the non-uniformity of the dopant distribution (either of P-donors or B-acceptors) which can make the local doping concentration much lower than designed. It is also possible that subsequent thermal processing of the sample can affect the doping concentrations further (e.g., for P-doping, one should take into account that B-doping is carried out later at the same thermal-budget conditions, while for B-doping, a short thermal-oxidation process for surface passivation may induce B-acceptor segregation into the oxide). Finally, strong

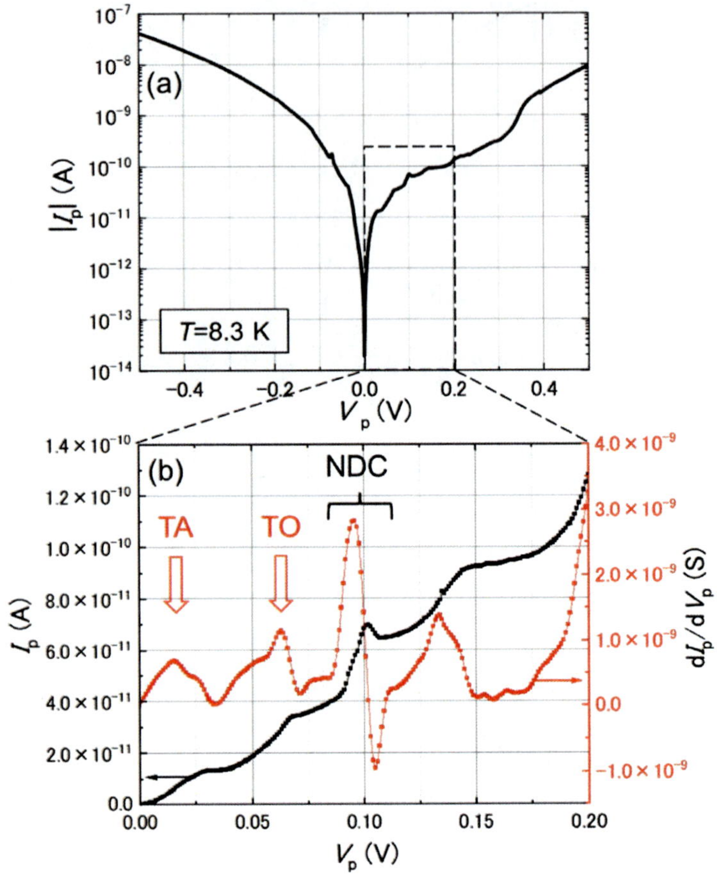

Fig. 4. (a) Low-temperature ($T \approx 8.3$ K) $I_p - V_p$ characteristics of a device with a co-doped region designed to be 500 nm and a width designed to be 7000 nm. (b) Forward-bias $I_p - V_p$ and $dI_p/dV_p - V_p$ (first derivative) characteristics. Features that can be ascribed to TA (transverse acoustic) and TO (transverse optical) phonon-assisted BTBT are marked.

compensation between P-donors and B-acceptors, especially in the overlap region, may result in an effective doping concentration much lower than N_D or N_A taken separately (i.e., in a much higher resistance). Such challenges require further study to clarify the key processes to optimize for high-yield nanowire Esaki diode fabrication [13].

The findings obtained from this work imply that, in order to fabricate nanowire Esaki diodes with full doping in such structures, the nano-patterning process should be optimized in the future to ensure full and homogeneous doping of such structures.

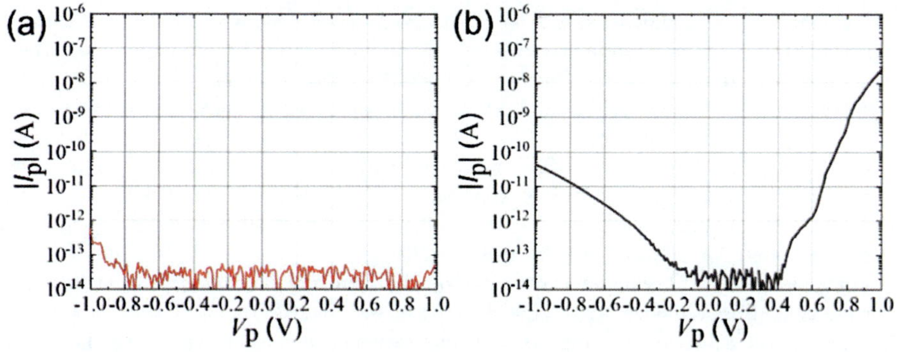

Fig. 5. (a)–(b) Low-temperature ($T \approx 8.3$ K) $I_P - V_P$ characteristics for two devices with nano-patterned central regions (designed width W < 200 nm) containing an overlap region, exhibiting no features resembling tunnel-diode features, but with different behavior: (a) practically no current in a wide range of bias; (b) current starting to flow at higher bias, beyond a no-current low-bias region.

4 Conclusions

Based on the measurements of devices with a width of 7000 nm (without nano-patterning), it can be concluded that we have successfully fabricated tunnel (Esaki) diodes. However, current flow was absent or limited in our target nanoscale devices. A most likely reason is the removal of the highly-doped edges of the central region by the nano-patterning process. This problem can be improved by doping after nano-patterning, although further research will be needed for optimizing other processes, as well, for fabrication of Si nanowire highly-doped tunnel diodes.

Acknowledgments. This work was partially supported by JSPS KAKENHI (no. 19K04529 and no. 22K04216). The authors thank K. Yamaguchi, Y. Iwasaki, and Y. Tamura for their contributions to initial device fabrication steps.

References

1. Waldrop, M.M.: The chips are down for Moore's law. Nat. News **530**, 144–147 (2016)
2. Ionescu, A.M., Riel, H.: Tunnel field-effect transistors as energy-efficient electronic switches. Nature **479**, 329–337 (2011)
3. Seabaugh, A.C., Zhang, Q.: Low-voltage tunnel transistors for beyond CMOS logic. Proc. IEEE **98**, 2095–2110 (2010)
4. Esaki, L.: New phenomenon in narrow germanium $p - n$ junctions. Phys. Rev. **109**, 603 (1958)
5. Chynoweth, A.G., Feldman, W.L., Logan, R.A.: Excess tunnel current in silicon Esaki junctions. Phys. Rev. **121**, 684 (1961)
6. Chynoweth, A.G., Logan, R.A., Thomas, D.E.: Phonon-assisted tunneling in silicon and germanium Esaki junctions. Phys. Rev. **125**, 877 (1962)
7. Schmid, H., Bessire, C., Björk, M.T., Schenk, A., Riel, H.: Silicon nanowire Esaki diodes. Nano Lett. **12**, 699–703 (2012)

8. Janardhanam, V., Boldbaatar, S., Jyothi, I., Kim, D.-H., Shim, K.-W.: Rapid-thermal-annealing-induced microstructural evolution of Au/Ni/β-Ga_2O_3 Schottky diodes correlated with their electrical properties. J. Alloy. Compd. **918**, 165622 (2022)
9. Alburaih, H.A., Albalawi, H., Henini, M.: Effect of rapid thermal annealing on the electrical properties of dilute GaAsPN based diodes. Semicond. Sci. Technol. **34**, 105009 (2019)
10. Gad, K.M., et al.: Improved Si/SiO_x interface passivation by ultra-thin tunneling oxide layers prepared by rapid thermal oxidation. Appl. Surf. Sci. **353**, 1269–1276 (2015)
11. Tabe, M., et al.: Atomistic nature in band-to-band tunneling in two-dimensional silicon pn tunnel diodes. Appl. Phys. Lett. **108**, 093502 (2016)
12. Prabhudesai, G., et al.: Single-charge band-to-band tunneling via multiple-dopant clusters in nanoscale Si Esaki diodes. Appl. Phys. Lett. **114**, 243502 (2019)
13. Moraru, D.: Challenges and progress in the fabrication of silicon nanowire tunnel diodes. Int. J. Electr. Comput. Biomed. Eng. **1**, 57–65 (2023)

Plasma Physics and Biotechnology

Interaction of Low Voltage Microplasma Discharge with Stratum Corneum

Jaroslav Kristof[1] (✉), Marium Maliha[2], Sadia Afrin Rimi[3], Abubakar Hamza Sadiq[3], Alam Md Jahangir[4], Yoshinori Mizuno[3,5], and Kazuo Shimizu[1,2,3,4]

[1] Organization for Innovative and Social Collaboration, Shizuoka University, Hamamatsu 432-8011, Japan
jaroslav.kristof@gmail.com

[2] Graduate School of Integrated Science and Technology, Shizuoka University, Hamamatsu 432-8011, Japan

[3] Graduate School of Science and Technology, Shizuoka University, Hamamatsu 432-8011, Japan

[4] Graduate School of Medical Photonics, Shizuoka University, Hamamatsu 432-8011, Japan

[5] Research and Development Center, Asahikogyosha Co. Ltd., Narashino 275-0001, Japan

Abstract. Number of transdermal patches is growing ever year. However, not every drug is possible deliver through the skin effectively. This creates pressure on searching a new method able to deliver drugs transdermal pathway. One of such methods is treatment of the skin by device producing plasma affecting barrier function of the skin. In our study, permeability of sodium fluorescein through the most upper part of the skin, stratum corneum was demonstrated. Permeability of stratum corneum was enhanced by argon microplasma treatment or air microplasma treatment. Lipid oxidation inside stratum corneum was proved by presence of lipid peroxides using fluorescent dye LiperFluo by performing a frozen section of the full thickness skin. FTIR measurement showed correlation between lipid disordered structures in lipid matrix of stratum corneum and presence of lipid peroxides.

Keywords: stratum corneum · microplasma · permeability

1 Introduction

Number of studies dedicated to transdermal delivery using plasma is increasing [1–4]. Permeability of several drugs through the skin has been already tested [5–8]. However, the principle of this drug delivery method is not fully understood, yet. Molecules delivered into a body have to pass through the two main barriers of the skin. The first barrier is stratum corneum, composed of dead corneocytes with lipid matrix in between them. This lipid matrix creates the first layer of skin defense. The second layer of the defense presents stratum granulosum with its tight junctions. Previous research focused on study of the second barrier defense showed that tight junctions are "opened" temporally for approximately three hours [9, 10]. Previously was also shown that plasma can lead to

Y. Ono and J. Kondoh (Eds.): Inter-Academia 2023, LNNS 939, pp. 91–97, 2024.
https://doi.org/10.1007/978-3-031-54450-7_12

disordered lipid structure in stratum corneum [11, 12] or their oxidation [13, 14]. In this study, we will focus on stratum corneum and its permeability depending on treatment method by argon or air microplasma. Presence of lipid peroxides will be shown depending treatment method will be shown, too.

2 Experiment

2.1 Microplasma Treatment

Dielectric barrier discharge was generated by a thin-film electrode. Atmospheric argon or air microplasma was maintained at frequency of 24 kHz (in the case of argon) and 35 kHz (in the case of air). A sinus-shape function of the voltage was set by a function generator (Tektronix, AFG3102) and amplified by high voltage amplifier (Trek, model 5/80). Power delivered to discharge was 2.5 W. The flow of argon or air was set at 5 L/min by a flow meter (Yamato). The skin treatment lasted for 5 min. The distance between the electrode and the skin was 1 mm. Experimental set-up is shown in Fig. 1.

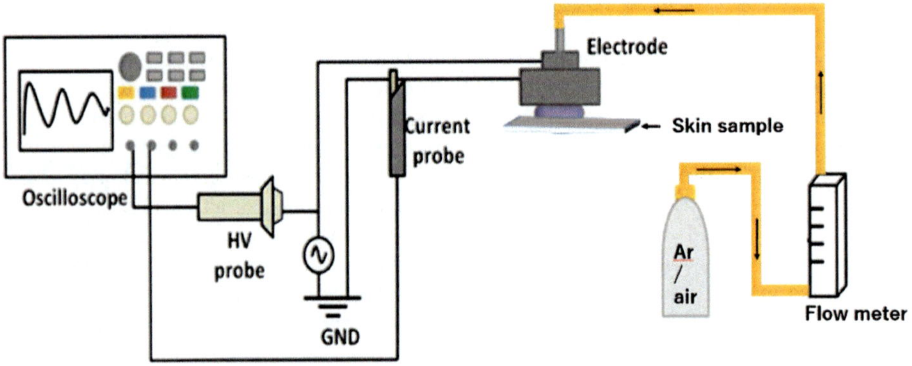

Fig. 1. Experimental set-up.

2.2 Penetration Study

Sample Preparation. The hairless Yucatan micropig's skin was ordered from Charles River Japan, Inc. (Yokohama, Japan). The skin samples were stored at −80 °C prior the experiment. In the day of experiment, the fat layer was removed by knife before cutting the skin into pieces of dimensions 2 cm × 2 cm. And then the skin samples were placed in 60 °C distilled water for 1 min after 3 h of soaking of the cut pieces at 4 °C in phosphate buffered saline (PBS). Finally, the epidermal layer was peeled off. Epidermis was placed on filter paper inside a culture dish and soaked by 0.5% (w/v) of Trypsin in 5.3 mmol/L EDTA·4Na solution for 2 h. Stratum corneum was separated by washing out in distilled water.

Franz Diffusion Cell. Franz diffusion cells were used to investigate the permeability of stratum corneum. Stratum corneum was used as a membrane between the receptor and the donor compartment. The diffusion area was equal to 1.65 cm^2. The donor compartment was filled with 1 ml of Sodium fluorescein solution dissolved in water at a concentration of 0.15 mg/ml. The receptor compartment with a volume equal to 10 ml was filled with PBS. The Franz cell was kept in a water bath at 37 °C with constant stirring.

2.3 ATR-FTIR Study

ATR-FTIR can tell us about the structures of lipids in stratum corneum. Lipids of the stratum corneum were characterized by the bandwidth of the methylene symmetric stretching vibration at 2,850 cm^{-1}. The bandwidth of a vibration is connected to the structural changes of the lipids. Usually, an increase in bandwidth means an increase in the interaction between functional groups. The increased interaction can be connected to disordering of lipid structure in stratum corneum. ATR-FTIR (Jasco FT/IR 6300 with ATR PRO610P-S) with a diamond prism was used to observe the upper layer of stratum corneum of full thickness pig skin with dimensions of 1 cm $\times$ 1 cm. Spectra were recorded with a resolution of 4 cm^{-1} and by accumulating 150 scans.

2.4 Frozen Section

The full thickness of the skin sample without the fat layer were cut down to smaller pieces (approximately 2.5 cm $\times$ 2.5 cm). The skin was cut to pieces of 5 mm wide and frozen in white tissue-coat at −80 °C for microscopic observation (skin sectioning). A frozen section of 10 μm thick were done by a microtome (NX70, Cryostar). Sectioned skin was placed on microslide glass for microscopic observation. Frozen section was dried for 10 min, then 1 ml of LiperFluo (Dojindo) was applied on skin section for 1 h. After 1 h, dye was washed in tap water and cover glass was mounted. Solution of distilled water and glycerol (50:50) was used as a mounting medium. Slide was measured by Nanozoomer S60, Hamamatsu Photonics. LipierFluo is fluorescent dye, selective for the detection of lipid peroxides. The excitation and emission wavelengths of the oxidized Liperfluo are 524 nm and 535 nm, respectively.

3 Results

3.1 Penetration Study

Figure 2 shows amount of Sodium fluorescein permeated through stratum corneum after 24 h.

Control represents non-treated stratum corneum. After argon plasma treatment, Sodium fluorescein permeation was increased by 48% in comparison to control stratum corneum. Using air microplasma treatment increased permeation of Sodium fluorescein by 117% in comparison to control stratum corneum. These results demonstrate that plasma affects most probably stratum corneum lipids as it was demonstrated in Fig. 2.

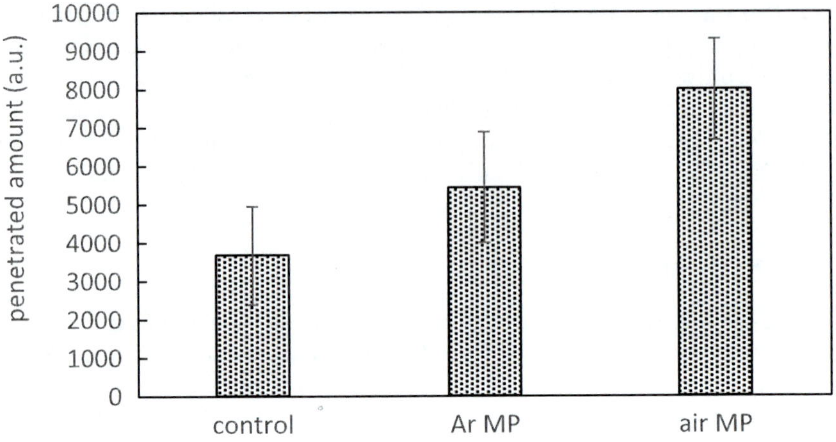

Fig. 2. Amount of permeated Sodium fluorescein through the non-treated (control), argon microplasma treated (Ar MP) and air microplasma treated (air MP) stratum corneum.

3.2 ATR-FTIR Study

We observed changes in the symmetric methylene stretching band in the stratum corneum after 5 min of treatment (Fig. 3 A, B). Symmetric CH_2 stretching band is component of several vibrational transitions. All these transitions were fitted by gaussian function. Bandwidth of symmetric CH_2 was after finding the best fit (Fig. 3A). Bandwidth of non-treated sample (control) was $14.8 \pm 0.3\ cm^{-1}$ (Fig. 3B). After argon microplasma treatment (Ar MP of Fig. 3B) of stratum corneum, the bandwidth increased to $15.6 \pm 0.2\ cm^{-1}$. When stratum corneum was treated by air microplasma (air MP of Fig. 3B), the bandwidth increased to $16.0 \pm 0.2\ cm^{-1}$. Observation of increasing of bandwidth after plasma treatment indicates increased disordering of stratum corneum lipids. This disordering of lipid structure is more effective after air microplasma treatment that argon microplasma treatment.

3.3 Frozen Section Study

Microplasma treatment of stratum corneum can lead to lipid peroxidation of lipid barrier and creation of lipid peroxides. Fluorescent dye LiperFluo was used to detect presence of lipid peroxides as a proof of lipid peroxidation. Figure 4 demonstrates presence of lipid peroxides in stratum corneum. Non-treated skin sample (Fig. 4A) does not indicate presence of any lipid peroxides. Argon microplasma treated skin sample (Fig. 4B) shows presence of lipid peroxides in upper part of stratum corneum. Air microplasma treated skin sample (Fig. 4C) display presence of lipid peroxides also in deeper layers when compare to argon microplasma treated skin. These results demonstrate that lipid peroxidation can be reason of higher permeability of stratum corneum after air microplasma treated stratum corneum.

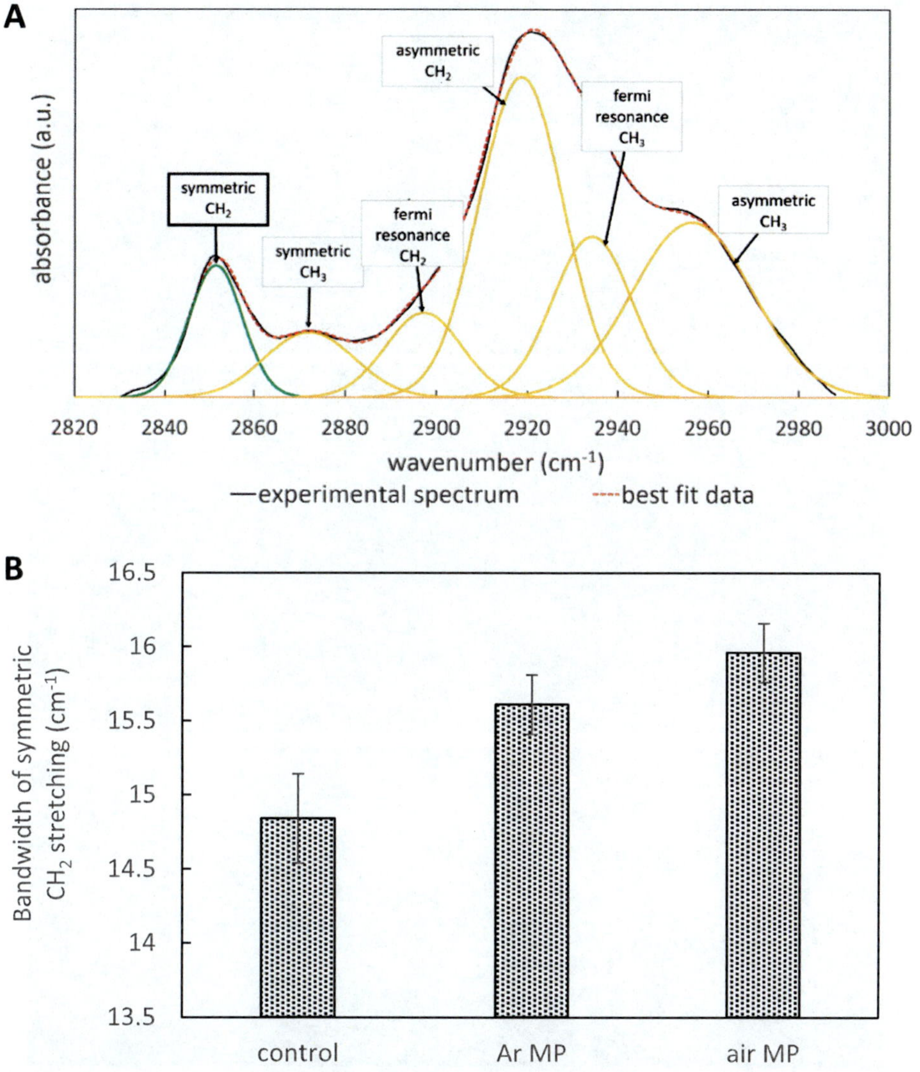

Fig. 3. A: Example of the best fit of experimental FTIR-ATR spectrum (symmetric CH_2 stretching depicted as green and other transitions depicted as orange) of vibrational transitions. B: Bandwidth of symmetric stretching band of CH_2 of the non-treated (control), argon microplasma treated (Ar MP) and air microplasma treated (air MP) stratum corneum.

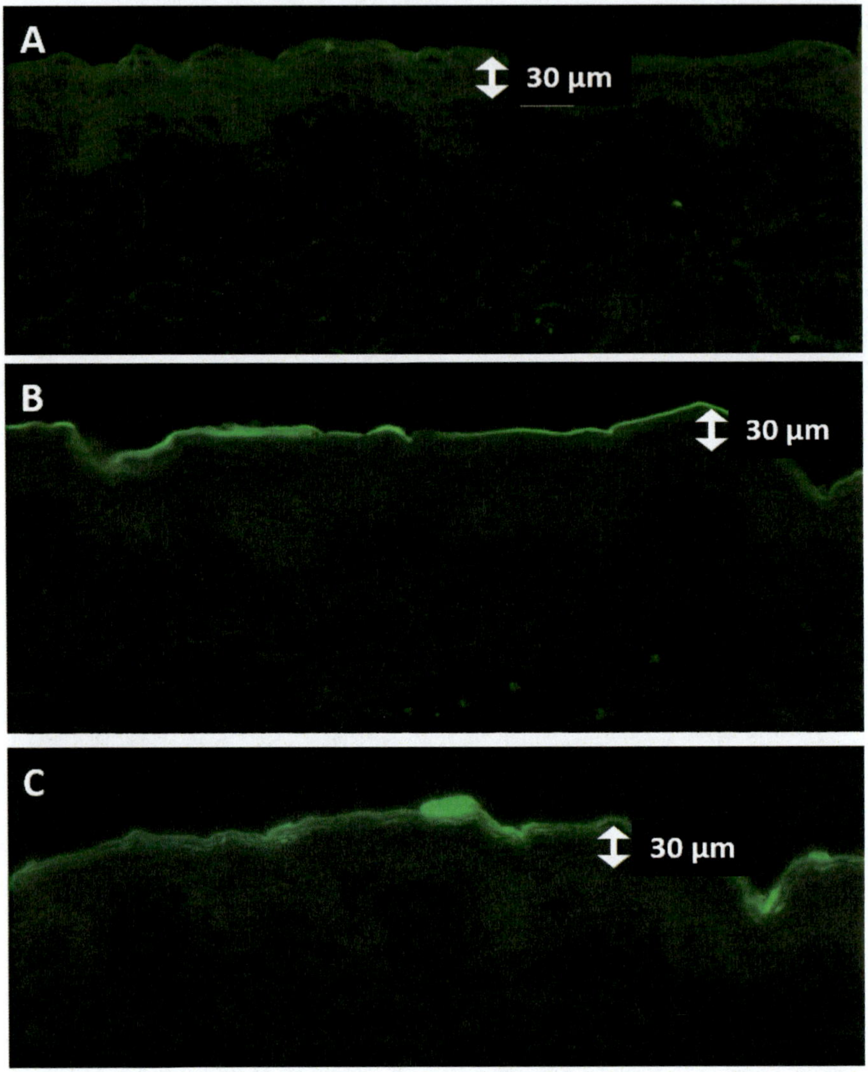

Fig. 4. Skin sections stained by LiperFluo. A: non-treated skin B: argon microplasma treated skin C: air microplasma treated skin.

4 Conclusion

Permeability of Microplasma treatment of stratum corneum was investigated. Comparison amount of Sodium fluorescein permeated through non-treated, argon microplasma treated and air microplasma treated stratum corneum showed that microplasma enhanced stratum corneum permeability. Air microplasma increased stratum corneum pemeability most effectively. ATR-FTIR confirmed lipid disordering after microplasma treatment of stratum corneum with higher disordering in the case of air microplasma treatment. Skin sectioning of treated skin revealed presence of lipid peroxides. Lipid peroxides were observed in deeper layer of stratum corneum in air microplasma treatment when compare to argon microplasma treatment. This observation can be possible reason that air microplasma treatment was more effective.

References

1. Xin, X.W., Hamblin, M.R., Yue, X.Y.: Applications of cold atmospheric plasma for transdermal drug delivery: a review. Drug Deliv. Transl. Res. **11**, 741–747 (2021). https://doi.org/10.1007/s13346-020-00808-2
2. Vijayarangan, V., Delalande, A., Dozias, S., Pouvesle, J.M., Robert, E., Pichon, C.: New insights on molecular internalization and drug delivery following plasma jet exposures. Int. J. Pharm. **589**, 119874 (2020)
3. Shimizu, K., Kristof, J.: Microplasma drug delivery. In: Tutar, Y., Tutar, L. (eds.) Plasma Medicine: Concepts and Clinical Applications, pp. 101–120. IntechOpen, London (2018)
4. Shimizu, K., Kristof, J.: Enhancement of percutaneous absorption on skin by plasma drug delivery method. In: Maiti, S., Sen, K.K. (eds.) Advanced Technology for Delivering Therapeutics, pp. 111–136. InTech, Rijeka (2017)
5. Shimizu, K., Tran, A.N., Kristof, J., Blajan, M.: Investigation of atmospheric microplasma for improving skin permeability. In: Proceedings of the 2016 Electrostatics Joint Conference, p. 9 (2016). [cited 2021 Nov17]. www.electrostatics.org/images/I4.pdf
6. Lv, Y., Nie, L., Duan, J., Li, Z., Lu, X.: Cold atmospheric plasma jet array for transdermal drug delivery. Plasma Process. Polym. **18**, 2000180 (2020)
7. Kristof, J., Miyamoto, H., Tran, A.N., Blajan, M., Shimizu, K.: Feasibility of transdermal delivery of Cyclosporine A using plasma discharges. Biointerphases **12**, 02B402 (2017)
8. Connolly, R.J., Lopez, G.A., Hoff, A.M., Jaroszeski, M.J.: Plasma facilitated delivery of DNA to skin. Biotechnol. Bioeng. **104**, 1034–1040 (2009)
9. Lee, H.Y., Choi, J.H., Hong, J.W., Kim, G., Lee, H.J.: Comparative study of the Ar and He atmospheric pressure plasmas on E-cadherin protein regulation for plasma-mediated transdermal drug delivery. J. Phys. D Appl. Phys. **51**, 215401 (2018)
10. Choi, J.H., et al.: Treatment with low-temperature atmospheric pressure plasma enhances cutaneous delivery of epidermal growth factor by regulating E-cadherin-mediated cell junctions. Arch. Dermatol. Res. **306**, 635–643 (2014). https://doi.org/10.1007/s00403-014-1463-9
11. Yahaya, A.G., Kristof, J., Mustafa, F., Blajan, M., Shimizu, K.: Effect of plasma discharge on structure of epidermal layer of pig skin. Plasma Med. **11**, 1–13 (2021)
12. Kristof, J., Tran, A.N., Blajan, M.G., Shimizu, K.: A Study of the influence of plasma particles for transdermal drug delivery. In: Jabłoński, R., Szewczyk, R. (eds.) Recent Global Research and Education: Technological Challenges. AISC, vol. 519, pp. 167–173. Springer, Cham (2017). https://doi.org/10.1007/978-3-319-46490-9_24
13. Striesow, J., et al.: Oxidative modification of skin lipids by cold atmospheric plasma (CAP): a standardizable approach using RP-LC/MS2 and DI-ESI/MS2. Chem. Phys. Lipids **226**, 104786 (2020)
14. Hirschberg, J., Loewenthal, L., Krupp, A., Emmert, S., Viöl, W.: Plasma induced changes in human lipid composition as revealed through XPS-analysis. Nat. sci. **8**, 125–137 (2016)

Brain-Drug Delivery Through Intercellular Junction of Blood Brain Barrier (BBB) Using Cold Atmospheric Plasma

Alam Md Jahangir[1](✉), Jaroslav Kristof[2], Abubakar Hamza Sadiq[3], Sadia Afrin Rimi[3], Takumi Okada[4], and Kazuo Shimizu[1,2,3,4]

[1] Graduate School of Medical Photonics, Shizuoka University, Hamamatsu 432-8011, Japan
alamm5408@gmail.com

[2] Organization for Innovative and Social Collaboration, Shizuoka University, Hamamatsu 432-8011, Japan

[3] Graduate School of Science and Technology, Shizuoka University, Hamamatsu 432-8011, Japan

[4] Graduate School of Integrated Science and Technology, Shizuoka University, Hamamatsu 432-8011, Japan

Abstract. To address the problems with brain drug delivery we aim to deliver the large molecular weight drug into the brain by opening intercellular junctions (e. g. tight junction) of blood brain barrier (BBB) using cold atmospheric plasma (CAP). The CAP produces reactive oxygen and nitrogen species (RONS) that include hydroxyl ($HO^{\bullet}$), superoxide ($O_2^{\bullet -}$), peroxyl ($ROO^{\bullet}$), nitric oxide ($NO^{\bullet}$), ozone (O_3), hydrogen peroxide (H_2O_2), and singlet oxygen ($^1O_2$). In this study, BBB cell model containing endothelial cells, pericytes and astrocytes was purchased from pharmaco-cell company Ltd. (Japan) and cultured using DMEM medium on 24-transwell plate. Upon confluency of the cells, fluorescein isothiocyanate dextran (FD-4) drug was added, and plasma was applied to the culture. The plasma Jet was prepared as follow (carrier gas: Argon, gas flow: 3 L/min, irradiation time: 60 s, irradiation distance: 20 mm, voltage: 2.2 k Pp., frequency: 10 kHz). After 1 h incubation at 37 °C with 5% CO_2, trans-endothelial electrical resistance (TEER) was measured. The TEER value was reduced in plasma-treated cells compared to non-treated cells. It indicates that this plasma condition produced reactive species (RONS) that are involved in breaking proteins of intercellular junction. Because the lower the TEER value, the more permeability in the intercellular junction. FD-4 drug was added to the apical part (insert) of trans-well plate, and after plasma treatment with 1 h incubation, florescence intensity was measured from the basal side. The higher florescence intensity was detected in plasma treated cells compared to non-treated cells. This result may provide a valuable insight into a new strategy for brain drug delivery.

Keywords: Blood-brain barrier · Plasma Jet · Permeability · Brain drug delivery

Y. Ono and J. Kondoh (Eds.): Inter-Academia 2023, LNNS 939, pp. 98–103, 2024.
https://doi.org/10.1007/978-3-031-54450-7_13

1 Introduction

The blood-brain barrier (BBB) is a protective barrier in the brain that prevents most drugs from entering brain [1, 2]. This barrier is made up of tight junctions in the endothelial cells of brain capillaries [3]. Its main job is to keep the brain's internal environment stable by controlling what goes in and out, preventing it from harmful agents [4].

The BBB is composed of endothelial cells, pericyte, and astrocyte [5]. The continuous layer of endothelial cells is held together by tight junctions, adherent junctions, and gap junctions. Among these, tight junctions are the most important for maintaining the barrier's resistance and controlling the passage of drugs [6].

Recently, cold atmospheric plasma (CAP) has become popular in medical science. It has various uses, such as treating wounds [7], drug delivery [8], sterilizing surfaces, and even cancer therapy [9]. CAP generates reactive oxygen and nitrogen species (RONS) like superoxide radical, proximities anion, and nitric oxide radical [10].

In our research, we want to use CAP-generated reactive species (RONS), especially nitric oxide (NO) to disrupt the cell membrane and tight junctions of the BBB, as well as other intercellular junctions like adherents junctions [11]. This will help drugs to pass through the BBB and reach the brain more effectively. By leveraging CAP's ability to produce these reactive species, we aim to improve drug delivery to the brain and enhance medical treatments.

2 Experiments

2.1 Cell Culture Conditions

In this study, BBB cell model containing endothelial cells, pericytes and astrocytes was purchased from pharmaco-cell company Ltd. (Nagasaki, Japan, http://www.pharmacocell.co.jp/, BBB kit (RBT-24H)) Japan) and cultured using DMEM medium on 24-transwell plate at 37 ℃ with 5% CO_2. The BBB kit was prepared by co-cultivating primary Wistar rat BCECs, brain pericytes, and astrocytes on trans-well membranes. This kit was stored at −80 °C and have a one-month shelf life. To prepare the BBB kit for use, it must be thawed and activated as per the protocol. The medium should be thawed by immersing it in a 37 °C water bath and then promptly added to both the brain-side and blood-side chambers. Subsequently, the BBB kit should be placed in an incubator, with daily medium changes.

2.2 Trans-Endothelial Electrical Resistance (TEER) Measurement

For checking the growth, trans-endothelial electrical resistance-TEER (millicell ERS-2 voltohmmeter)) was measured everyday. Each of two electrodes of the EVOM voltammeter were placed on insert (top side) and well (bottom side) of the trans-wells. Along with cell culture on the transwell plate, one transwell plate was filled with only medium. The TEER value was calculated using the following formula:

$$\text{TEER}\left(\Omega \times \text{cm}^2\right) = (\text{Total R} - \text{Blank R}) \times 0.33.$$

Here, total R denotes the relative resistance value of each cell culture, blank R signifies the resistance of the medium alone, and 0.33 cm^2 is the growth area of the cells on insert. When the TEER reaches 150 Ω-cm^2, it indicates the confluency of the cells and strong barrier of intercellular tight junction.

2.3 Florescence Intensity Measurement

Upon confluency of the cells, fluorescein isothiocyanate dextran (FD-4) drug was added to the cells with the concentration of 3 μM. The molecular weight of this drug was 4.4 kDa. The drug was added to the insert (apical/top side) and florescence intensity was measured from the top side.

2.4 Plasma Irradiation

When the TEER reached 150 Ω-cm^2 or more and FD-4 drug was added to cells, the cells were incubated at 37 °C with 5% CO_2. Then plasma was irradiated. For this, plasma Jet was prepared with Argon (Ar) gas with the flow of 3 L/min. The voltage and frequency were set as 2.2 kV peak-to-peak and 10 kHz respectively. The duration of plasma irradiation was 60 s, and the irradiation distance between plasma discharge to cell was 20 mm (Fig. 1).

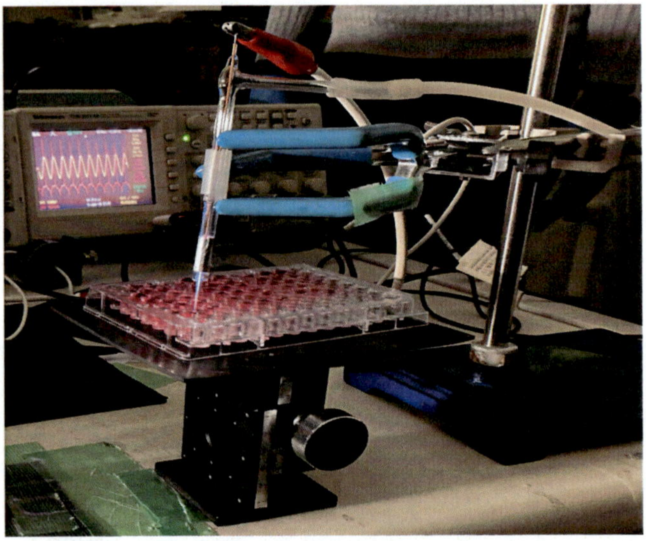

Fig. 1. Plasma Jet application on cells.

3 Results and Discussions

3.1 Permeability Assay

After plasma irradiation, the cells were incubated again for 60 min to allow the plasma components (e.g., reactive oxygen and nitrogen species etc.) to interact with intercellular tight junction. The TEER was measured again to check if the tight junction was broken. The TEER value was reduced in plasma-treated cells compared to non-treated cells. It indicates that this plasma condition produced reactive species (RONS) that are involved in breaking proteins of intercellular junction. Because the lower the TEER value, the more permeability in the intercellular junction (Fig. 1).

The experiment was run in 3 set of cells. First is the control cell where no drug was applied. In second cells and third cells, only FD-4 drug was applied. Since drug was added in top side of the trans-well plate, florescence intensity was measured from top side. Therefore, these two cells show higher florescence compared to control.

After plasma irradiation, the higher florescence intensity was measured from the bottom side of the plate. Since the plasma irradiation breaks the tight junction barrier, the drug may pass from the insert (top) to the well (bottom). The higher florescence intensity was detected in plasma treated cells compared to non-treated cells. The higher florescence intensity indicates the higher amount of drug in basal sites of the plates (Fig. 2).

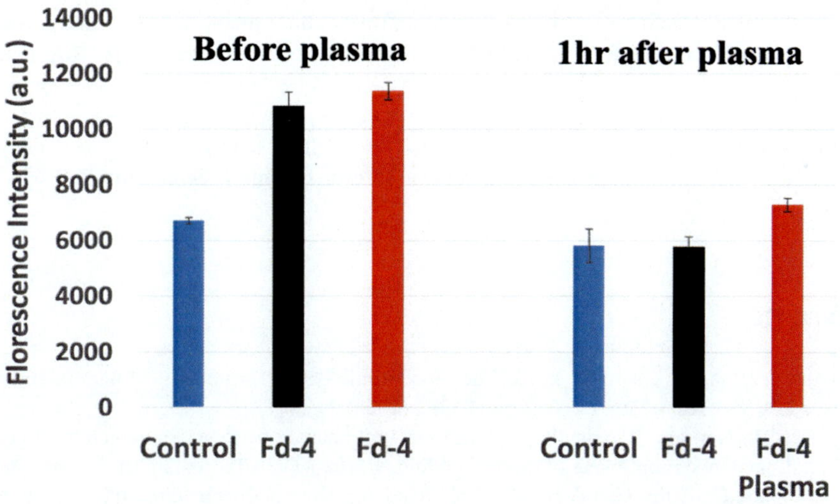

Fig. 2. Florescence Intensity before plasma (left) and after plasma (right)

On the other hand, TEER was measured every day. On day 4 when it reached more than 250 Ω-cm^2, plasma was applied and incubated for 1 h. Then TEER was measured again, and the TEER value was reduced in the plasma-treated cell compared to control and FD-4 treated cells. The lower TEER value in the plasma-treated cell indicates more permeability in the intercellular junction (Fig. 3).

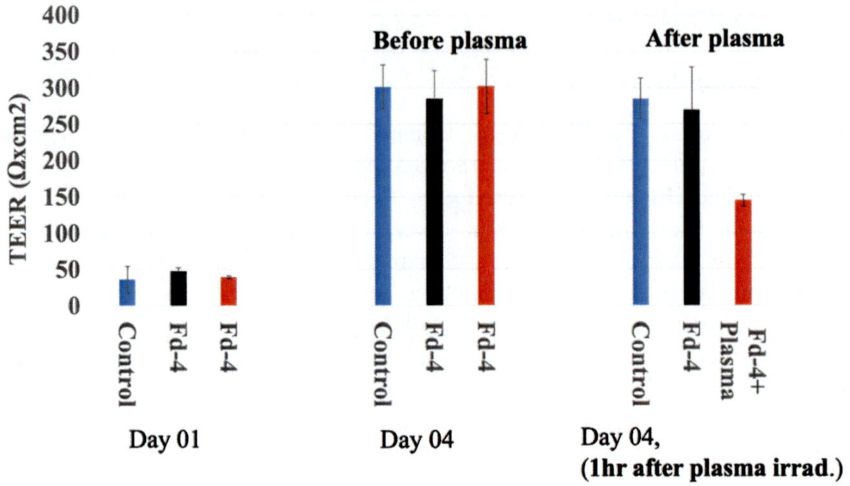

Fig. 3. TEER measurement before plasma (left and middle) and after plasma (right)

4 Conclusion

We aimed to deliver the drug through the tight junction of blood brain barrier (BBB). To confirm it, we measured TEER and florescence intensity. The reduced TEER and increased florescence intensity compared to control cell confirms successful drug delivery. This result may provide a valuable insight into a new strategy for brain drug delivery.

Acknowledgement. This work was supported by **JSPS KANKENHI** grant numbers JP23K17473 and by **the Shizuoka University Interdisciplinary Fellowship for Medical Photonics.**

References

1. Wu, D., Chen, Q., Chen, X., et al.: The blood–brain barrier: structure, regulation, and drug delivery. Sig. Transduct. Target. Ther. **8**, 217 (2023)
2. Kadry, H., Noorani, B., Cucullo, L.: A blood–brain barrier overview on structure, function, impairment, and biomarkers of integrity. Fluids Barriers CNS **17**, 69 (2020)
3. Luissint, A.C., Artus, C., Glacial, F., et al.: Tight junctions at the blood brain barrier: physiological architecture and disease-associated dysregulation. Fluids Barriers CNS **9**, 23 (2012). https://doi.org/10.1186/2045-8118-9-23
4. Pardridge, W.M.: Drug transport across the blood-brain barrier. J. Cereb. Blood Flow Metab. **32**(11), 1959–1972 (2012)
5. Yao, Y., Chen, Z.-L., Norris, E.H., Strickland, S.: Astrocytic laminin regulates pericyte differentiation and maintains blood brain barrier integrity. Nat. Commun. **5**, 3413 (2014)
6. Brightman, M.W., Reese, T.S.: Junctions between intimately apposed cell membranes in the vertebrate brain. J. Cell Biol. **40**, 648–677 (1969)

7. Lloyd, G., Friedman, G., Jafri, S., Schultz, G., Fridman, A., Harding, K.: Gas plasma: medical uses and developments in wound care. Plasma Process. Polym. **7**, 194–211 (2010)
8. Shimizu, K., Hayashida, K., Blajan, M.: Novel method to improve transdermal drugdelivery by atmospheric microplasma irradiation. Biointerphases **10**, 029517 (2015)
9. Shimizu, K.: Biological effects and enhancement of percutaneous absorption on skin by atmospheric microplasma irradiation. Plasma Med. **5**(2–4), 205–221 (2016)
10. Kim, S., Chung, T.: Cold atmospheric plasma jet-generated RONS and their selective effects on normal and carcinoma cells. Sci. Rep. **6**, 20332 (2016)
11. Thiel, V.E., Audus, K.L.: Nitric oxide and blood-brain barrier integrity. Antioxid. Redox Signal. **3**(2), 273–278 (2001)

Oak Aging and Ultrasound Treatment for Improving the Sensory Profile of Sauvignon Blanc Wines

Camelia Elena Luchian[1(✉)], Elena Cornelia Focea[1], Elena Cristina Scutarașu[1,2], Iuliana Motrescu[1], Ana Maria Vlase[3], Laurian Vlase[3], Lucia Cintia Colibaba[1], and Valeriu Cotea[1,2]

[1] Ion Ionescu de la Brad Iasi University of Life Sciences, 3rd M. Sadoveanu Alley, 700490 Iași, Romania
kamelia_luchian@yahoo.com

[2] Research Center of Oenology, Romanian Academy, Iași Branch, 9th M. Sadoveanu Alley, 700505 Iași, Romania

[3] Faculty of Pharmacy, Iuliu Hațieganu University of Medicine and Pharmacy, 8th Victor Babes, Street, 400012 Cluj-Napoca, Romania

Abstract. Aging and maturation with oak contacts improves color stability and gives smoothness, roundness and flavor to wine. These wines are preferred by many consumers due to their complex sensory profile but it usually have a higher price, traditional maturation being expensive. The tendency of consumers to focus on innovative products with a complex, rich, interesting sensory profile, but at reduced prices. On this line, this paper focuses on reducing maturation period and costs using ultrasounds waves (35 kHz) and different French oak fragments (with varying degrees of roasting) to improve wines sensory profile. Considering that most studies focus on the influence of wood on the quality of red wines, white wines obtained from the Sauvignon blanc variety (from Iasi-Romania) are used in this study. The results highlight a significant impact on wine sensory profile in a shorter time. A woody and spicy note was better expressed when oak chips were used. Variables such as the shape of the oak fragment (granulated vs. chips), the used dosage (1 vs. 2 gL^{-1}) and the contact time (15 min. With ultrasounds vs. 10 and 20 days in traditional maturation) have a major effect on the quality of the wines. The results contribute to the optimization of winemaking process.

Keywords: Optimization · Ultrasounds · Oak Fragments · White Wines

1 Introduction

Contemporary winemaking has continuously evolved by embracing diverse techniques and technologies aimed at enhancing wine quality [1]. The aging of wine represents an extended temporal trajectory, occasionally contrasting with the exigencies of consumers, consequently posing substantial financial burdens on wineries. To mitigate these

Y. Ono and J. Kondoh (Eds.): Inter-Academia 2023, LNNS 939, pp. 104–121, 2024.
https://doi.org/10.1007/978-3-031-54450-7_14

challenges, strategic interventions necessitate consideration to expedite vinification processes while upholding the requisite standards of wine quality [2]. Wine aging in contact with wood leads to assortments diversification by obtaining wines with a complex sensory profile. Oak fragments are often used in winemaking for wine enrichment in phenolic and aroma compounds. Concurrently, controlled oxidation and the alteration of tannins and anthocyanins are recognized for enhancing the intensity, stability, and structural attributes of wines. Consequently, wines subjected to barrel aging are commonly perceived to possess superior quality, consequently commanding higher market valuations. Although maturing and aging in barrels involves a long time and high costs, oak chips are an advantageous alternative. The oak species usually used for wine aging belong to the *Quercus genus*, as follows: *Quercus alba* (American white oak), *Quercus petraea*, and *Quercus robur* (French oak). The characteristics of the used chips (the species used, origin, porosity, granularity, permeability, toasting degree) and contact time significantly influence the quality of the final product. Several papers have studied the evolution of oak-derived volatiles and organoleptic properties in wine, but few of them refer to white wines [3–6]. The exploration of innovative techniques for wine aging presents the potential to reduce the customary aging duration, enabling expedited availability of wines in the market. Further research endeavors are imperative to conclusively ascertain the viability and efficacy of ultrasonic, electric field, and gamma irradiation applications as substitutes for conventional wine aging processes [1]. Some studies have also explored combinations with micro-oxygenation, trying to imitate maturation conditions akin to traditional barrel aging [7–9].

Ultrasound represents an economical and environmentally sustainable technological approach applicable across various domains within food technology. Despite the abundance of research in this domain, the translation of findings to industrial-scale operations has been limited, primarily due to the absence of standardized design protocols and scalable strategies [2]. Recently, the International Organization of Vine and Wine has officially allowed the application of ultrasound for expediting the extraction of compounds from grapes in the pre-fermentation maceration phase, following destemming and crushing [10]. A critical consideration in advancing ultrasound technology is ensuring seamless translation from laboratory-scale experimentation to industrial application—a fundamental aspect within the realm of sonochemistry that has intrigued researchers for a considerable duration. Sonication within the low-frequency range (20–100 kHz) has exhibited advantageous impacts on a multitude of food products, particularly those that are fermented. Numerous comprehensive studies have consistently demonstrated that ultrasound exhibits the potential to enhance the polysaccharide composition of wine while concurrently reducing the requisite aging period. In a broader context, ultrasound showcases its efficacy in facilitating a spectrum of crucial chemical reactions pivotal to the maturation and aging processes of fermented foods. These encompass oxidation, esterification, and proteolysis. Furthermore, ultrasound significantly contributes to elevating the overall quality of such food products by fostering favorable attributes concerning texture, chromatic characteristics, flavor, and taste [2]. For instance, Mărgean & Pădureanu [11] conducted an assessment of ultrasound treatment on must fermentation, reporting a notable increase in phenolic compounds content. Likewise, Ruiz-Rodriguez et al. [12] provided empirical evidence illustrating that ultrasound treatments facilitate the extraction of volatile compounds in wines. Additionally, Gracin et al. [13] proposed the utilization of high-power ultrasound treatment and thermosonication for the inactivation of *Brettanomyces bruxellensis* in red wines. Moreover, Xie et al. [14] documented a significant augmentation in antioxidant activity within wines treated with ultrasound in combination with reduced temperatures.

Phenolic compounds manifest a major role in wine sensory characteristics and aging process. Non-enzymatic oxidative reactions play an essential role in inducing significant sensory changes during the aging process. This involves the transfer of an electron (or hydrogen atom) from the oxidized compound to oxygen or another acceptor. In the context of bottled wines, reactions involving molecular oxygen proceed slowly and depend on oxygen content and the catalyzers-substratum ratio. The cavitation phenomenon leads to temperature increases, accompanied by a slight pH elevation, and the phenolic content plays a substantial role in influencing oxidative potential. Ultrasound has the potential to facilitate various reactions during aging, including both oxidative reactions (with or without molecular oxygen) and reductive reactions. A key transformation during aging is the gradual enhancement and stabilization of color, attributed to copigment anthocyanin complexes, the development of new pigments, and the progressive formation of tannin–tannin and anthocyanin–tannin complexes [15].

On this line, this study examines the influence of oak aging and ultrasound treatment on the sensory profile of Sauvignon blanc white wines. Furthermore, the research focuses on finding the optimal convergence of wood variety and aging methodologies in order to produce a high-grade wine enriched with innovative sensory dimensions, all the while preserving the distinctive primary and secondary aromas intrinsic to individual grape varietals.

2 Materials and Methods

2.1 Materials

For this experiment, French granular oak fragments and French chips oak fragments (*Quercus robus* and *Quercus sessiflora*), light toasted, fresh and medium toasted were purchased from Lamothe-Abiet (Bordeaux, France). For the determination of phenolic compounds, the following standards were used: epicatechin, catechin, syringic acid, gallic acid, protocatechuic acid, procyanidin B1, procyanidin B2, procyanidin B3, procyanidin B4, esculetin, syringaldehyde, caftaric acid, caffeic acid, *p*-coumaric acid, *trans*-resveratrol, *cis*-resveratrol (from Merck, Darmstadt, Germany).

2.2 Samples Production

Sauvignon blanc grapes from Iasi vineyard (Romania) were used. The grapes were manually harvested (with 240 g/L total sugar) in autumn of 2021 and then processed. A pre-fermentative maceration of the grape juice was applied (for 5 h). After the must clarification, the sediment was removed and *Saccharomyces cerevisiae* yeasts were inoculated into the must (20 g/hL). Alcoholic fermentation was carried out at 14 °C. The resulted wine was stabilized (sulphited) and divided in 38 aliquots (S0–S37), as follows:

S0 – control sample, conventional aging; S25 – control sample, ultrasound treatment; S1–S37 – work samples. All samples were obtained in triplicate. Variables such as oak shapes (granular vs. chips), toasting degree (light vs. fresh vs. medium degree), dosage (1 g/L vs. 2 g/L) and contact period (10 days vs. 20 days) were established. Oak fragments presented different technological characteristics. For example, granular fresh oak usually helps in decreasing vegetal nose; light variant gives roundness and sweetness; medium granular oak improves complexity, length in the mouth. On the other hand, fresh oak chips gives structure to wine, medium chips gives vanilla, coconut and sweetness, while medium toasted contribute to the intense caramel notes.

The ultrasound treatments were conducted employing a 35 kHz ultrasonic frequency at a temperature of 20 °C. The analyzed samples displayed an ethanol level ranging between 13.7% vol. Alc. And 14.1% vol. Alc., and a density varying from 0.9912 to 0.9921. The total acidity ranged from 7.65 to 8.10 g/L tartaric acid (in S0 and S9, respectively), while volatile acidity ranged from 0.32 g/L acetic acid (S0) to 0.46 g/L acetic acid.

2.3 Phenolic Compounds

The main phenolic compounds were identified using an HPLC system. The following apparatus were used: 204 Sigma Centrifuge (Osterode am Harz, Germany); Analytical Plus Balance (Mettler-Toledo, Switzerland); Ultrasonic bath Elma Transsonic 700/H (Singen, Germany). The HPLC system used was an 1100 series Agilent Technologies model (Darmstadt, Germany) consisting of a G1312A binary pump, an in-line G1379A degasser, an G1329A autosampler, a G1316A column thermostat and an Agilent Ion Trap Detector 1100 SL. The calibration curves were linear for all analytes in range of 0.1–50 ug/mL. The samples were analyzed in triplicates and the results represent de arithmetic average of the obtained data.

2.4 Sensory Analysis

The sensory analysis session involved a panel of 30 experienced tasters, comprising 18 men and 12 women, with ages ranging from 25 to 65. Tasters evaluated each analyzed descriptor on a scale of 1 to 10. A pre-established tasting sheet encompassing a total of 22 sensory attributes, including descriptors such as vegetal, hay, vanilla, sweet tobacco, coconut, and coffee spices, was employed for the evaluation process [16–18].

2.5 Statistical Analysis

Data processing including Anova was made with XLSTAT package for Excel (Luminevo, Denver, USA). For the visual representation of phenolic and sensory profiles, the online version of Heatmapper (http://www.heatmapper.ca, accessed on 23 November 2023, Wishart Research Group, University of Alberta, Canada) was used (Table 1).

Table 1. Sample codification

Granular oak fragments				Chips oak fragments			
Sample	T.D.	Dosage (g/L)	Time	Sample	T.D.	Dosage (g/L)	Time
S0	Control sample, conventional maturation						
S1	L	1	10 days	S4	L	1	10 days
S2	F			S5	F		
S3	M			S6	M		
S7	L		20 days	S10	L		20 days
S8	F			S11	F		
S9	M			S12	M		
S13	L	2	10 days	S16	L	2	10 days
S14	F			S17	F		
S15	M			S18	M		
S19	L		20 days	S22	L		20 days
S20	F			S23	F		
S21	M			S24	M		
S25	Control sample, ultrasound treatments						
S26	L	1	15 min	S29	L	1	15 min
S27	F			S30	F		
S28	M			S31	M		
S32	L	2		S35	L	2	
S33	F			S36	F		
S34	M			S37	M		

T.D.-Toasting degree; L-light; F-fresh; M-medium; min.-minutes

3 Results and Discussions

3.1 Phenolic Compounds

The results obtained for the determination of phenolic profile are presented in Table 2 and Fig. 1. The samples presented different amounts of phenolic compounds, depending on the applied treatment. In general, samples in which wood fragments were applied have been characterized by higher values of the identified phenolic compounds. Comparative values were obtained between the samples obtained by conventional aging and those treated with ultrasound. Therefore, these procedures enable us to obtain similar results in a shorter time and with minimum costs.

As can be seen in Fig. 1, caftaric acid was predominant in all variants, but the highest value was obtained in S27 (granular form, fresh, 1 g/L, ultrasounds, 15 min) - 27.213 μg/mL. An important increase in gallic acid levels (the second predominant phenolic compound) was registered in sample S20 - 4.294 μg/mL (granular form, fresh, 2 g/L, 20 days) compared to the others.

Caffeic acid was found in higher amounts in S12 (chips, medium roasted, 1 g/L, 20 days), S29 (chips, light roasted, 1 g/L, 15 min. Ultrasound), and S33 (granular, fresh, 2 g/L, 15 min. Ultrasound). We can also mention that ultrasounds enabled us to obtain the same values again in a shorter time and with minimum costs.

According to Table 2, the higher amounts of syringaldehyde were identified in S17 - 1.429 27.213 μg/mL (granular form, fresh, 1 g/L, ultrasounds, 15 min.). The use of ultrasound allowed the reduction of both the aging time and the amount of added wood for syringicaldehyde.

Significant differences can be observed on phenolic compounds such as gallic acid levels, procyanidin B1, esculetin, syringaldehyde and caftaric acid. The introduced variables did not have a significant impact on the other analysed phenolic compounds. French oak are usually rich in those phenolic compounds. According to Glabasnia and Hofmann [19] gallic acid and caffeic acid are found in higher amounts in *Q. robur* French oak compared to *Q. alba*. Prida et al. [20] postulated that European oak is rich in syringldehyde. Cadahía et al. [21] also reported high levels of gallic acid, esculetin and syringaldehyde in French oak. In most cases, the granular form of the wood allowed the extraction of larger amounts of phenolic compounds. Also, the fresh toasted products permitted a better synthesis of the phenolic compounds. These results are in concordance with data presented by Cerdán et al. [22].

The identified compounds manifest an important impact on sensory perception of wine. According to Sterneder et al. [23], gallic acid, catechin and procyanidin B2 usually give bitter taste and astringency in wines. Canas et al. [24] postulated that esculetin can cause neutral or acidic taste in wine.

Table 2. Phenolic profile of the analysed samples

Category	1	2	3	4	5	6	7	8	9	10	11	12	13	14	15	16
S0	0.881 a	0.656 a	0.000 a	1.481 i	0.874 a	0.876 ab	0.197 a	0.160 a	0.077 a	0.003 def	0.000 k	24.800 l	1.679 a	0.166 a	0.161 a	0.439 a
SD	0.05	0.35	0	0.1	0.09	0.05	0.02	0.01	0.01	0	0	0.01	0.04	0.03	0.03	0.12
S1	0.880 a	0.675 a	0.043 a	2.746 cdefghi	1.027 a	0.807 ab	0.174 a	0.156 a	0.073 a	0.005 bcde	0.294 fghijk	26.619 abc	1.723 a	0.164 a	0.162 a	0.482 a
SD	0.13	0.18	0.01	0.09	0.1	0.01	0.04	0.09	0.04	0.001	0.2	0.45	0.06	0.11	0.01	0.2
S2	0.829 a	0.651 a	0.021 a	3.612 abcdef	1.024 a	0.793 ab	0.139 a	0.143 a	0.055 a	0.005 bcde	0.000 k	25.749 fghij	1.657 a	0.216 a	0.202 a	0.363 a
SD	0.05	0.1	0.01	0.45	0.05	0.25	0.01	0.02	0.03	0	0	0.35	0.01	0.02	0.01	0.15
S3	0.849 a	0.663 a	0.062 a	2.266 efghi	0.877 a	0.654 b	0.126 a	0.140 a	0.079 a	0.003 def	0.694 cde	25.828 fghij	1.636 a	0.197 a	0.267 a	0.520 a
SD	0.2	0.3	0.55	0.05	0.25	0.45	0.03	0.01	0.01	0.001	0.2	0.05	0.15	0.18	0.01	0.05
S4	0.908 a	0.696 a	0.028 a	2.567 cdefghi	0.983 a	0.906 ab	0.157 a	0.151 a	0.091 a	0.003 def	0.158 hijk	25.987 cdefghij	1.723 a	0.197 a	0.156 a	0.399 a
SD	0.35	0.03	0.04	0.2	0.02	0.4	0.1	0.01	0.02	0.001	0.01	0.35	0.02	0.01	0.01	0.03
S5	0.858 a	0.659 a	0.026 a	2.911 abcdefg	0.959 a	0.766 ab	0.129 a	0.164 a	0.072 a	0.006 bcd	0.000 k	25.749 fghij	1.679 a	0.245 a	0.207 a	0.351 a
SD	0.12	0.05	0	0.01	0.4	0.2	0.01	0.09	0.01	0.001	0	0.35	0.04	0.05	0.01	0.12
S6	0.873 a	0.688 a	0.056 a	2.283 efghi	1.018 a	0.787 ab	0.140 a	0.143 a	0.073 a	0.001 f	0.506 defghi	25.789 fghij	1.679 a	0.237 a	0.229 a	0.383 a
SD	0.18	0.09	0.01	0.1	0.05	0.01	0.05	0.02	0.45	0	0.03	0.11	0.14	0.04	0.05	0.15
S7	0.776 a	0.599 a	0.062 a	3.683 abcd	0.998 a	0.669 b	0.129 a	0.128 a	0.066 a	0.005 bcde	0.632 cdefg	25.591 fghij	1.657 a	0.150 a	0.204 a	0.467 a
SD	0.25	0.1	0.01	0.35	0.09	0.02	0.05	0.02	0.01	0	0.24	0.01	0.02	0.01	0.05	0.03
S8	0.851 a	0.670 a	0.026 a	4.195 ab	0.952 a	0.773 ab	0.166 a	0.167 a	0.065 a	0.006 bcd	0.103 jk	25.433 hijkl	1.614 a	0.197 a	0.189 a	0.381 a
SD	0.1	0.18	0.01	0.2	0.05	0.05	0.1	0.05	0.02	0.001	0.05	0.05	0.15	0.09	0.22	0.15
S9	0.660 a	0.528 a	0.117 a	2.685 cdefghi	0.975 a	0.513 b	0.099 a	0.125 a	0.062 a	0.004 cdef	1.183 ab	25.314 jkl	1.701 a	0.224 a	0.171 a	0.484 a
SD	0.45	0.01	0.05	0.1	0.02	0.45	0.04	0.01	0.06	0.001	0.15	0.01	0.04	0.15	0.03	0.15
S10	0.886 a	0.659 a	0.029 a	3.620 abcde	0.892 a	0.827 ab	0.168 a	0.156 a	0.091 a	0.007 bc	0.167 hijk	25.433 hijkl	1.723 a	0.224 a	0.097 a	0.464 a
SD	0.15	0.15	0.25	0.035	0.02	0.32	0.05	0.05	0.01	0.002	0.02	0.05	0.45	0.2	0.04	0.18

(continued)

Table 2. *(continued)*

S11	0.868 a	0.699 a	0.025 a	3.272 abcdefg	0.956 a	0.965 ab	0.155 a	0.142 a	0.070 a	0.008 ab	0.000 k	25.551 fghij	1.636 a	0.173 a	0.180 a	0.436 a
SD	0.3	0.1	0.45	0.05	0.09	0.05	0.1	0.1	0.01	0	0	0.09	0.02	0.04	0.02	0.05
S12	0.920 a	0.647 a	0.111 a	3.004 abcdefg	1.110 a	0.745 ab	0.142 a	0.105 a	0.094 a	0.007 bc	1.172 ab	26.145 bcdefg	1.745 a	0.225 a	0.213 a	0.445 a
SD	0.32	0.15	0.04	0.2	0.45	0.05	0.01	0.05	0.02	0	0.01	0.45	0.23	0.11	0.1	0.05
S13	0.878 a	0.710 a	0.038 a	2.859 bcdefghi	1.021 a	0.880 ab	0.137 a	0.157 a	0.088 a	0.003 def	0.354 efghijk	26.026 cdefghi	1.723 a	0.164 a	0.194 a	0.446 a
SD	0.45	0.18	0.01	0.09	0.02	0.01	0.02	0.04	0.02	0	0.05	0.06	0.15	0.01	0.01	0.03
S14	0.862 a	0.666 a	0.025 a	3.732 abc	0.868 a	0.755 ab	0.128 a	0.161 a	0.084 a	0.006 bcd	0.000 k	25.710 fghij	1.679 a	0.199 a	0.158 a	0.517 a
SD	0.3	0.1	0.02	0.45	0.02	0.03	0.01	0.01	0.02	0.001	0	0.02	0.01	0.02	0.05	0.05
S15	0.856 a	0.679 a	0.068 a	2.318 defghi	1.036 a	0.754 ab	0.161 a	0.131 a	0.080 a	0.002 ef	0.664 cdef	25.868 efghij	1.657 a	0.190 a	0.196 a	0.372 a
SD	0.15	0.45	0.03	0.45	0.01	0.02	0.05	0.01	0.01	0	0.2	0.05	0.06	0.13	0.1	0.04
S16	0.872 a	0.706 a	0.021 a	2.883 bcdefgh	1.026 a	0.742 ab	0.156 a	0.148 a	0.088 a	0.004 cdef	0.165 hijk	25.907 defghij	1.679 a	0.221 a	0.226 a	0.410 a
SD	0.45	0.1	0.01	0.1	0.09	0.05	0.05	0.02	0.05	0	0.11	0.18	0.01	0.01	0.04	0.04
S17	0.887 a	0.672 a	0.021 a	3.118 abcdefg	0.896 a	0.875 ab	0.142 a	0.169 a	0.078 a	0.004 cdef	0.052 k	25.631 fghij	1.679 a	0.224 a	0.229 a	0.377 a
SD	0.05	0.05	0.02	0.55	0.18	0.04	0.03	0.05	0.03	0.001	0.03	0.09	0.01	0.55	0.04	0.25
S18	0.825 a	0.686 a	0.055 a	2.414 cdefghi	1.019 a	0.681 ab	0.145 a	0.151 a	0.067 a	0.004 cdef	0.666 cdef	26.580 abcd	1.701 a	0.163 a	0.224 a	0.564 a
SD	0.08	0.45	0.01	0.25	0.45	0.035	0.02	0.03	0.02	0	0.05	0.01	0.05	0.15	0.2	0.05
S19	0.834 a	0.593 a	0.051 a	2.471 cdefghi	1.028 a	0.587 b	0.129 a	0.113 a	0.094 a	0.005 bcde	0.695 cde	25.472 ghijkl	1.570 a	0.190 a	0.185 a	0.440 a
SD	0.1	0.01	0.01	0.01	0.01	0.01	0.05	0.04	0.04	0.001	0.05	0.02	0.04	0.05	0.11	0.12
S20	0.802 a	0.647 a	0.031 a	4.294 a	1.014 a	0.684 ab	0.123 a	0.118 a	0.105 a	0.005 bcde	0.105 jk	25.907 defghij	1.636 a	0.160 a	0.192 a	0.521 a
SD	0.2	0.25	0.02	0.05	0.02	0.02	0.06	0.02	0.1	0.004	0.01	0.1	0.01	0.02	0.01	0.15
S21	0.738 a	0.606 a	0.112 a	2.816 bcdefghi	1.079 a	0.559 b	0.116 a	0.107 a	0.091 a	0.005 bcde	1.236 ab	25.512 ghijk	1.657 a	0.171 a	0.181 a	0.463 a
SD	0.45	0.32	0.07	0.45	0.5	0.04	0.05	0.03	0.02	0.001	0.5	0.02	0.15	0.03	0.02	0.3
S22	0.815 a	0.666 a	0.038 a	3.774 abc	0.973 a	0.664 b	0.139 a	0.125 a	0.078 a	0.007 bc	0.248 ghijk	25.947 cdefghij	1.701 a	0.159 a	0.200 a	0.418 a

(continued)

Table 2. *(continued)*

SD	0.1	0.05	0.01	0.35	0.02	0.03	0.11	0.02	0.05	0.001	0.01	0.05	0.15	0.1	0.1	0.01
S23	0.929 a	0.655 a	0.028 a	3.729 abc	0.974 a	0.872 ab	0.162 a	0.159 a	0.090 a	0.011 a	0.084 jk	26.540 abcde	1.657 a	0.190 a	0.204 a	0.450 a
SD	0.2	0.05	0.01	0.09	0.01	0.01	0.15	0.03	0.05	0	0.03	0.4	0.27	0.1	0.01	0.05
S24	0.765 a	0.577 a	0.147 a	3.268 abcdefg	1.004 a	0.597 b	0.143 a	0.114 a	0.061 a	0.004 cdef	1.429 a	26.224 bcdef	1.614 a	0.221 a	0.227 a	0.547 a
SD	0.25	0.02	0.03	0.4	0.02	0.05	0.01	0.3	0.41	0.001	0.01	0.03	0.04	0.01	0.05	0.11
S25	1.023 a	0.733 a	0.000 a	1.509 hi	0.871 a	1.316 a	0.212 a	0.171 a	0.082 a	0.004 cdef	0.000 k	24.839 kl	1.636 a	0.221 a	0.252 a	0.365 a
SD	0.35	0.05	0	0.25	0.4	0.02	0.02	0.01	0.01	0	0	0.55	0.1	0.12	0.14	0.11
S26	0.836 a	0.651 a	0.041 a	2.219 fghi	0.869 a	0.769 ab	0.159 a	0.136 a	0.082 a	0.005 bcde	0.341 efghijk	26.105 bcdefgh	1.723 a	0.160 a	0.244 a	0.442 a
SD	0.4	0.01	0.02	0.01	0.35	0.2	0.05	0.02	0.01	0	0.12	0.4	0.3	0.1	0.2	0.01
S27	0.863 a	0.650 a	0.013 a	2.578 cdefghi	0.936 a	0.711 ab	0.155 a	0.107 a	0.065 a	0.006 bcd	0.072 k	27.213 a	1.657 a	0.221 a	0.201 a	0.415 a
SD	0.2	0.01	0.001	0.09	0.04	0.02	0.11	0.05	0.05	0.001	0.01	0.09	0.11	0.14	0.04	0.04
S28	0.722 a	0.506 a	0.042 a	2.050 ghi	0.978 a	0.456 b	0.095 a	0.130 a	0.068 a	0.006 bcd	0.545 defgh	26.778 ab	1.592 a	0.222 a	0.078 a	0.404 a
SD	0.1	0.02	0.02	0.01	0.03	0.03	0.03	0.02	0.02	0	0.02	0.35	0.1	0.02	0.01	0.01
S29	0.967 a	0.690 a	0.030 a	2.214 ghi	1.039 a	0.941 ab	0.163 a	0.189 a	0.102 a	0.002 ef	0.115 ijk	26.224 bcdef	1.745 a	0.194 a	0.166 a	0.417 a
SD	0.15	0.45	0.01	0.4	0.01	0.45	0.12	0.01	0.02	0	0.1	0.05	0.1	0.15	0.05	0.04
S30	0.977 a	0.738 a	0.017 a	2.147 ghi	0.977 a	0.995 ab	0.189 a	0.162 a	0.096 a	0.003 def	0.000 k	25.907 defghij	1.701 a	0.160 a	0.220 a	0.412 a
SD	0.55	0.35	0.01	0.4	0.05	0.02	0.05	0.02	0.05	0.001	0	0.08	0.11	0.1	0.15	0.2
S31	0.879 a	0.681 a	0.044 a	2.101 ghi	1.048 a	0.714 ab	0.155 a	0.143 a	0.091 a	0.002 ef	0.334 efghijk	26.026 cdefghi	1.679 a	0.160 a	0.210 a	0.371 a
SD	0.45	0.4	0.01	0.01	0.35	0.15	0.02	0.18	0.05	0.001	0.01	0.12	0.4	0.01	0.15	0.35
S32	0.902 a	0.625 a	0.059 a	3.137 abcdefg	0.986 a	0.763 ab	0.130 a	0.160 a	0.092 a	0.003 def	0.476 defghij	26.105 bcdefgh	1.636 a	0.154 a	0.175 a	0.354 a
SD	0.2	0.5	0.02	0.09	0.18	0.55	0.01	0.01	0.01	0.001	0.18	0.09	0.18	0.05	0.01	0.05
S33	0.881 a	0.690 a	0.012 a	3.423 abcdefg	0.989 a	0.832 ab	0.168 a	0.140 a	0.079 a	0.008 ab	0.079 jk	25.512 ghijk	1.745 a	0.170 a	0.226 a	0.532 a
SD	0.45	0.01	0.01	0.02	0.3	0.02	0.15	0.2	0.05	0.002	0.02	0.05	0.09	0.12	0.04	0.15

(continued)

Table 2. *(continued)*

S34	0.900 a	0.640 a	0.096 a	2.547 cdefghi	0.989 a	0.748 ab	0.150 a	0.162 a	0.082 a	0.003 def	1.028 bc	25.749 fghij	1.592 a	0.193 a	0.174 a	0.450 a
SD	0.32	0.01	0.01	0.01	0.01	0.04	0.01	0.005	0.01	0	0.15	0.15	0.04	0.01	0.11	0.01
S35	0.882 a	0.654 a	0.031 a	2.879 bcdefgh	1.056 a	0.807 ab	0.161 a	0.156 a	0.082 a	0.008 ab	0.211 hijk	25.472 ghijkl	1.657 a	0.160 a	0.228 a	0.406 a
SD	0.1	0.5	0.01	0.4	0.1	0.15	0.03	0.01	0.03	0.001	0.04	0.01	0.18	0.01	0.05	0.4
S36	0.918 a	0.665 a	0.021 a	2.942 abcdefg	0.977 a	0.962 ab	0.175 a	0.151 a	0.073 a	0.006 bcd	0.000 k	25.749 fghij	1.614 a	0.221 a	0.190 a	0.411 a
SD	0.15	0.2	0.01	0.01	0.09	0.2	0.02	0.02	0.01	0	0	0.02	0.02	0.03	0.02	0.01
S37	0.856 a	0.646 a	0.066 a	2.284 efghi	0.905 a	0.762 ab	0.156 a	0.153 a	0.086 a	0.007 bc	0.757 cd	25.354 ijkl	1.548 a	0.190 a	0.213 a	0.399 a
SD	0.2	0.02	0.01	0.01	0.4	0.02	0.15	0.15	0.01	0	0.01	0.04	0.55	0.01	0.2	0.15
Pr > F(Model)	**1.000**	**1.000**	**1.000**	**<0.0001**	**1.000**	**0.012**	**0.999**	**1.000**	**1.000**	**<0.0001**	**<0.0001**	**<0.0001**	**1.000**	**1.000**	**0.989**	**0.996**
Significant	**No**	**No**	**No**	**Yes**	**No**	**Yes**	**No**	**No**	**No**	**Yes**	**Yes**	**Yes**	**No**	**No**	**No**	**No**

1-epicatechin, 2-catechin, 3-syringic acid, 4-gallic acid, 5-protocatechuic acid, 6-procyanidin B1, 7-procyanidin B2, 8-procyanidin B3, 9-procyanidin B4, 10-esculetin, 11-syringaldehyde, 12-caftaric acid, 13-caffeic acid, 14-*p*-coumaric acid, 15-*trans*-resveratrol, 16-*cis*-resveratrol.

S0 - control sample, conventional aging; S1 -1 g/L granular oak fragments, 10 days of maturation; S2 - 1 g/L granular oak fragments, 10 days of maturation; S3 - 1 g/L granular oak frag-ments, 10 days of maturation; S4 - 1 g/L chips oak fragments, 10 days of maturation; S5 - 1 g/L chips oak fragments, 10 days of maturation; S6 - 1 g/L chips oak fragments, 10 days of maturation; S7 - 1 g/L granular oak fragments, 20 days of maturation; S8 - 1 g/L granular oak fragments, 20 days of maturation; S9 - 1 g/L granular oak fragments, 20 days of maturation; S10 - 1 g/L chips oak fragments, 20 days of maturation; S11 - 1 g/L chips oak fragments, 20 days of maturation; S12 - 1 g/L chips oak fragments, 20 days of maturation; S13 - 2 g/L granular oak fragments, 10 days of maturation; S14 - 2 g/L granular oak fragments, 10 days of maturation; S15 - 2 g/L granular oak fragments, 10 days of maturation; S16 - 2 g/L chips oak fragments, 10 days of maturation; S17 - 2 g/L chips oak fragments, 10 days of maturation; S18 - 2 g/L chips oak fragments, 10 days of maturation; S19 - 2 g/L granular oak fragments, 20 days of maturation; S20 - 2 g/L granular oak fragments, 20 days of maturation; S21 - 2 g/L granular oak fragments, 20 days of maturation; S22 - 2 g/L chips oak fragments, 20 days of maturation; S23 - 2 g/L chips oak fragments, 20 days of maturation; S24 - 2 g/L chips oak fragments, 20 days of maturation; S25 - control sample, ultrasound treatment; S26 - 1 g/L granular oak fragments, ultrasound treatment, 15 min.; S27 - 1 g/L granular oak fragments, ultrasound treatment, 15 min.; S28 - 1 g/L granular oak fragments, ultrasound treatment, 15 min.; S29 - 1 g/L chips oak fragments, ultrasound treatment, 15 min.; S30 - 1 g/L chips oak fragments, ultrasound treatment, 15 min.; S31 - 1 g/L chips oak fragments, ultrasound treatment, 15 min.; S32 - 2 g/L granular oak fragments, ultrasound treatment, 15 min.; S33 - 2 g/L granular oak fragments, ultrasound treatment, 15 min.; S34 - 2 g/L granular oak fragments, ultrasound treatment, 15 min.; S35 - 2 g/L chips oak fragments, ultrasound treatment; S36 - 2 g/L chips oak fragments, ultrasound treatment; S37 - 2 g/L chips oak fragments, ultrasound treatment.

The samples were analysed in triplicate. The presented data represent the arithmetic mean of the obtained results and the calculated standard deviation. Different letters indicate homogeneous groups.

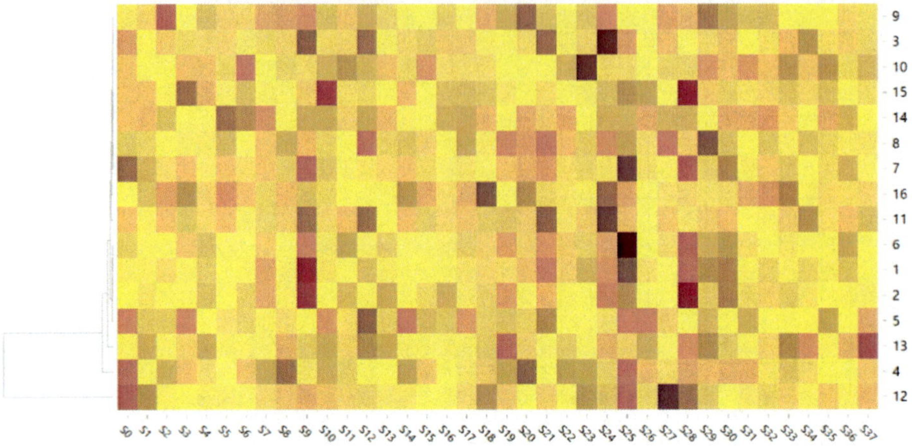

Fig. 1. Heatmap on sensory characteristics of wines. S0 - control sample, conventional aging; S1 - 1 g/L granular oak fragments, 10 days of maturation; S2 - 1 g/L granular oak fragments, 10 days of maturation; S3 - 1 g/L granular oak fragments, 10 days of maturation; S4 - 1 g/L chips oak fragments, 10 days of maturation; S5 - 1 g/L chips oak fragments, 10 days of maturation; S6 - 1 g/L chips oak fragments, 10 days of maturation; S7 - 1 g/L granular oak fragments, 20 days of maturation; S8 - 1 g/L granular oak fragments, 20 days of maturation; S9 - 1 g/L granular oak fragments, 20 days of maturation; S10 - 1 g/L chips oak fragments, 20 days of maturation; S11 - 1 g/L chips oak fragments, 20 days of maturation; S12 - 1 g/L chips oak fragments, 20 days of maturation; S13 - 2 g/L granular oak fragments, 10 days of maturation; S14 - 2 g/L granular oak fragments, 10 days of maturation; S15 - 2 g/L granular oak fragments, 10 days of maturation; S16 - 2 g/L chips oak fragments, 10 days of maturation; S17 - 2 g/L chips oak fragments, 10 days of maturation; S18 - 2 g/L chips oak fragments, 10 days of maturation; S19 - 2 g/L granular oak fragments, 20 days of maturation; S20 - 2 g/L granular oak fragments, 20 days of maturation; S21 - 2 g/L granular oak fragments, 20 days of maturation; S22 - 2 g/L chips oak fragments, 20 days of maturation; S23 - 2 g/L chips oak fragments, 20 days of maturation; S24 - 2 g/L chips oak fragments, 20 days of maturation; S25 - control sample, ultrasound treatment; S26 - 1 g/L granular oak fragments, ultrasound treatment, 15 min.; S27 - 1 g/L granular oak fragments, ultrasound treatment, 15 min.; S28 - 1 g/L granular oak fragments, ultrasound treatment, 15 min.; S29 - 1 g/L chips oak fragments, ultrasound treatment, 15 min.; S30 - 1 g/L chips oak fragments, ultrasound treatment, 15 min.; S31 - 1 g/L chips oak fragments, ultrasound treatment, 15 min.; S32 - 2 g/L granular oak fragments, ultrasound treatment, 15 min.; S33 - 2 g/L granular oak fragments, ultrasound treatment, 15 min.; S34 - 2 g/L granular oak fragments, ultrasound treatment, 15 min.; S35 - 2 g/L chips oak fragments, ultrasound treatment; S36 - 2 g/L chips oak fragments, ultrasound treatment; S37 - 2 g/L chips oak fragments, ultrasound treatment. 1-epicatechin; 2-catechin; 3-syringic acid; 4-gallic acid; 5-protocatechuic acid; 6-procyanidin B1; 7-procyanidin B2; 8-procyanidin B3; 9-procyanidin B4; 10-esculetin; 11-syringaldehide; 12-caftaric acid; 13-caffeic acid; 14-p-coumaric acid; 15-*trans*-resveratrol; 16-*cis*-resveratrol.

3.2 Sensory Analysis

The results showed an important influence of the oak fragment type on the sensory profile (Figs. 2, 3 and 4). The heatmap from Fig. 5 highlights the intensity of sensory descriptors in the analyzed samples. It can be observed that all samples were perceived as having a good persistence and texture, but also a high acidity.

In general, wines aged with oak exhibited heightened sweet-taste intensity and garnered elevated ratings concerning texture in comparison to control samples. A notable divergence was observed, whereby the oak-aged wines displayed diminished levels of acidity and astringency. This disparity may be ascribed to the mitigating effect of sweetness on the perception of astringency. Concurrently, according to the results presented by Yu et al. [25], tannins exhibited a progressive softening throughout the maturation process, consequently leading to a gradual reduction in astringency.

The samples in which 1 g/L wood fragments (for 10 days) were administered in the form of chips revealed a richer olfactory and taste profile compared to the samples in which the granulated form was used, with more intense notes of exotic fruits, vanilla, but also a persistent and better texture.

Increasing the contact time leads to the accentuation of the spicy sensation and vanilla taste. From this category, S3 sample (granular, medium toasting, dosage - 1 g/L, 10 days of maturation) was characterized by more intense spicy aromas, while S28 (granular, medium toasting, dosage - 1 g/L, 15 min with ultrasounds) and S27 (granular, fresh toasting, dosage - 1 g/L, 15 min with ultrasounds) by vanilla notes. According to Yu et al. [25], the manifestation of vanilla notes in the wines is purportedly linked to an augmented presence of 1-hexanol, a compound recognized for its association with a vanilla olfactory perception. Additionally, compounds such as 3-hexen-1-ol and trans-2-hexenyl-acetate are acknowledged for their contribution to the manifestation of vanilla aroma in wines [25].

After 20 days of contact with wood fragments, vanilla, coconut and sweet tobacco aromas were more intense in samples treated with granular forms. Variants in which oak chips were used were defined by almond and coconut notes. According to Perreira et al. [26], furans are responsible for the roasted almond notes in wines, while lactones contribute to the coconut aroma.

Major differences were obtained between the control sample (without oak fragments) and the oaked wines. The application of ultrasounds conducted to more intense spicy and woody notes, but also a longer persistence. The utilization of oak fragments for aging wine augmented its complexity, mitigated certain bitter and astringent attributes, although at the expense of attenuated fruit aroma. Notably, wines aged with oak chips exhibited a more prominent fruity aroma. The manifestation of fruity aroma in the wines was more pronounced in those subjected to the aging process with oak fragments compared to the control samples. As presented by Yu et al. [20], oak-chip-treated wines typically possess heightened concentrations of ethyl lactate and isopropyl acetate, imparting an elegant fruity and creamy flavor profile.

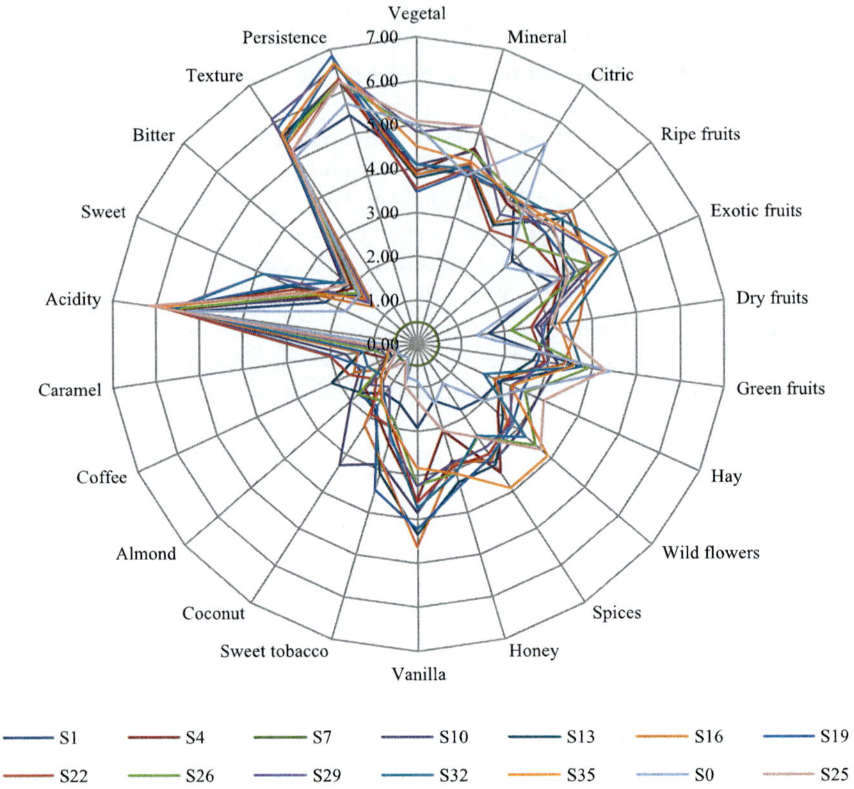

Fig. 2. Sensory profile of samples obtained with light toasting degree comparing to control samples. S1 - 1 g/L granular oak fragments, 10 days of maturation; S4 - 1 g/L chips oak fragments, 10 days of maturation; S7 - 1 g/L granular oak fragments, 20 days of maturation; S10 - 1 g/L chips oak fragments, 20 days of maturation; S13 - 2 g/L granular oak fragments, 10 days of maturation; S16 - 2 g/L chips oak fragments, 10 days of maturation; S19 - 2 g/L granular oak fragments, 20 days of maturation; S22 - 2 g/L chips oak fragments, 20 days of maturation; S26 - 1 g/L granular oak fragments, ultrasound treatment, 15 min.; S29 - 1 g/L chips oak fragments, ultrasound treatment, 15 min.; S32 - 2 g/L granular oak fragments, ultrasound treatment, 15 min.; S35 - 2 g/L chips oak fragments, ultrasound treatment; S0 – control sample, conventional aging; S25 - control sample, ultrasound treatment.

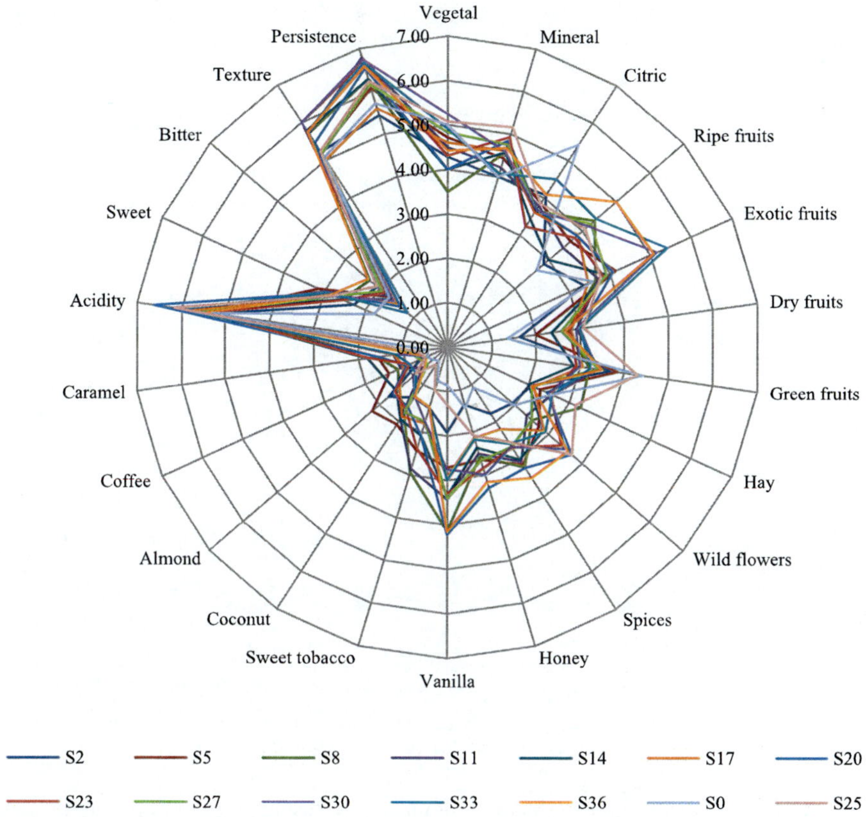

Fig. 3. Sensory profile of samples obtained with fresh toasting degree comparing to control samples. S2 - 1 g/L granular oak fragments, 10 days of maturation; S5 - 1 g/L chips oak fragments, 10 days of maturation; S8 - 1 g/L granular oak fragments, 20 days of maturation; S11 - 1 g/L chips oak fragments, 20 days of maturation; S14 - 2 g/L granular oak fragments, 10 days of maturation; S17 - 2 g/L chips oak fragments, 10 days of maturation; S20 - 2 g/L granular oak fragments, 20 days of maturation; S23 - 2 g/L chips oak fragments, 20 days of maturation; S27 - 1 g/L granular oak fragments, ultrasound treatment, 15 min.; S30 - 1 g/L chips oak fragments, ultrasound treatment, 15 min.; S33 - 2 g/L granular oak fragments, ultrasound treatment, 15 min.; S36 - 2 g/L chips oak fragments, ultrasound treatment; S0 – control sample, conventional aging; S25 - control sample, ultrasound treatment.

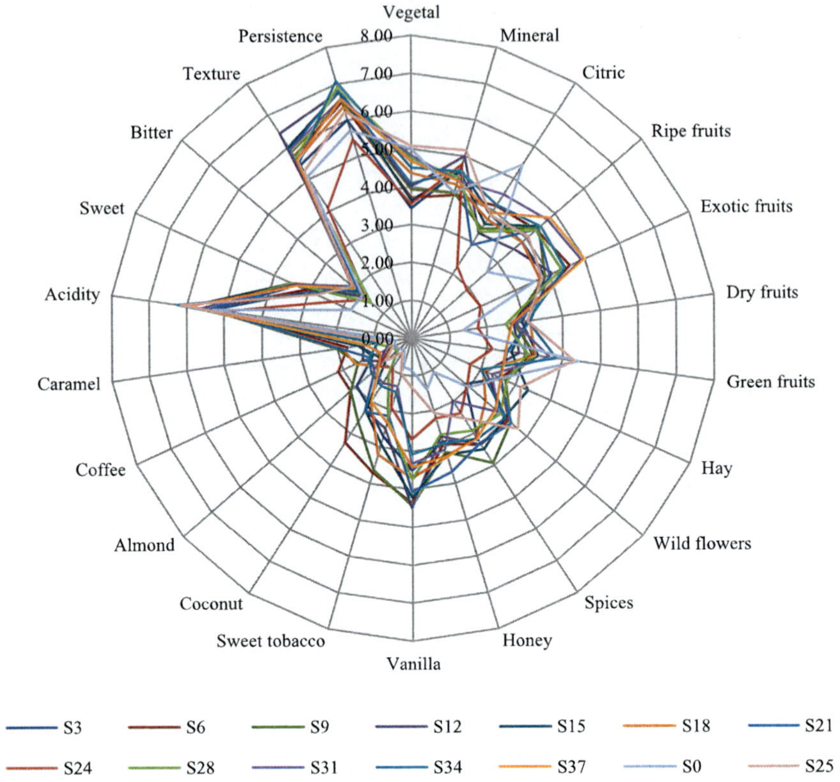

Fig. 4. Sensory profile of samples obtained with medium toasting degree comparing to control samples. S3 - 1 g/L granular oak fragments, 10 days of maturation; S6 - 1 g/L chips oak fragments, 10 days of maturation; S9 - 1 g/L granular oak fragments, 20 days of maturation; S12 - 1 g/L chips oak fragments, 20 days of maturation; S15 - 2 g/L granular oak fragments, 10 days of maturation; S18 - 2 g/L chips oak fragments, 10 days of maturation; S21 - 2 g/L granular oak fragments, 20 days of maturation; S24 - 2 g/L chips oak fragments, 20 days of maturation; S28 - 1 g/L granular oak fragments, ultrasound treatment, 15 min.; S31 - 1 g/L chips oak fragments, ultrasound treatment, 15 min.; S34 - 2 g/L granular oak fragments, ultrasound treatment, 15 min.; S37 - 2 g/L chips oak fragments, ultrasound treatment; S0 – control sample, conventional aging; S25 - control sample, ultrasound treatment.

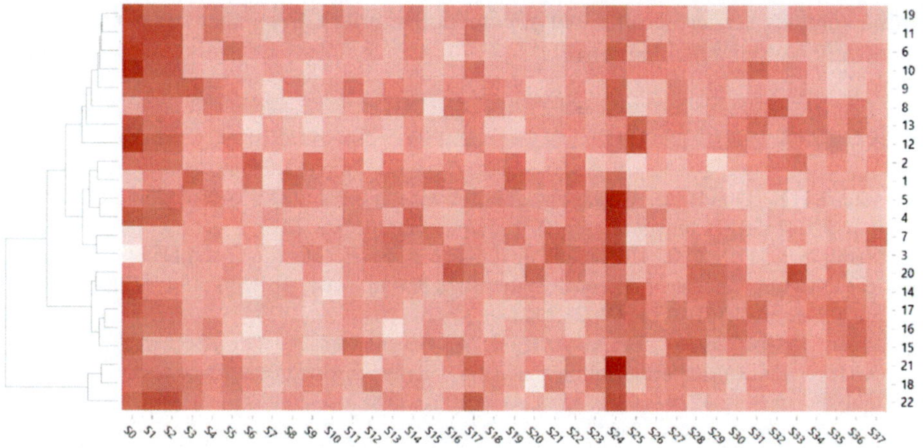

Fig. 5. Heatmap on sensory characteristics of wines. S0 - control sample, conventional aging; S1 - 1 g/L granular oak fragments, 10 days of maturation; S2 - 1 g/L granular oak fragments, 10 days of maturation; S3 - 1 g/L granular oak fragments, 10 days of maturation; S4 - 1 g/L chips oak fragments, 10 days of maturation; S5 - 1 g/L chips oak fragments, 10 days of maturation; S6 - 1 g/L chips oak fragments, 10 days of maturation; S7 - 1 g/L granular oak fragments, 20 days of maturation; S8 - 1 g/L granular oak fragments, 20 days of maturation; S9 - 1 g/L granular oak fragments, 20 days of maturation; S10 - 1 g/L chips oak fragments, 20 days of maturation; S11 - 1 g/L chips oak fragments, 20 days of maturation; S12 - 1 g/L chips oak fragments, 20 days of maturation; S13 - 2 g/L granular oak fragments, 10 days of maturation; S14 - 2 g/L granular oak fragments, 10 days of maturation; S15 - 2 g/L granular oak fragments, 10 days of maturation; S16 - 2 g/L chips oak fragments, 10 days of maturation; S17 - 2 g/L chips oak fragments, 10 days of maturation; S18 - 2 g/L chips oak fragments, 10 days of maturation; S19 - 2 g/L granular oak fragments, 20 days of maturation; S20 - 2 g/L granular oak fragments, 20 days of maturation; S21 - 2 g/L granular oak fragments, 20 days of maturation; S22 - 2 g/L chips oak fragments, 20 days of maturation; S23 - 2 g/L chips oak fragments, 20 days of maturation; S24 - 2 g/L chips oak fragments, 20 days of maturation; S25 - control sample, ultrasound treatment; S26 - 1 g/L granular oak fragments, ultrasound treatment, 15 min.; S27 - 1 g/L granular oak fragments, ultrasound treatment, 15 min.; S28 - 1 g/L granular oak fragments, ultrasound treatment, 15 min.; S29 - 1 g/L chips oak fragments, ultrasound treatment, 15 min.; S30 - 1 g/L chips oak fragments, ultrasound treatment, 15 min.; S31 - 1 g/L chips oak fragments, ultrasound treatment, 15 min.; S32 - 2 g/L granular oak fragments, ultrasound treatment, 15 min.; S33 - 2 g/L granular oak fragments, ultrasound treatment, 15 min.; S34 - 2 g/L granular oak fragments, ultrasound treatment, 15 min.; S35 - 2 g/L chips oak fragments, ultrasound treatment; S36 - 2 g/L chips oak fragments, ultrasound treatment; S37 - 2 g/L chips oak fragments, ultrasound treatment.

4 Conclusions

Ageing wine in contact with wood is a complex process, which depends on numerous variables. Oak aging and ultrasound treatment present a significant impact on wine phenolic and sensory profile, in correlation with a shorter time of contact. The samples obtained with oak chips were characterized by more intense fruity, but also woody and spicy notes. The results are influenced by the type of the oak fragment, the used dosage, and the contact time. Considering the tendency of consumers to turn to wines with a complex sensory profile, but at competitive prices, this paper contributes to the optimization of winemaking process.

References

1. Yildiri, H.K., Dündar, E.: New techniques for wine aging. In: BIO Web of Conferences, vol. 9, p. 02012 (2017). https://doi.org/10.1051/bioconf/20170902012
2. Natolino, A., Celotti, E.: Ultrasound treatment of red wine: effect on polyphenols, mathematical modeling, and scale-up considerations. Lebensm.-Wiss. Technol. **154**, 112843 (2022). https://doi.org/10.1016/j.lwt.2021.112843
3. Botha, A., Du Toit, W., Brand, J., Kidd, M., Groenewald, N.: The effect of different oak products used during fermentation and ageing on the sensory properties of a white wine over time. Foods **9**(9), 1220 (2020). https://doi.org/10.3390/foods9091220
4. Sánchez-Gómez, R., Alamo-Sanza, M., Martínez-Gil, A.M., Nevares, I.: Red wine micro-oxygenation systems and oak wood-effects on anthocyanins, copigmentation and color evolution. Processes **8**(10), 1250 (2020). https://doi.org/10.3390/pr8101250
5. Rubio-Bretón, P., Garde-Cerdán, T., Martínez, J.: Use of oak fragments during the aging of red wines. Effect on the phenolic, aromatic, and sensory composition of wines as a function of the contact time with the wood. Beverages **4**(4), 102 (2018). https://doi.org/10.3390/beverages4040102
6. Crump, A.M., Johnson, T.E., Wilkinson, K.L., Bastian, S.E.P.: Influence of oak maturation regimen on composition, sensory properties, quality, and consumer acceptability of Cabernet Sauvignon wines. J. Agric. Food Chem. **63**, 1593–1600 (2015). https://doi.org/10.1021/jf5044025
7. Bozalongo, R., Carillo, J.D., Torroba, M.A.F., Tena, M.T.: Analysis of French and American oak chips with different toasting degrees by headspace solid-phase microextraction-gas chromatography-mass spectrometry. J. Chromatogr. A **1173**, 10–17 (2017)
8. Arfelli, G., Sartini, E., Corzani, C., Fabiani, A.: Chips, lees, and micro-oxygenation: influence on some flavours and sensory profile of a bottles red Sangiovese wine. Eur. Food Res. Technol. **233**, 1–10 (2011). https://doi.org/10.1007/s00217-011-1480-2
9. Focea, E., et al.: Application of ultrasounds to improve oak aging of white wines. In: BIO Web of Conferences, vol. 56, p. 02021 (2023). https://doi.org/10.1051/bioconf/20235602021
10. Natrella, G., Noviello, M., Trani, A., Faccia, M., Gabacorta, G.: The effect of ultrasound treatment in winemaking on the volatile compounds of Aglianico, Nero di Troia, and Primitivo red wines. Foods **12**(3), 648 (2023). https://doi.org/10.3390/foods12030648
11. Mărgean, A., Pădureanu, V.: The influence of ultrasound treatment on must fermentation process. Bull. Transilvania Univ. Brașov **11**(60), 127–132 (2018)
12. Ruiz-Rodríguez, A., Carrera, C., Palma, M., Barroso, C.G.: Ultrasonic treatments during the alcoholic fermentation of red wines: effects on "Syrah" wines. Vitis: J. Grapevine Res. **58**(5), 83–88 (2019). https://doi.org/10.5073/vitis.2019.58.special-issue.83-88
13. Gracin, L., et al.: Monitoring the influence of high power ultrasound treatment and thermosonication on inactivation of Brettanomyces bruxellensis in red wine. Croatian J. Food Technol. Biotechnol. Nutr. **12**(3–4), 107–112 (2017)
14. Xie, Q., et al.: Combined ultrasound and low temperature pretreatment improve the content of anthocyanins, phenols and volatile substance of Merlot red wine. Ultrason. Sonochem. (2023). https://doi.org/10.2139/ssrn.4542039
15. Ferraretto, P., Celotti, E.: Preliminary study of the effects of ultrasound on red wine polyphenols. Cyta-J. Food **14**(4), 529–535 (2016). https://doi.org/10.1080/19476337.2016.1149520

16. ISO 8589:2007; Sensory Analysis: General Guidance for the Design of Test Room. ISO, Geneva (2007)
17. ISO 3591:1977; Sensory Analysis: Apparatus. Wine-Tasting Glass. ISO, Geneva (1977)
18. International Organization of Wine and Vine. Review Document on Sensory Analysis of Wine. International Organization of Wine and Vine, Paris (2015). https://www.oiv.int/public/medias/3307/review-on-sensory-analysis-of-wine.pdf. Accessed 21 July 2023
19. Glabasnia, A., Hofmann, T.: Sensory-directed identification of taste-active ellagitannins in American (*Quercus alba L.*) and European oak wood (*Quercus robur L.*) and quantitative analysis in bourbon whiskey and oak-matured red wines. J. Agric. Food Chem. **54**, 3380–3390 (2006). https://doi.org/10.1021/jf052617b
20. Prida, A., Boulet, J.C., Ducousso, A., Nepveu, G., Puech, J.L.: Effect of species and ecological conditions on ellagitannin content in oak wood from an even-aged and mixed stand of *Quercus robur L.* and *Quercus petraea* Liebl. Ann. For. Sci. **63**, 415–424 (2006). https://doi.org/10.1051/forest:2006021
21. Cadahía, E., Muñoz, L., Fernández de Simón, B., García-Vallejo, M.C.: Changes in low molecular weight phenolic compounds in Spanish, French, and American oak woods during natural seasoning and toasting. J. Agric. Food Chem. **49**, 1790–1798 (2001). https://doi.org/10.1021/jf0006168
22. Cerdán, T.G., Mozaz, S.R., Azpilicueta, C.A.: Volatile composition of aged wine in used barrels of French oak and of American oak. Food Res. Int. **35**(7), 603–610 (2002). https://doi.org/10.1016/s0963-9969(01)00151-x
23. Sterneder, S., et al.: Astringent gallic acid in red wine regulates mechanisms of gastric acid secretion via activation of bitter taste sensing receptor TAS2R4. J. Agric. Food Chem. **69**(36), 10550–10561 (2021). https://doi.org/10.1021/acs.jafc.1c03061
24. Canas, S.: Phenolic compounds and related properties of aged wine spirits: influence of barrel characteristics. A review. Beverages. **3**, 4–55 (2017). https://doi.org/10.3390/beverages3040055
25. Yu, Y., et al.: Impact of different oak chips' aging on the volatile compounds and sensory characteristics of *Vitis amurensis* wines. Foods **11**(8), 1126 (2022). https://doi.org/10.3390/foods11081126
26. Pereira, A.G., et al.: Management of wine aroma compounds: principal basis and future perspectives. In: Cosme, F., Fernando, M., Nunes, M., Ribeiro, L.F. (eds.) Chemistry and Biochemistry of Winemaking, Wine Stabilization and Aging. Intechopen eBooks (2021). https://doi.org/10.5772/intechopen.92973

Evaluation of the Nutritional Quality of Some Fruits Grown in Romania

Elena Cristina Scutarașu[1], Camelia Elena Luchian[1](✉), Bogdan Ionel Cioroiu[2], Marius Niculaua[2], Ana Maria Vlase[3], Valeriu Cotea[1], and Laurian Vlase[3]

[1] Ion Ionescu de la Brad Iasi University of Life Sciences, 3rd M. Sadoveanu Alley, 700490 Iași, Romania
kamelia_luchian@yahoo.com

[2] Research Center of Oenology, Romanian Academy, Iași Branch, 9th M. Sadoveanu Alley, 700505 Iași, Romania

[3] Faculty of Pharmacy, Iuliu Hațieganu University of Medicine and Pharmacy, 8th Victor Babes Street, 400012 Cluj-Napoca, Romania

Abstract. Many species of fruit are grown in Romania but their production is limited to a certain period of the year. However, there is still no habit of the population to consume fresh fruit, but a growing tendency of consumers to choose food with a high nutritional value was observed. Fresh fruits can contribute to a balanced diet and good development of metabolic processes, being valuable sources of bioactive compounds, vitamins, and energy. Few studies were focused on the nutritional potential of fruits cultivated in Romania and no paper refers to the antioxidant capacity of fruits from Romania in correlation with melatonin content. This work aims to promote the introduction of local fruits into the daily diet, as functional foods. In this regard, 19 samples such as cherries, sour cherries, blackberries, raspberries, strawberries, and table grapes were evaluated. On this line, the antioxidant activity of the samples in relation to their melatonin content and the presence of phenolic compounds were analyzed. In general, the highest concentrations of melatonin were obtained in table grapes and cherries - Amar de Maxut. Among the analyzed species, table grapes presented the highest antioxidant activity, in accordance with their melatonin level.

Keywords: Melatonin · Phenolic Compounds · Antioxidant Activity · Bioactive Compounds · Balanced-Diet

1 Introduction

Melatonin is an indoleamine, usually produced from tryptophan [1] but also from norepinephrine, resulting from the conversion of serotonin. Its production is favoured by a darker environment [2–4] and limited by hydroxyindole-O-methyltransferase activity [5]. In plants, melatonin synthesis ensues from the conversion of tryptophan into serotonin, a process that may unfold through diverse pathways. Nevertheless, noteworthy

Y. Ono and J. Kondoh (Eds.): Inter-Academia 2023, LNNS 939, pp. 122–139, 2024.
https://doi.org/10.1007/978-3-031-54450-7_15

findings indicate a reversal in the sequence of steps governing serotonin biosynthesis from tryptophan within plants. Contrary to the conventional enzymatic conversion yielding 5-hydroxytryptophan, the initial product manifests as tryptamine, catalyzed by tryptophan decarboxylase. Subsequently, this tryptamine is subjected to catalysis by tryptamine 5-hydroxylase, ultimately yielding serotonin. Beyond its role in melatonin and serotonin biosynthesis, tryptophan in plants serves as the precursor for other phytohormones, such as indole-3-acetic acid and various auxins, including indole-3-butyric and p-hydroxyphenylacetic acids. The synthesis of indole-3-acetic acid in plants exhibits versatility through several l-tryptophan-dependent pathways involving intermediate products such as (a) indole-3-pyruvic acid and indole-3-acetaldehyde, (b) indole-3-acetamide, or (c) indole-3-acetonitrile. Moreover, indole-3-acetic acid could derive from tryptamine, catalyzed by tryptamine deaminase - a pathway directly intersecting with serotonin and melatonin biosynthesis [6].

In the last few years, the mentioned indoleamine has been identified in many vegetable substrates, fruits and microorganisms. A correlation between yeasts and melatonin production has been reported [7–9]. Its concentration in food depends on many factors, such as plant species, variety, maturation degree, soil characteristics and climate conditions [10]. Melatonin was identified in high amounts in seeds and cereals, such as soybean [11], rice [12, 13], barley [14], or wheat [15]. Also, important concentrations can be found in many fruit species, including apples [16, 17], pears [18], peaches [19], cherries [20], bananas [21], strawberries [22], and grapes [23]. Current studies suggest that melatonin is implicated in the plant and fruit development processes and protects against environmental stress from heavy metals, UV radiation, temperature variations, and drought [10].

Beverages are important sources of essential nutrients and manifest a major contribution not only to human health and nutrition but also to economic affairs, culture, and social relations [24]. Beverages like fermented apple and orange juices, coffee, black tea, wine [23–27], and beer [28, 29] were tested for melatonin. The occurrence of this new bioactive compound in fermented drinks is usually low (pg/mL to ng/mL) [30], but it presents important benefits for consumers health. Many authors reported its functional roles being often linked to its antioxidant capacity by reducing oxidative stress [2, 30], with positive effects on the regulation of circadian rhythm, reproduction [31, 32], and cardiovascular function [33], with an important effect against cancer cells [8, 34, 35] and with anti-aging properties [36], manifest neurodegenerative diseases protection [37, 38], helping in improving sleep quality [39–43], with an important anti-inflammatory [29] and anti-depression effect [44], inhibitory action on thyroid growth and secretion [45], positive effect on scoliosis [46], and with thermoregulatory properties [47].

As there is an increased interest in discovering natural sources of bioactive compounds, existing studies do not provide sufficient information on the melatonin content of fermented fruits and beverages. Considering the many positive actions of melatonin on the health of consumers as well as the current trends towards functional foods, the importance of the topic addressed is explained, as is the need to enrich drinks with bioactive compounds. The study is interdisciplinary, representing a special interest in the field of research. The nutritional potential of table grapes grown in Romania is little known.

Most studies focus on wine varieties. No studies regarding the antioxidant capacity in correlation to melatonin levels of fruits from Romanian cultivars were found.

2 Materials and Methods

The evaluation of table grapes quality is still a relatively underexplored area of research in Romania, with the majority of existing studies focusing on grapevine varieties intended for winemaking. No study was found on the antioxidant capacity of fruits grown in Romania in relation to their melatonin content.

In order to fulfill the main purpose, a comprehensive series of determinations were made. These determinations were centered on the quantification of melatonin content by LC-MS. Additionally, the investigation encompassed the quantification of anthocyanin levels using LC-MS and total anthocyanin, total polyphenols, and phenolic compounds with reducing properties through spectrophotometric methods.

2.1 Reagents

For these experiments, standard solutions (e.g. melatonin, L-tryptophan, cyanidin-3-glucoside, cyanidin, malvidin-3-glucoside, petunidin 3-glucoside, cyanidin 3-rutinoside, delphinidin 3-glucoside, gallic acid – GAE, 2,2'-Diphenyl-1- picrylhydrazyl – DPPH), and different reagents (Folin-Ciocâlteu reagent, acetonitrile, methanol, isopropanol, acetone, formic acid, triethylamine) were purchased from Merck (Darmstadt, Germany).

2.2 Sample Preparation

In order to fulfill the proposed purpose, 19 fruit samples were analyzed, as shown in Table 1. The analyzed fruits were harvested at maturity for consumption by local producers from the Iasi area, Romania (47°10′ north latitude, 27°35′ east longitude).

The samples were frozen until the analysis. A medium aliquot of each sample was crushed with a pestle mortar. The work protocol involved organic solvent extraction of melatonin (0.6 mL methanol and 0,05 mL internal standard solution was mixed with 1 g of fruit), homogenization (10 min with a vortex, 15 min in a ultrasonic bath, at room temperature), centrifugation (5 min, at 15000 rpm) and analysis.

2.3 Laboratory Analysis

The total polyphenolic content of the extracts was quantified using the D280 index. The measurements were made using a UV-VIS spectrometer Analytik Jena Specord 200, and the absorbance was measured at 280 nm. Moreover, Folin-Ciocâlteu assay was used for the evaluation of **phenolic compounds with reducing properties**. Measurements were conducted using a UV-VIS spectrometer, at a wavelength of 750 nm. The determinations were made in triplicate, the results representing the arithmetic mean of the three determinations. The results were expressed in mg of gallic acid/100 g fruits.

The antioxidant capacity of the analyzed samples was evaluated using 1,1-diphenyl-2-picryl hydrazyl (DPPH) assay, with absorbance measurements carried out

Table 1. The analysed fruit samples

No.	Variety	Species
1.	Stella	Cherry
2.	Boambe de Cotnari	Cherry
3.	Germersdorf	Cherry
4.	Amar de Maxut	Cherry
5.	Van	Cherry
6.	Botoşani	Sour cherry
7.	Thornfree	Blackberry
8.	Elsanta	Strawberry
9.	Red Giant	Rasberry
10.	Muscat timpuriu de Bucureşti	Table grapes
11.	Paula	Table grapes
12.	Coarnă neagră selecţionată	Table grapes
13.	Augusta	Table grapes
14.	Victoria	Table grapes
15.	Timpuriu de Pietroasa	Table grapes
16.	Gelu	Table grapes
17.	Muscat Hamburg	Table grapes
18.	Napoca	Table grapes
19.	Cardinal	Table grapes

at 515 nm. A calibration curve was constructed using gallic acid, and the results were reported as IC50 values. The determinations were made in triplicate, the results representing the arithmetic mean of the three determinations.

Identification and quantification of melatonin and internal standard were performed on a Transcend XT Ultimate 3000 UHPLC system (Thermo Scientific TM, Waltham, Massachusetts, USA) equipped with quaternary pump RPG-3400RS (TX), binary pump TM LPG-3400RS (LX), injection system PAL LX, thermostated column compartment TCC-3000 RS and valve interface module (Tee). For detection, a TSQ Access Max mass spectrometer was used. Sample cleanup was achieved on Cyclone P – polymeric column (0,5 mm ID and 50 mm length). Elution of compounds was performed on Agilent Poroshell C18 (4.6 mm ID, 100 mm length and 1.8 μm particle size). In sample cleanup, water was used to retain the compounds on the column TX polymeric column (TA). Transfer was realized with water:methanol (30% (TA):70%(TD) (v/v) mixture (in loop). TX column was cleaned with acetonitrile:isopropanol:acetone (45%:45%:10% v/v/v) (TB) and then with 0.05% triethylamine in acetonitrile (TC). A mobile phase with 0,1% formic acid in water (LA) and 0,1% formic acid in methanol

(LB) was used for separation of melatonin and internal standard. Total chromatographic time was 4.5 min and the elution program was established according to the Table 2:

Table 2. Elution program applied for the chromatographic system

Sec	Flow-rate	%TA	%TB	%TC	%TD	Tee	Loop	Flow-rate	%LA	%LB
60	1.5	100				==	Out	0.2	80	20
30	0.7	100				==	Out	0.1	80	20
60	2.5		100			T	In	0.2	50	50
60	2.5			100		==	Out	0.2	50	50
30	2.0	30			70	==	In	0.2	80	20
30	1.5	100				==	Out	0.2	80	20

In the mass spectrometry analysis, a Heated Electrospray Ionization source operated in positive mode was utilized for ionization, while a collision cell within Q2 was employed for fragmenting and separating specific ions to enable accurate compound identification. The ionization conditions included an ionization potential of 3 kV, an ionization source temperature of 350 °C, a nebulization gas pressure of 35 psi, and an auxiliary gas pressure of 10 psi. The capillary tube was maintained at a temperature of 350 °C, and the polarity was set to positive. The determinations were made in triplicate, the results representing the arithmetic mean of the three determinations. The results were expressed in ng melatonin/g of fruit (Fig. 1).

The anthocyanins have been quantified by Agilent Technologies 1100 HPLC Series system (Agilent, Santa Clara, CA, USA) coupled with a mass spectrometer from Agilent, model with Ion Trap 1100 SL (LC/MSD Ion Trap VL, Agilent, Santa Clara, CA, USA). The results were expressed as μg/mL extract. The analytical procedure followed was in accordance with the methodology detailed by Vlase et al. [48]. The analytes subjected to separation and subsequent quantification included the following compounds: 1 - delphinidin 3-glucoside, 2 - cyanidin 3-glucoside, 3 - petunidin 3-glucoside, and 4 - malvidin 3-glucoside. The determinations were made in triplicate, the results representing the arithmetic mean of the three determinations (Fig. 2).

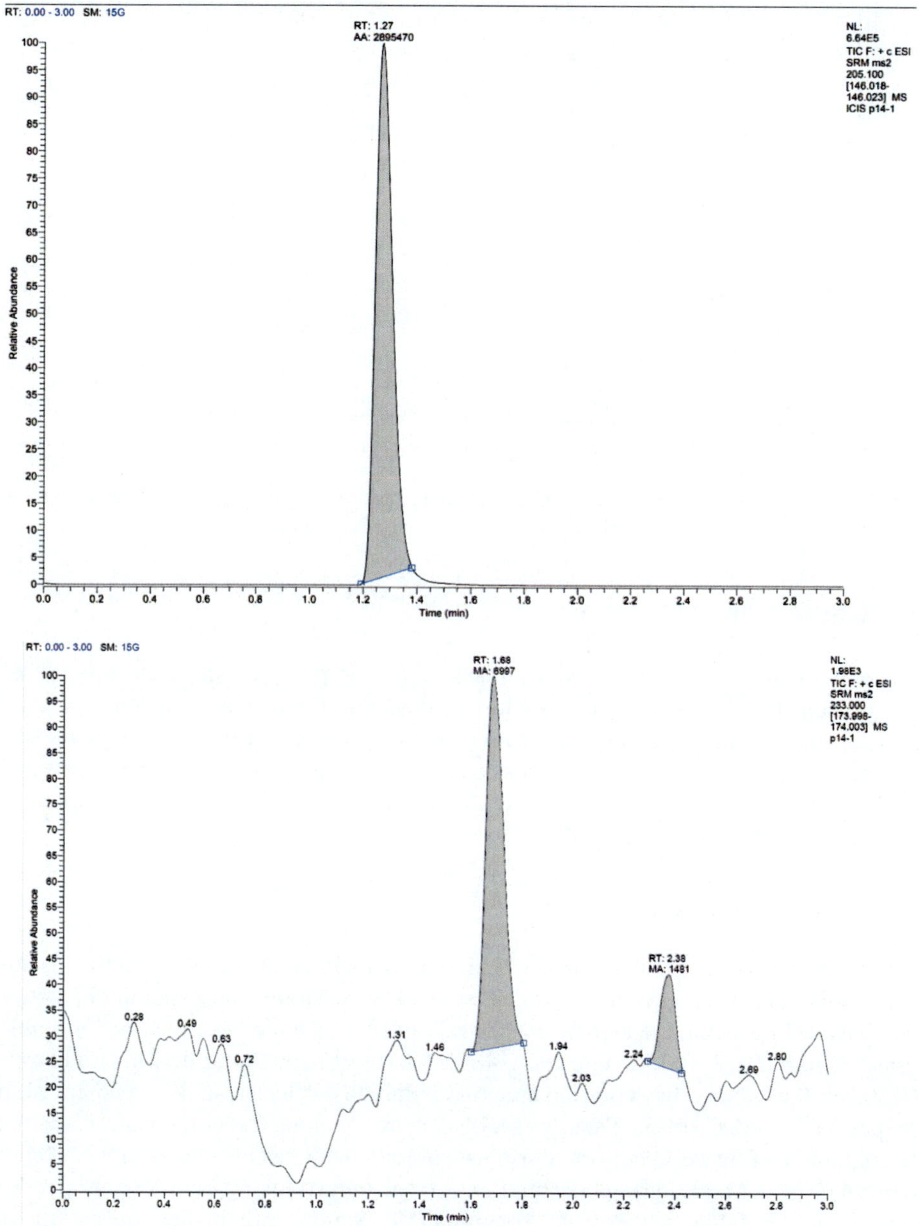

Fig. 1. Example of chromatogram: Evaluation of melatonin – Victoria grapes (internal standard solution RT = 1.27 min. - upper part; melatonin RT = 2.38 min. - lower part).

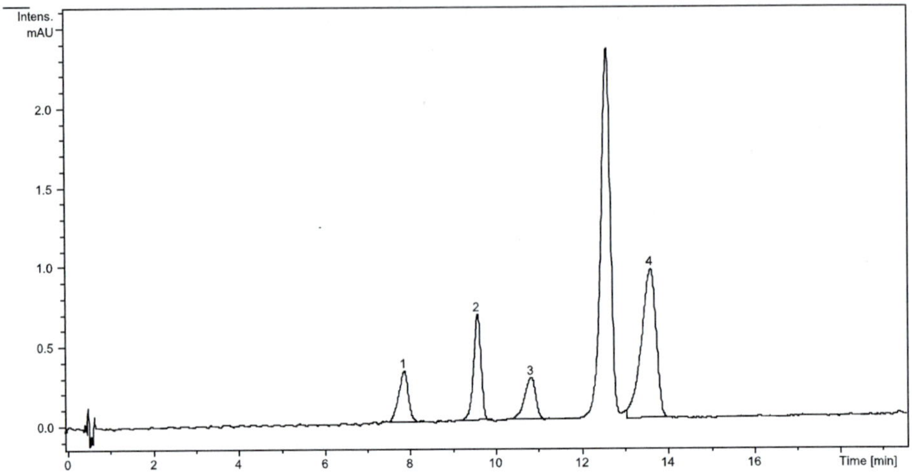

Fig. 2. Example of chromatogram – Cardinal table grapes (1 - delphinidin 3-glucoside; 2 - cyanidin 3-glucoside; 3 - petunidin 3-glucoside; 4 - malvidin 3-glucoside)

3 Results and Discussions

The antioxidant activity of fruits is influenced by the presence and concentration of various compounds, including melatonin, anthocyanins, total anthocyanin content, total polyphenols, and phenol compounds with reduction properties. Fruits with higher levels of these compounds generally exhibit stronger antioxidant activity, which is beneficial for human health. It's important to note that the combined effect of these compounds contributes to the overall antioxidant capacity of fruits.

3.1 Melatonin Content

Table 3 shows the results from the analysis of melatonin content in the examined samples. Indigenous fruit serve as significant sources of melatonin, an important bioactive compound. In general, the highest concentrations of melatonin were obtained for table grapes (from 0.079 ± 0.01 ng/g to 1.148 ± 0.01 ng/g) and Amar de Maxut cherries (0.302 ± 0.01 ng/g). In 2006, melatonin was initially detected in *Vitis vinifera* skin grapes (Nebbiolo, Croatina, Sangiovese, Merlot, Marzemino, Cabernet Franc, Cabernet Sauvignon, and Barbera) cultivated in Treviso, Italy, with levels ranging from 0.005 to 0.965 ng/g [49]. Much higher concentrations of melatonin (ranging from 9.3 to 17.5 ng/g) were found in Malbec grapes from Argentina [50]. Significantly higher concentrations of melatonin were observed in whole berries (ranging from 100 to 150 mg/g), depending on the phenological stage, for Merlot cultivars grown in Canada [51]. However, only 1.2 and 1.5 ng/g of melatonin was observed in Sangiovese and Albana grapes, respectively [52]. Nevertheless, in general, most studies report melatonin content in grapes to be in the range of ng/g, which is consistent with the data obtained in this experiment.

Regarding strawberry samples, Stürtz et al. [53] reported values approximately ten times lower for the Camarosa, Candoga, Festival, and Primoris varieties. Similarly,

González-Goméz et al. [54] identified values ranging from 0.01 ng/g (Van) to 0.05 ng/g (Pico Colorado). These discrepancies can be attributed primarily to variations related to species, *terroir*, and the specific applied processing techniques. Samples displaying high concentrations of melatonin typically exhibited correspondingly higher antioxidant activity. Table grapes notably demonstrated the highest levels of antioxidant activity, consistent with their melatonin content. In a study published by González-Goméz et al. [54], melatonin values of up to 4.3 ng/mL were reported for cherries. The observed differences can be ascribed to the fact that they utilized fresh fruit without previous freezing. On this line, Arnao et al. [55] postulated that some variables such as the plant variety, agronomic cultivation conditions, extraction protocols, and the techniques employed for measurement purposes are influencing the melatonin concentrations in plants.

Table 3. Melatonin determination

Variety	Concentration (ng/g)	Species
Stella	nd	Cherry
Boambe de Cotnari	nd	Cherry
Germersdorf	0.163 ± 0.03	Cherry
Amar de Maxut	0.302 ± 0.01	Cherry
Van	0.151 ± 0.02	Cherry
Botoşani	0.109 ± 0.01	Sour cherry
Thornfree	0.268 ± 0.02	Blackberry
Elsanta	0.12 ± 0.03	Strawberry
Red Giant	0.095 ± 0.02	Rasberry
Muscat timpuriu de Bucureşti	0.132 ± 0.02	Table grapes
Paula	0.626 ± 0.02	Table grapes
Coarnă neagră selecţionată	0.305 ± 0.01	Table grapes
Augusta	0.079 ± 0.01	Table grapes
Victoria	0.329 ± 0.03	Table grapes
Timpuriu de Pietroasa	1.148 ± 0.01	Table grapes
Gelu	0.084 ± 0.02	Table grapes
Muscat Hamburg	0.119 ± 0.03	Table grapes
Napoca	0.282 ± 0.01	Table grapes
Cardinal	0.207 ± 0.02	Table grapes

3.2 Phenolic Compound Content

Polyphenols are a large group of compounds found in plants that have antioxidant properties. Total polyphenols contribute to the overall antioxidant capacity of fruits. These compounds can scavenge free radicals and reduce oxidative damage. Anthocyanins are a specific type of polyphenol, and their contribution to the total polyphenol content enhances the overall antioxidant capacity [56].

Phenolic compounds with reduction properties refer to compounds that can donate electrons to reduce oxidative stress. Many phenolic compounds in fruits possess reduction properties, allowing them to act as antioxidants. They can counteract the harmful effects of free radicals by donating electrons [56].

According to Table 4, Coarnă neagră selecționată (542.85 ± 0.00 mg GAE/100 g fruit) and Timpuriu de Pietroasa (401.14 ± 0.00 mg GAE/100 g fruit) grapes, followed by Elsanta (458.48 ± 0.0 mg GAE/100 g fruit) strawberries and Amar de Maxut (391.01 ± 0.00 mg GAE/100 g fruit) cherries presented the highest values of total polyphenolic content (D280).

Consistent with the total polyphenolic content, Timpuriu de Pietroasa grapes (292.87 ± 0.00 mg GAE/100 g fruit) and Coarnă neagră selecționată grapes (220.82 ± 0.02 mg GAE/100 g fruit), followed by Elsanta strawberries, displayed the highest values for the Folin-Ciocâlteu index (Table 5). Cherries such as Van (98.87 ± 0.01 mg GAE/100 g fruit) and Stella (109.74 ± 0.01 mg GAE/100 g fruit) registered the lowest values for total phenolic content with reducing properties. In line with these findings, Du et al. [57] reported values ranging from 115.8 mg GAE/100 g in the Red Globe variety to 219.5 mg/100 g in Cabernet Sauvignon.

In general, the samples with higher concentration of melatonin were also characterised by high content in phenolic compounds. This can be attributed to the protective role of melatonin other antioxidants.

Table 4. Total polyphenolic content (D280 index)

Variety	mg GAE/100 g fruit	Species
Stella	387.03 ± 0.02	Cherry
Boambe de Cotnari	294.41 ± 0.03	Cherry
Germersdorf	260.01 ± 0.04	Cherry
Amar de Maxut	391.01 ± 0.01	Cherry
Van	250.30 ± 0.02	Cherry
Botoșani	350.42 ± 0.01	Sour cherry
Thornfree	371.01 ± 0.02	Blackberry
Elsanta	458.48 ± 0.03	Strawberry
Red Giant	317.34 ± 0.04	Rasberry
Muscat timpuriu de București	276.32 ± 0.01	Table grapes
Paula	276.32 ± 0.04	Table grapes
Coarnă neagră selecționată	542.85 ± 0.03	Table grapes

(continued)

Table 4. (*continued*)

Variety	mg GAE/100 g fruit	Species
Augusta	230.90 ± 0.02	Table grapes
Victoria	271.03 ± 0.01	Table grapes
Timpuriu de Pietroasa	401.14 ± 0.01	Table grapes
Gelu	323.08 ± 0.03	Table grapes
Muscat de Hamburg	352.18 ± 0.02	Table grapes
Napoca	380.41 ± 0.01	Table grapes
Cardinal	369.83 ± 0.01	Table grapes

Table 5. Phenolic compounds with reducing properties (Folin-Ciocâteu assay)

Variety	mg GAE/100 g fruit	Species
Stella	109.74 ± 0.04	Cherry
Boambe de Cotnari	116.32 ± 0.03	Cherry
Germersdorf	110.85 ± 0.02	Cherry
Amar de Maxut	212.67 ± 0.01	Cherry
Van	98.87 ± 0.04	Cherry
Botoșani	207.11 ± 0.02	Sour cherry
Thornfree	217.14 ± 0.03	Blackberry
Elsanta	232.68 ± 0.02	Strawberry
Red Giant	127.73 ± 0.03	Rasberry
Muscat timpuriu de București	146.03 ± 0.03	Table grapes
Paula	122.80 ± 0.01	Table grapes
Coarnă neagră selecționată	220.82 ± 0.02	Table grapes
Augusta	140.38 ± 0.02	Table grapes
Victoria	145.97 ± 0.02	Table grapes
Timpuriu de Pietroasa	292.87 ± 0.04	Table grapes
Gelu	201.99 ± 0.01	Table grapes
Muscat de Hamburg	239.20 ± 0.02	Table grapes
Napoca	186.75 ± 0.03	Table grapes
Cardinal	238.04 ± 0.01	Table grapes

3.3 Anthocyanin Content

Anthocyanins are known for their strong antioxidant properties. They can neutralize free radicals and help prevent oxidative stress in cells. The total anthocyanin content in a fruit is an indicator of the overall concentration of these antioxidant compounds. Higher levels of total anthocyanins often correlate with increased antioxidant activity [58].

Table 6 illustrates the results obtained from the determination of total anthocyanin content in the analyzed samples. Thornfree blackberries presented the highest total anthocyanin content (69.51 ± 0.03 mg malvidin-3-glucoside/100 g fruit), followed by Amar de Maxut cherries (51.85 ± 0.05 mg malvidin-3- glucoside/100 g fruit). Also, the autochthonous varieties of table grapes were highlighted by high values of total anthocyanins (from 3.22 ± 0.01 mg malvidin-3-glucoside/100 g of fruit to 47.77 ± 0.03 mg malvidin-3-glucoside/100 g fruit). Du et al. [57] reported comparable values for this parameter, depending on the variability of the analyzed variety (5.0 mg cyanidin 3-glucosides/100 g in the Red Globe variety, 13.7 mg cyanidin 3-glucosides/100 g in Muscat, 164, 2 mg cyanidin 3-glucosides/100 g in Cabernet Gernischt, 68.9 mg cyanidin 3-glucosides/100 g in Merlot, 17.4 mg cyanidin 3-glucosides/100 g in Cabernet Franc, 67.8 mg cyanidin 3-glucosides/100 g for Cabernet Sauvignon).

Table 6. Total anthocyanin content

Variety	mg malvidin 3-glucoside/100 g of fruit	Species
Stella	0.76 ± 0.04	Cherry
Boambe de Cotnari	1.70 ± 0.02	Cherry
Germersdorf	29.29 ± 0.04	Cherry
Amar de Maxut	51.85 ± 0.05	Cherry
Van	8.37 ± 0.04	Cherry
Botoșani	25.07 ± 0.02	Sour cherry
Thornfree	69.51 ± 0.03	Blackberry
Elsanta	11.47 ± 0.04	Strawberry
Red Giant	9.29 ± 0.05	Rasberry
Muscat timpuriu de București	18.10 ± 0.03	Table grapes
Paula	3.22 ± 0.04	Table grapes
Coarnă neagră selecționată	41.31 ± 0.02	Table grapes
Augusta	15.56 ± 0.03	Table grapes
Victoria	4.39 ± 0.02	Table grapes
Timpuriu de Pietroasa	47.77 ± 0.03	Table grapes
Gelu	15.96 ± 0.04	Table grapes
Muscat Hamburg	35.24 ± 0.01	Table grapes
Napoca	18.68 ± 0.02	Table grapes
Cardinal	47.75 ± 0.02	Table grapes

Table 7 shows the results of the identification and quantification of anthocyanins in the studied samples. The compounds cyanidin 3-glucoside, cyanidin 3-rutinoside, delphinidin 3-glucoside, petunidin 3-glucoside, and malvidin 3-glucoside are observed. Their presence varies depending on the species studied and the diversity of the variety. Specifically, in Thornfree blackberries, high concentrations of cyanidin 3-glucosides (170.9 μg/mL) were identified. In Amar de Maxut and Germersdorf cherries, the predominant compound is cyanidin 3-rutinoside (109.5 μg/mL and respectively, 57.2 μg/mL). Timpuriu de Pietroasa grapes were characterized by an increased content of malvidin 3-glucosides (55 μg/mL), and delphinidin 3-rutinoside (8.656 μg/mL). The anthocyanin profile of fruits exhibits variability and is influenced by the harvesting moment, being in continuous changing and degradation during the processing methods and storage. Acylated anthocyanins are usually reported as minor components in fruits [59]. These compounds possess the potential to play a significant role in fruit authentication. According to Avula et al. [59], malvidin and delphinidin aglycones represent the main anthocyanin glycosides in the analysed fruits. For the red fruits, these groups serve as markers or major components. The findings from Table 7 emphasize that cyanidin aglycone is the main anthocyanin in the analysed samples. This reaffirms the profound impact of *terroir* and varietal diversity, highlighting their potential to provide valuable insights into the origins of the fruits.

Table 7. Identification and quantification of the main anthocyanins

Specie	Variety	Anthocyanin	Concentrations (μg/mL)
Cherry	Stella	Cyanidin 3-glucoside	0.490 ± 0.04
		Cyanidin 3-rutinoside	11.164 ± 0.01
Cherry	Boambe de Cotnari	Cyanidin 3-glucoside	0.372 ± 0.03
		Cyanidin 3-rutinoside	6.837 ± 0.02
Cherry	Germersdorf	Cyanidin 3-glucoside	2.656 ± 0.03
		Cyanidin 3-rutinoside	57.246 ± 0.02
Cherry	Amar de Maxut	Cyanidin 3-glucoside	3.482 ± 0.01
		Cyanidin 3-rutinoside	109.509 ± 0.01
Cherry	Van	Cyanidin 3-glucoside	0.890 ± 0.01
		Cyanidin 3-rutinoside	33.643 ± 0.03
Sour cherry	Botoșani	Cyanidin 3-glucoside	1.351 ± 0.03
		Cyanidin 3-rutinoside	15.660 ± 0.02
Blackberry	Thornfree	Cyanidin 3-glucoside	170.945 ± 0.02
		Cyanidin 3-rutinoside	10.602 ± 0.02
Strawberry	Elsanta	Cyanidin 3-glucoside	0.921 ± 0.01
Rasberry	Red Giant	Cyanidin 3-glucoside	4.787 ± 0.03

(*continued*)

Table 7. (*continued*)

Specie	Variety	Anthocyanin	Concentrations (μg/mL)
		Cyanidin 3-rutinoside	7.231 ± 0.02
Table grapes	Coarnă neagră selecționată	Delphinidin 3-glucoside	2.401 ± 0.01
		Cyanidin 3-glucoside	3.539 ± 0.03
		Petunidin 3-glucoside	2.365 ± 0.03
		Malvidin 3-glucoside	10.062 ± 0.01
Table grapes	Timpuriu de Pietroasa	Delphinidin 3-glucoside	8.656 ± 0.02
		Cyanidin 3-glucoside	1.743 ± 0.01
		Petunidin 3-glucoside	7.355 ± 0.02
		Malvidin 3-glucoside	55.074 ± 0.02
Table grapes	Muscat de Hamburg	Delphinidin 3-glucoside	0.970 ± 0.03
		Cyanidin 3-glucoside	2.308 ± 0.01
		Petunidin 3-glucoside	1.440 ± 0.01
		Malvidin 3-glucoside	8.409 ± 0.03
Table grapes	Napoca	Delphinidin 3-glucoside	0.912 ± 0.01
		Cyanidin 3-glucoside	0.916 ± 0.02
		Petunidin 3-glucoside	1.454 ± 0.01
		Malvidin 3-glucoside	13.783 ± 0.03
Table grapes	Cardinal	Delphinidin 3-glucoside	0.433 ± 0.04
		Cyanidin 3-glucoside	0.847 ± 0.02
		Petunidin 3-glucoside	0.539 ± 0.02
		Malvidin 3-glucoside	8.776 ± 0.01

3.4 Antioxidant Activity

In general, table grapes stand out with the highest antioxidant activity, consistent with their level of melatonin (Table 8). Thus, the selected Timpuriu de Pietroasa (0.49 ± 0.00) and Coarnă neagră selecționată (0.57 ± 0.00) varieties, but also the Germersdorf (0.51 ± 0.01) and Amar de Maxut cherries (0.67 ± 0.01) showed increased antioxidant activity. It is thus promoted that indigenous fruit varieties are valuable sources of bioactive compounds.

Melatonin serves to stimulate the activity of antioxidant enzymes, including peroxidases, glutathione reductase, superoxide dismutase, and catalase. Concomitantly, it acts as an inhibitor of prooxidative enzymes such as nitric oxide synthase, myeloperoxidase, and eosinophil peroxidase. Melatonin undergoes enzymatic or nonenzymatic transformations, yielding biologically active metabolites such as 5-methoxytryptamine, cyclic

Table 8. Antioxidant activity of the samples

Variety	IC50 mg/mL	Species
Stella	3.23 ± 0.03	Cherry
Boambe de Cotnari	2.13 ± 0.02	Cherry
Germersdorf	0.51 ± 0.02	Cherry
Amar de Maxut	0.67 ± 0.03	Cherry
Van	2.23 ± 0.02	Cherry
Botoșani	2.25 ± 0.01	Sour cherry
Thornfree	1.32 ± 0.01	Blackberry
Elsanta	3.00 ± 0.03	Strawberry
Red Giant	0.72 ± 0.04	Rasberry
Muscat timpuriu de București	2.61 ± 0.02	Table grapes
Paula	3.92 ± 0.03	Table grapes
Coarnă neagră selecționată	0.57 ± 0.02	Table grapes
Augusta	4.12 ± 0.01	Table grapes
Victoria	3.85 ± 0.01	Table grapes
Timpuriu de Pietroasa	0.49 ± 0.04	Table grapes
Gelu	3.40 ± 0.02	Table grapes
Muscat de Hamburg	1.80 ± 0.02	Table grapes
Napoca	3.21 ± 0.02	Table grapes
Cardinal	2.31 ± 0.01	Table grapes

3-hydroxymelatonin, N1-acetyl-N2-formyl-5-methoxykynuramine, and N1-acetyl-5-methoxykynuramine. These metabolites, similar to melatonin, employ analogous mechanisms to confer protection against oxidative damage [6]. Melatonin has been reported to influence enzymes involved in the browning reactions of fruits, such as polyphenol oxidase. Inhibiting enzymatic browning can help preserve the quality and appearance of fruits by preventing the breakdown of anthocyanins and polyphenols. Xu et al. [60] found that the presence of high levels of melatonin increases the content of total anthocyanins and phenolic compounds in fruits, corroborated to an increased antioxidant capacity, which is in accordance to the results presented in this paper.

The samples analysed in this paper were collected at various times of the day. The results published by Boccalandro et al. [61] support an antioxidant role of melatonin and reveal its inherent fluctuations in fruits throughout the day, implying that its functional significance is more pronounced during the nocturnal and early morning periods. Consequently, products derived from fruits harvested during these specific times may exhibit elevated nutraceutical properties. In recent years, considerable attention has been directed towards the incorporation of plant foods characterized by elevated melatonin levels as a dietary supplement. This approach aims to augment melatonin concentrations in human blood plasma, thereby enhancing its scavenging capabilities and antioxidant activity [62]. This fact highlights a significant contribution of the presented results to the enrichment of literature data, as well as to the promotion of Romanian fruits as important nutritional sources.

4 Conclusions

Melatonin, anthocyanins and phenolic compounds play a significant role in defining the antioxidant activity of the evaluated samples. Samples with higher values of melatonin presented the highest values for total anthocyanins and total phenolic compounds. Melatonin's antioxidant properties contribute to the stability of anthocyanins and phenolic compounds, helping maintain their concentration and biological activity in fruits. The results obtained for the melatonin content are in concordance with the antioxidant activity of the analysed fruits. Between the analysed fruits, grapes are the most important sources of bioactive compounds, presenting the highest values for most of the analysed indicators. Even if the existing studies mainly focus on wine grapes of *Vitis vinifera*, Romanian table grape varieties (especially Coarnă neagră selecționată) are important sources of natural antioxidants. Given the numerous health benefits they offer to consumers, a more comprehensive study on melatonin content in fruits is necessary.

References

1. Lerner, A.B.: Melatonin, without the hype. In: Olcese, J. (ed.) Melatonin After Four Decades: An Assessment of Its Potential, vol. 460. Kluwer Academic Publishers, New York (2002)
2. Reiter, R.J., Tan, D.X., Cabrera, J., Arpa, D.D.: Melatonin and tryptophan derivatives as free radical scavengers and antioxidants. In: Huether, G., Kochen, W., Simat, T.J., Steinhart, H. (eds.) Tryptophan, Serotonine and Melatonin, Basic Aspects and Applications. Kluwer Academic/Plenum Publishers, New York (1999)
3. Korf, H.W.: Evolution of melatonin – producing pinealocytes. In: Olcese, J. (ed.) Melatonin After Four Decades: An Assessment of Its Potential, vol. 460. Kluwer Academic Publishers, New York (2002)
4. Zawilska, J.B., Rosiak, J., Nowak, J.Z.: Ultraviolet light suppresses melatonin biosynthesis in chick pineal gland. In: Olcese, J. (ed.) Melatonin After Four Decades: An Assessment of Its Potential, vol. 460. Kluwer Academic Publishers, New York (2002)
5. Klein, D.C.: Serotonin *N*-acetyltransferase: a personal historical perspective. In: Olcese, J. (ed.) Melatonin After Four Decades: An Assessment of Its Potential, vol. 460. Kluwer Academic Publishers, New York (2002)
6. Martín-Cabrejas, M.A., Aguilera, Y., Benítez, V., Reiter, R.J.: Melatonin synthesis in fermented foods. In: Frias, J., Martinez-Villaluenga, C., Penãs, E. (eds.) Fermented Foods in Health and Disease Prevention, pp. 105–129. Elsevier, Amsterdam (2017)
7. Vásquez, J., Grillitsch, K., Daum, G., Mas, A., Torija, M.J., Beltran, G.: Melatonin minimizes the impact of oxidative stress induced by hydrogen peroxide in Saccharomyces and non-conventional yeast. Front. Microbiol. **9**, 1933 (2018)
8. Morcillo-Parra, M.Á., Beltran, G., Mas, A., Torija, M.J.: Determination of melatonin by a whole cell bioassay in fermented beverages. Sci. Rep. **9**, 9120 (2019)
9. Vigentini, I., et al.: Yeast contribution to melatonin, melatonin isomers and tryptophan ethyl ester during alcoholic fermentation of grape musts. J. Pineal Res. **58**, 388–396 (2015)

10. Feng, X., Wang, M., Zhao, Y., Han, P., Dai, Y.: Melatonin from different fruit sources, functional roles, and analytical methods. Trends Food Sci. Technol. **37**, 21–31 (2014)
11. Wei, W., et al.: Melatonin enhances plant growth and abiotic stress tolerance in soybean plants. J. Exp. Bot. **66**, 695–707 (2014)
12. Setyaningsih, W., Saputro, I.E., Barbero, G.F., Palma, M., Barroso, G.: Determination of melatonin in rice (*Oryza sativa*) grains by pressurized liquid extraction. J. Agric. Food Chem. **63**, 1107–1115 (2015)
13. Yan, F., et al.: Melatonin improves K^+ and Na^+ homeostasis in rice under salt stress by mediated nitric oxide. Ecotoxicol. Environ. Saf. **206**, 111358 (2020)
14. Arnao, M.B., Hernández-Ruiz, J.: Chemical stress by different agents affects the melatonin content of barley roots. J. Pineal Res. **46**, 295–299 (2009)
15. Buttar, Z.A., Wu, S.N., Arnao, M.B., Wang, C., Ullah, I., Wang, C.: Melatonin suppressed the heat stress-induced damage in wheat seedlings by modulating the antioxidant machinery. Plants **9**, 809 (2020)
16. Lei, Q., et al.: Identification of genes for melatonin synthetic enzymes in 'Red Fuji' apple (*Malus domestica Borkh.cv.Red*) and their expression and melatonin production during fruit development. J. Pineal Res. **55**, 443–451 (2013)
17. Zhang, H., et al.: Melatonin in apples and juice: inhibition of browning and microorganism growth in apple juice. Molecules **23**, 521 (2018)
18. Liu, J., et al.: Effects of melatonin treatment of postharvest pear fruit on aromatic volatile biosynthesis. Molecules **24**, 4233 (2019)
19. Hui, G., et al.: Melatonin treatment reduces chilling injury in peach fruit through its regulation of membrane fatty acid contents and phenolic metabolism. Food Chem. **245**, 659–666 (2018)
20. Schor, J.: Cherry juice supplies melatonin and improve sleep. Nat. Med. J. **4** (2012)
21. Salehi, B., et al.: Melatonin in medicinal and food plants: occurrence, bioavailability, and health potential for humans. Cells **8**, 681 (2019)
22. Liu, C., Zheng, H., Sheng, K., Liu, W., Zheng, L.: Effects of melatonin treatment on the postharvest quality of strawberry fruit. Postharvest Biol. Technol. **139**, 47–55 (2018)
23. Meng, J.F., Shi, T., Song, S., Zhang, Z.W., Fang, Y.L.: Melatonin in grapes and grape-related foodstuffs. Food Chem. **231**, 185–191 (2017)
24. Bulman, A., D'Cunha, N.M., Marx, W., McKune, A.J., Jani, R., Naumovski, N.: Nutraceuticals as potential targets for the development of a functional beverage for improving sleep quality. Beverages **7**, 33 (2021)
25. Fernández-Cruz, E., Álvarez-Fernández, M.A., Valero, E., Troncoso, A.M., García-Parrilla, M.C.: Melatonin and derived L-tryptophan metabolites produced during alcoholic fermentation by different wine yeast strains. Food Chem. **217**, 431–437 (2017)
26. Guerrini, S., Mangani, S., Romboli, Y., Luti, S., Pazzagli, L., Granchi, L.: Impact of *Saccharomyces cerevisiae* strains on health-promoting compounds in wine. Fermentation **4**, 26 (2018)
27. Albu, C., Radu, L.E., Radu, G.L.: Assessment of melatonin and its precursors content by a HPLC-MS/MS method from different Romanian wines. AMC Omega **5**, 27254–27260 (2020)
28. Maldonado, M.D., Moreno, H., Calvo, J.R.: Melatonin present in beer contributes to increase the levels of melatonin and antioxidant capacity of the human serum. Clin. Nutr. **28**(2), 188–191 (2009)
29. Marhuenda, J., Villaño, D., Arcusa, R., Zafrilla, P.: Melatonin in wine and beer: beneficial effects. Molecules **26**, 343 (2021)
30. Muñiz-Calvo, S., Bisquert, R., Guillamón, J.M.: Melatonin in yeast and fermented beverages: analytical tools for detection physiological role and biosynthesis. Melatonin Res. **3**, 144–160 (2020)
31. Galano, A., Tan, D.X., Reiter, R.J.: Melatonin as a natural ally against oxidative stress: a physicochemical examination. J. Pineal Res. **51**, 1–16 (2011)

32. Ianăs, O., Manda, D., Câmpean, D., Ionescu, M., Soare, G.: Effects of melatonin and its relation to the hypothalamic-hypophyseal-gonadal axis. In: Olcese, J. (ed.) Melatonin After Four Decades: An Assessment of Its Potential, vol. 460. Kluwer Academic Publishers, New York (2002)
33. Krause, D.N., Greg, G.G., Doolen, S., Duckles, S.P.: Melatonin and cardiovascular function. In: Olcese, J. (ed.) Melatonin After Four Decades: An Assessment of Its Potential, vol. 460. Kluwer Academic Publishers, New York (2002)
34. Blask, D.E., Sauer, L.A., Dauchy, R.T., Holowachuk, E.W., Ruhoff, M.S.: New insights into melatonin regulation of cancer growth. In: Olcese, J. (ed.) Melatonin After Four Decades: An Assessment of Its Potential, vol. 460. Kluwer Academic Publishers, New York (2002)
35. Pawlikowski, M., Kunert-Radek, J., Winczyk, K., Melen-Mucha, G., Gruszka, A., Karasek, M.: The antiproliferative effects of melatonin on experimenta. In: Olcese, J. (ed.) Melatonin After Four Decades: An Assessment of Its Potential, vol. 460. Kluwer Academic Publishers, New York (2002)
36. Turek, F.W., Zee, P., Van Reeth, O.: Melatonin and aging. In: Olcese, J. (ed.) Melatonin After Four Decades: An Assessment of Its Potential, vol. 460. Kluwer Academic Publishers, New York (2002)
37. Franceschini, D., Skaper, S.D., Floreani, M., Borin, G., Giusti, P.: Further evidences for neuroprotective effects of melatonin. In: Huether, G., Kochen, W., Simat, T.J., Steinhart, H. (eds.) Tryptophan, Serotonin, and Melatonin, vol. 467, pp. 207–215. Springer, Boston (1999). https://doi.org/10.1007/978-1-4615-4709-9_27
38. Kopustinskiene, D.M., Bernatoniene. J: Molecular Mechanisms of Melatonin-Mediated Cell Protection and Signaling in Health and Disease. Pharmaceutics 13(2),129 (2021)
39. Pandi-Perumal, S.R., Srinivasan, V., Poeggeler, B., Hardeland, R., Cardinali, D.P.: Drug Insight: the use of melatonergic agonists for the treatment of insomnia-focus on ramelteon. Nat. Clin. Pract. Neurol. **3**, 221–228 (2007)
40. Meng, X., et al.: Dietary sources and bioactivities of melatonin. Nutrients **9**, 367 (2017)
41. Armstrong, S.M.: Melatonin as a chronobiotic for circadian insomnia: clinical observations and animal models. In: Olcese, J. (ed.) Melatonin After Four Decades: An Assessment of Its Potential, vol. 460. Kluwer Academic Publishers, New York (2002)
42. Skene, D.J., Lockley, S.W., Arendt, J.: Use of melatonin in the treatment of phase shift and sleep disorders. In: Huether, G., Kochen, W., Simat, T.J., Steinhart, H. (eds.) Tryptophan, Serotonine and Melatonin, Basic Aspects and Applications. Kluwer Academic/Plenum Publishers, New York (1999)
43. Rodenbeck, A., Huether, G., Riither, E., Hajak, G.: Nocturnal melatonin secretion and its modification by treatment in patients with sleep disorders. In: Huether, G., Kochen, W., Simat, T.J., Steinhart, H. (eds.) Tryptophan, Serotonine and Melatonin, Basic Aspects and Applications. Kluwer Academic/Plenum Publishers, New York (1999)
44. Putilov, A.A., Russkikh, G.S., Danilenko, K.V.: Phase of melatonin rhythm in winter depression. In: Olcese, J. (ed.) Melatonin After Four Decades: An Assessment of Its Potential, vol. 460. Kluwer Academic Publishers, New York (2002)
45. Lewinski, A.: Melatonin and the thyroid gland. In: Pandi-Perumal, S.R., Cardinali, D.P. (eds.) Melatonin: Biological Basis of Its Function. Eurekah.com (2006)
46. Bagnall, K.M., et al.: The role of melatonin in the development of scoliosis. In: Pandi-Perumal, S.R., Cardinali, D.P. (eds.) Melatonin: biological basis of its function. Eurekah.com (2006)
47. Gilbert, S.S., van den Heuvel, C.J., Dawson, D., Lushington, K.: The role of thermoregulation in the soporific effects of melatonin: a new perspective. In: Pandi-Perumal, S.R., Cardinali, D.P. (eds.) Melatonin: Biological Basis of Its Function. Eurekah.com (2006)
48. Vlase, A.M., et al.: Epilobium species: from optimization of the extraction process to evaluation of biological properties. Antioxidants **12**, 91 (2022)

49. Iriti, M., Rossoni, M., Faoro, F.: Melatonin content in grape: myth or panacea? J. Sci. Food Agric. **86**, 1432–1438 (2006)
50. Vitalini, S., Gardana, C., Zanzotto, A., Simonetti, P., Faoro, F., Fico, G., et al.: The presence of melatonin in grapevine (*Vitis vinifera* L.) berry tissues. J. Pineal Res. **51**, 331–337 (2011)
51. Murch, S.J., Hall, B.A., Le, C.H., Praveen, K.S.: Changes in the levels of indoleamine phytochemicals during veraison and ripening of wine grapes. J. Pineal Res. **49**, 95–100 (2010)
52. Mercolini, L., Mandrioli, R., Raggi, M.A.: Content of melatonin and other antioxidants in grape-related foodstuffs: measurement using a MEPS-HPLC-F method. J. Pineal Res. **53**(1), 21–28 (2011)
53. Stürtz, M., Cerezo, A.B., Cantos-Villar, E., Garcia-Parilla, M.C.: Determination of the melatonin content of different varieties of tomatoes (*Lycopersicon esculentum*) and strawberries (*Fragariaananassa*). Food Chem. **127**(3), 1329–1334 (2011)
54. González-Goméz, D., Lozano, M., Fernández-León, M.F., Ayuso, M.C., Bernalte, M.J., Rodríguez, A.B.: Detection and quantification of melatonin and serotonin in eight sweet cherry cultivars (*Prunus avium* L.). Eur. Food Res. Technol. **229**, 223–229 (2009). https://doi.org/10.1007/s00217-009-1042-z
55. Arnao, M.B., Hernández-Ruíz, J.: The physiological function of melatonin in plants. Plant Signal. Behav. **1**(3), 89–95 (2006)
56. Álvarez, R., Araya, H., Navarro-Lisboa, R., De Dicastillo, C.L.: Evaluation of polyphenols and antioxidant capacity of fruits and vegetables using a modified enzymatic extraction method. Food Technol. Biotechnol. **54**(4), 462–467 (2016)
57. Du, B., He, B.J., Shi, P.B., Li, F.Y., Li, J., Zhu, F.M.: Phenolic content and antioxidant activity of wine grapes and table grapes. J. Med. Plants Res. **6**(17), 3381–3387 (2012)
58. Tena, N., Martín, J., Asuero, A.G.: State of the art of anthocyanins: antioxidant activity, sources, bioavailability, and therapeutic effect in human health. Antioxidants **9**(5), 451 (2020)
59. Avula, B., et al.: Advances in the chemistry, analysis and adulteration of anthocyanin rich-berries and fruits: 2000–2022. Molecules **28**(2), 560 (2023)
60. Xu, L., Yue, Q., Bian, F., Sun, H., Zhai, H., Yao, Y.: Melatonin enhances phenolics accumulation partially via ethylene signaling and resulted in high antioxidant capacity in grape berries. Front. Plant Sci. **8**, 426 (2017)
61. Boccalandro, H.E., González, C.V., Wunderlin, D.A., Silva, M.F.: Melatonin levels, determined by LC-ESI-MS/MS, fluctuate during the day/night cycle in *Vitis vinifera* cv Malbec: evidence of its antioxidant role in fruits. J. Pineal Res. **51**(2), 226–232 (2011)
62. Aguilera, Y., et al.: Effect of Illumination on the content of melatonin, phenolic compounds, and antioxidant activity during germination of lentils (*Lens culinaris* L.) and kidney beans (*Phaseolus vulgaris* L.). J. Agric. Food Chem. **62**(44), 10736–10743 (2014)

Bactericidal Activity of Mesoporous SiO_2 Matrices Inlaid with Cu° and CuO Nanoparticles Against P. Aeruginosa

Marwan F. S. H. AL-Kamali[1], Andrei A. Boika[2], Dmitry V. Tapalski[3], Dmitry Kovalenko[4(✉)], and Hamdan A. S. AL-Shamiri[5]

[1] Industrial Electronics Department, Sukhoi State Technical University of Gomel, Prospect Octiabria, 48, 246029 Gomel, Republic of Belarus

[2] Technical Sciences, Sukhoi State Technical University of Gomel, Prospect Octiabria, 48, 246029 Gomel, Republic of Belarus
boiko@gstu.by

[3] Medical Sciences, Institute of Physiology of the National Academy of Sciences of Belarus, Akademicheskaya St., 28, 220012 Minsk, Republic of Belarus

[4] F. Skorina Gomel State University, Sovetskaya Str., 102, 246019 Gomel, Republic of Belarus
dkov@gsu.by

[5] Physics Department, Faculty of Science, University of Bish, Box 551, Bisha 61922, Saudi Arabia
hasultan@ub.edu.sa

Abstract. The paper presents the results of research into mesoporous SiO_2 matrices inlaid with nanoparticles Cu° and CuO to establish the bactericidal activity of powders of the SiO_2:Cu° and SiO_2:CuO composition against *pseudomonas aeruginosa* ATCC 27853. The Brunauer–Emmett–Teller method established the dependence of the specific surface of xerogels depending on the concentration of copper oxide and reduced copper. The structure of the resulting xerogels was studied by electron microscopy. It is shown that these xerogels can be used as biologically active substances for applications in creating antimicrobial agents. Reducing particle agglomerates to Cu° in SiO_2 matrix can effectively affect the stabilization of antibacterial properties in the system of the xerogel closed pores.

Keywords: Sol-gel · nanoparticle · copper ions · porosity · structure · bactericidal activity · antimicrobial properties

1 Introduction

Studies related to the development of new methods for synthesis of oxide functional materials in ultrafine form, based on low-temperature colloid-chemical processes, have recently become widespread, which make it possible to obtain materials of a given chemical composition, as well as dispersions of particle sizes and morphology [1–3]. It is associated with the promising use of mesoporous silica structures for drug delivery, removal of contaminants and organic compounds, since catalytic reactions are accelerated [4].

Y. Ono and J. Kondoh (Eds.): Inter-Academia 2023, LNNS 939, pp. 140–150, 2024.
https://doi.org/10.1007/978-3-031-54450-7_16

Metal nanoparticles have biological activity, including bacteriostatic and bactericidal effects [5]. The study of the regularities of biological activity of nanodispersed metals depending on the method of their production, shape, size and physicochemical properties is an urgent task.Copper nanoparticles are known to exhibit pronounced biological activity, including bacteriostatic and bactericidal actions. There are some examples of studying the bactericidal effect of iron and copper nanoparticles on E.coli, Staphylococcus aureus [6]. The presence of reduced copper particles in the porous structure of the SiO_2:Cu° xerogel can exhibit bacteriostatic and bactericidal effects. The suppressive ability to Pseudomonas aeruginosa and Staphylococcus aureus was studied to determine the possibility of using the resulting materials.

2 Materials and Methods

2.1 Materials

All chemicals given below were purchased and used in the form of aerosil A-380 nanopowders (SiO_2, 99.9%, the average size of primary particles from 5–15 nm, specific surface area 50–380 m^2/g, adsorption capacity from 100 to 340 g at 100 g of aerosil, bulk density 35–150 g/l), Copper nitrate tri-hydrate ($Cu(NO_3)_2 \times 3H_2O$, 99,6%) and Polyvinyl alcohol $(\text{-}C_2H_4O)_n$. These chemicals were obtained from the Vekton Company in Russia. The Sol of SiO_2 was made using distilled water. All compounds were utilized without being purified further.

2.2 Synthesis Approach

The SiO_2 powders were synthesized using the sol-gel method [7]. They contained copper nitrate, copper oxide, and reduced copper in their structure. During their synthesis, reagents of a chemical purity not lower than analytic grade (for copper nitrate and aerosil marks A-380) and "ACS" (for bidistilled water and hydrogen) were used. The first stage was accompanied by sol formation. Then a pure gel and a xerogel were formed on its basis. Inlaid's were introduced into the liquid sol in the form of water-soluble salts of a given concentration (in our case, copper nitrate was used). The sol gelation (pure and containing copper) was carried out in open plastic molds in air. The formed gels was dried in a ventilated oven at T = 80 ℃. Dry xerogel preforms inlaid with copper nitrate were subjected to successive heat treatment, first in air and then in a hydrogen environment, in order to obtain composite materials of SiO_2:Cu° composition containing separately localized micro- and nanoparticles Cu°. After that, the powders were subsequently grinded and sieved into fractions on a planetary mill "Pulvirisette 5" and a vibrating sieve screen "Annalizette 3", respectively.

Figure 1 shows the scheme of the technological stages of obtaining xerogel SiO_2 matrices inlaid with reduced metal nanoparticles of copper ions.

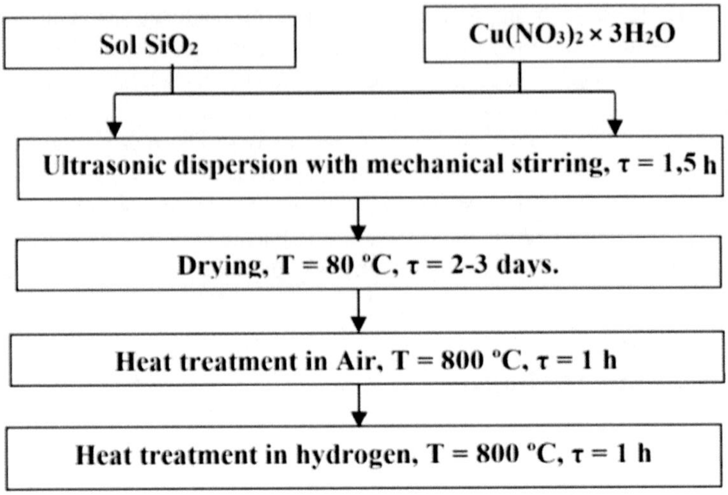

Fig. 1. Scheme of the technological stages of obtaining xerogel SiO_2 matrices inlaid with Cu°.

3 Characterization

The specific surface area (Ssp) of microspheres was assessed using a specific surface area and porosity analyzer "ASAP 2020 MP" (Micromeritics, SAS) and low-temperature nitrogen adsorption, according to the Brunauer–Emmett–Teller theory. The measurement error was 3%.

The samples were degassed for 3 h before to the experiments. The average size of aggregates was determined using the assumption that all particles are the same size and shape:

$$D_{av} = \frac{6}{\rho \times S}; \quad (1)$$

where ρ *- pycnometric density, kg/m*^{3}*;*
S - specific surface, m^2*/kg;*
D_{av} *- aggregate average diameter, m.*

The Brunauer–Emmett–Teller (BET) (A_{BET}) and Langmuir (AL) specific surface area measuring techniques were employed to determine the dimensional properties of the resultant powders of diverse compositions. The single-point approach was also utilized to determine adsorption and desorption pore volumes ($V_{sp\ ads}$ and $V_{sp\ des}$). The Barrett-Joyner-Halenda technique was used to compute the cumulative volumes of adsorption and desorption ($V_{BJH\ ads}$ and $V_{BJH\ des}$) of pores with sizes ranging from 1.7 to 300 nm. The average pore diameter D of the mesopore group is defined as (D = 4V/A), where (4V/A) is the volume to area ratio of the group walls. The Horvath-Cavaso technique was used by VHK and DHK to calculate pore volume and diameter. The DFT (density functional theory) model was utilized to determine the pore distribution structure in the acquired samples.

The morphology and structure of the obtained samples were investigated through a JEOL JSM-2100 transmission electron microscope (JEOL, Japan) (point resolution 2.3, line resolution 1.4, maximum accelerating voltage 200 kV, cathode acts as an electron source LaB_6, minimum beam diameter 20 nm, standard sample size in the plane of the TEM holder 3 mm in diameter) [8].

The morphology of the materials was investigated using an electron scanning microscope model S-4800 (Hitachi, Japan) using a 15 kV accelerating voltage and a resolution of 1 nm [9].

To evaluate the bactericidal activity of powders of the composition SiO_2:Cu° and SiO_2:CuO, tablets with an average thickness of 5 mm and an average diameter of 11 mm were formed by the uniaxial pressing method [10]. Similarly-sized tablets made from SiO_2 powder without copper inclusions were tested as controls.

The samples were sterilized using air method at 180 °C for 60 min. After sterilization, control and experimental samples were placed in containers with 50 ml of sterile distilled water and washed on an ES-20 shaker-incubator (Biosan, Latvia) at 35 °C for 24 h with orbital mixing 150 rpm.

Pseudomonas aeruginosa **ATCC 27853** was used as a reference culture. A suspension containing 10^8 CFU/ml (0.5 McFarland units, densitometer control) was prepared in the sterile isotonic sodium chloride solution.

Control and experimental samples (both intact and washed) were placed in test tubes with 10 ml of sterile isotonic sodium chloride solution and inoculated with a bacterial suspension in a volume of 10 μl. So, the calculated starting concentration of microbial cells was 10^5 cfu/ml After inoculation, the contents of the tubes were mixed by vortexing. The tubes were incubated for 2 h at 35 °C. Then, 50 μl of their contents were transferred to nutrient agar plates (HiMedia, India). Inoculation was performed using a spatula and a spiral inoculator for uniform distribution of bacteria over the surface of the nutrient medium. The cultures were incubated for 18 h at 35 °C. After that, the presence and growth pattern were assessed, and the number of grown colonies was counted, the studies were performed in triplicate.

4 Results and Discussion Synthesis

4.1 Surface Area Analyzer

The Brunauer–Emmett–Teller (BET) method determined the values of the specific surface (S_{sp}), and performed the analysis of the average particle size ($D_{av)}$ in the structure of the obtained xerogels. The data are presented in Table 1, the resulting powders were investigated for their possible use as antimicrobial agents with a high inhibitory ability against pseudomonas aeruginosa and staphylococcus aureus. The presence of reduced agglomerates of Cu° particles in the structure of SiO_2:Cu° compounds can effectively influence the stabilization of their antibacterial properties as a result of a decrease in the mechanisms of their natural oxidation during localization in the system of the xerogel closed pores.

The distribution of pores over effective radii shows that for pure aerosil and xerogel (heat-treated at 800 °C) the pores are in a narrow range with a maximum in the region

Table 1. Measurement values of the specific surface of nanopowders with compositions SiO_2, SiO_2:CuO and SiO_2:Cu°

Composition of powder	Sample weight, mg	S_{sp}, m^2/g	D_{av}, m^2/g	Notes
SiO_2 (aerosil brand A-380)	35.1	264	149	Method of determination is thermal desorption of argon (GOST 23401 - 90) The samples were dried at 120 °C for 1 h
	41.6	278		
SiO_2 annealed at (T = 800 °C)	34.8	167	174	
	75.4	180		
SiO_2:CuO (1:0.20)	51.5	150	149	
	49.7	148		
SiO_2:CuO (1:0.30)	52.8	124	126	
	51.7	127		
SiO_2:CuO (1:0.40)	54.7	119	119	
	54.8	118		
SiO_2:Cu° (1:0.20)	47.4	155	156	
	41.6	156		
SiO_2:Cu° (1:0.30)	50.1	152	151	
	52.2	149		
SiO_2:Cu° (1:0.40)	55.2	129	124	
	50.6	119		

of 27 and 55 nm, respectively (see Fig. 2a). In xerogels inlaid with copper ions, the distribution of pores along the effective radii shows a transitional character in the range from 5 to 40 nm, and their value slightly changes upwards with an increase in the concentration of copper oxide. When copper oxide is reduced to Cu°, the maximum of the pore distribution curve along the radii shifts downward with the expansion of the pore radius range to 50 nm (see Fig. 2b, c).

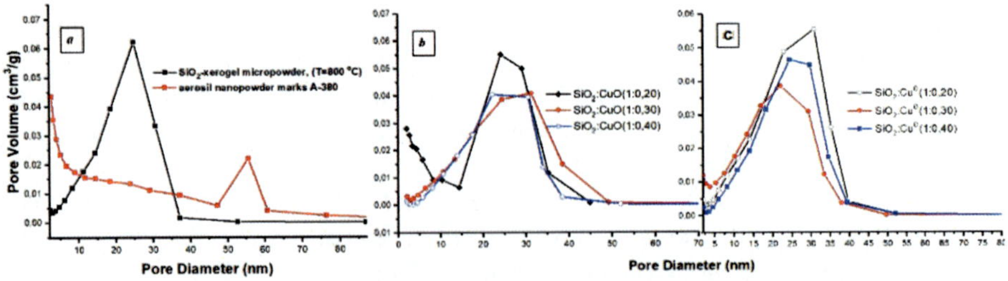

Fig. 2. Pore size distribution for the samples of nanopowders under study.

This is confirmed by the isotherms of nitrogen adsorption and desorption [11, 12], indicating a shift in the capillary-condensation part of the isotherms towards lower relative pressures (Fig. 3 for pure aerosil and xerogel heat-treated at 800 °C and Fig. 4 - inlaid with copper ions).

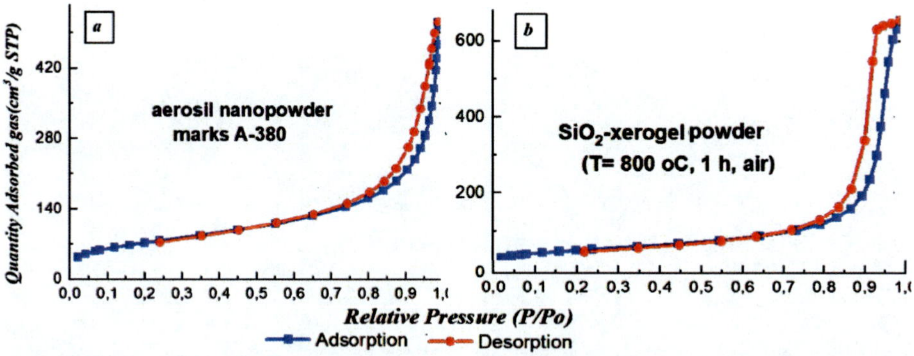

Fig. 3. Isotherms of low-temperature nitrogen adsorption-desorption for the samples of nano-powders under study

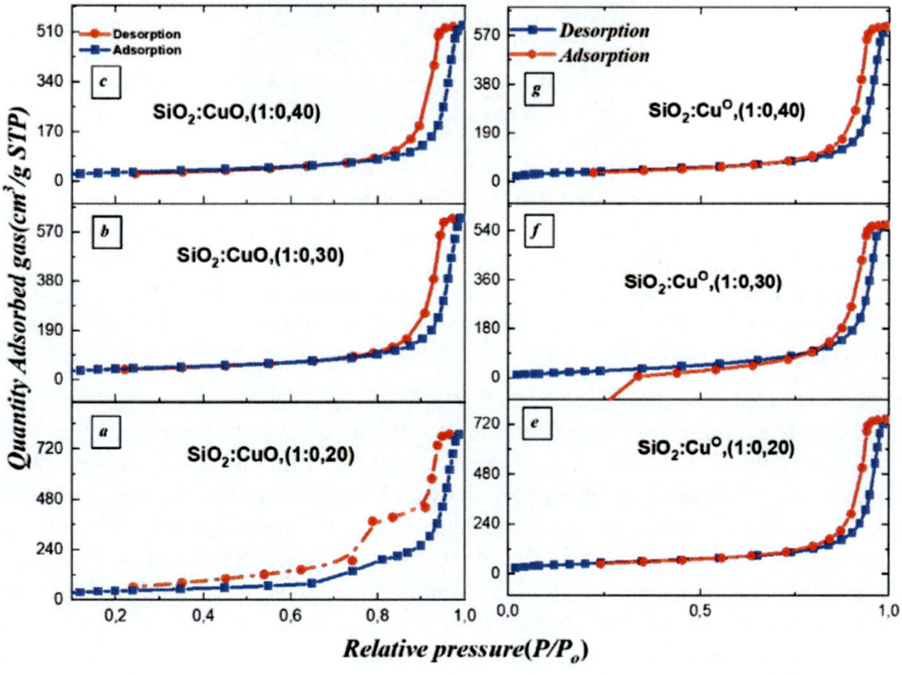

Fig. 4. Isotherms of low-temperature adsorption-desorption of nitrogen for the samples of powders under study.

4.2 Scanning Electron Microscopy

To study the nature of the interaction of inlaids with the surface of the SiO_2 - framework xerogel globules, three types of samples were prepared to analyze their surface morphology by scanning electron microscopy (SEM): SiO_2:$Cu(NO_3)_2 \times 3H_2O$; SiO_2:CuO and SiO_2:Cu° (see Fig. 5).

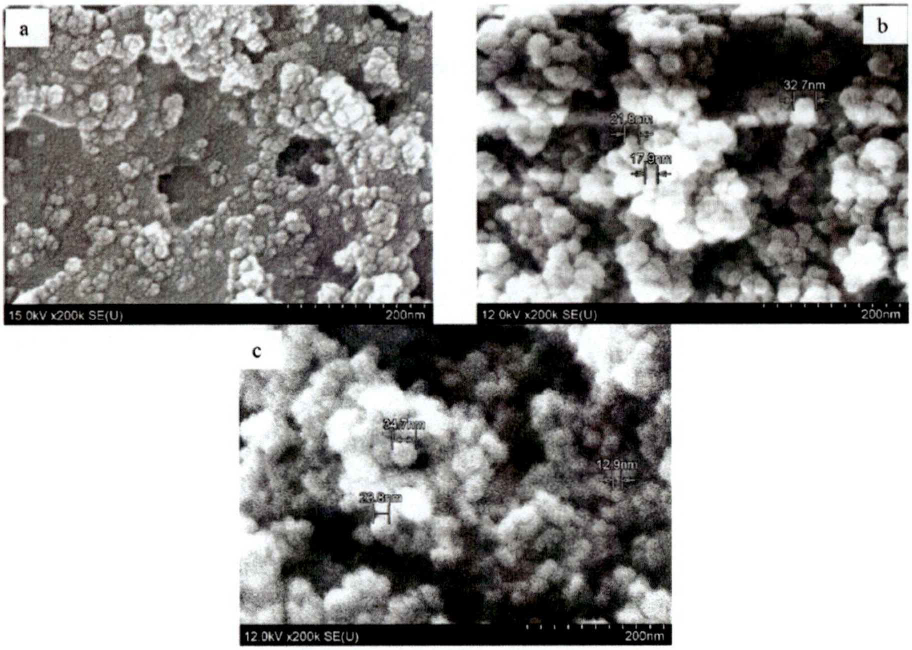

Fig. 5. SEM images of the xerogel surface formed on the basis of aqueous dispersion of aerosil A-380 and copper nitrate (Copper nitrate concentration in 79 ml of sol is 0.40 mol.%), a - drying for 7 days at T = 80 °C, The phase composition of the xerogel - SiO_2:Cu ($NO_3)_2 \times 3H_2O$, b - annealing in air at T = 800 °C for 1 h, The phase composition of the xerogel is SiO_2:CuO, c - annealing in hydrogen at T = 800 °C for 1 h, The phase composition of the xerogel is SiO_2:Cu^0

A clearly expressed globular structure of xerogels and a high homogeneity of their initial morphology are visible. Large globular aggregates that form the xerogel framework are formed by aerosil primary particles which have a much smaller size (for A-380 is about 5–15 nm, according to available data). Due to the high degree of hydration, the copper nitrate salt literally "clogs" the interglobular and interpore spaces of the xerogel (see Fig. 5a). The formation of SiO_2:CuO matrices based on such xerogels returns the "relief" to the xerogel structure, which indicates the mutual chemical resistance of the SiO_2 and CuO phases (see Fig. 5b). The transformation of copper oxide Cu (II) into the state of reduced metal "levels" the overall xerogel structure "(see Fig. 5c), micrograph for a magnification scale of 0.2 μm" [13, 14].

In the general case, a fairly uniform pattern of distribution of inlaid substances over the internal structure of the SiO_2 -framework of the xerogel was observed both for copper oxide and for the reduced-to-metal form. It should be noted that no separate formation of micro- and nanoparticles at the indicated resolution was found while studying the surface morphology of xerogels by SEM. However, the reduced copper probably forms a 2D coating over the entire inner surface of the xerogel. The authors of [15] demonstrated the possibility of controlling chemical bonds on the surface of noble metal nanoparticles (for Ag°). It is entirely possible that this coating may have a structure formed from agglomerates of nanometer-sized metal particles, which will be the aim of our subsequent studies.

4.3 Transmission Electron Microscopy

The hypothesis about the formation of the "core-shell" system is confirmed by transmission electron microscopy (TEM) data obtained for composite materials of $SiO_2{:}Cu(NO_3)_3 \times H_2O$, $SiO_2{:}CuO$ and $SiO_2{:}Cu^{o}$ composition (see Fig. 6). The final functional (sensory) properties of such materials, which are manifested, for example, for porous glasses containing silver nanoparticles are also of high interest [16].

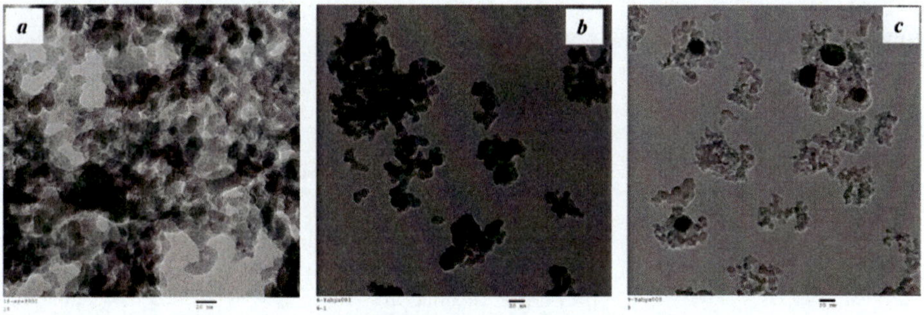

Fig. 6. TEM image of the surface of xerogels formed on the basis of aqueous dispersion of aerosil A-380, and copper nitrate by the sol-gel method (copper nitrate concentration is 0.15 mol.%); drying for 7 days at T = 80 °C. (a - $SiO_2{:}Cu(NO_3)_3 \times H_2O$, b - $SiO_2{:}CuO$, c-$SiO_2{:}Cu^{o}$)

4.4 Bactericidal Activity

The results of determining the antimicrobial properties of control and experimental samples are presented (see Fig. 7 and Table 2).

Control samples and samples containing copper oxide did not show antimicrobial activity. The bactericidal effect of xerogels with the inclusion of copper nitrate was concentration-dependent (maximum was for the samples with copper nitrate concentrations of 0.4 mol. and 0.5 mol. of the content of 1 mol of SiO_2 in the final sol). The antibacterial effect was partially eliminated in the samples washed for 24 h; its severity also depended on the copper nitrate concentration in the samples.

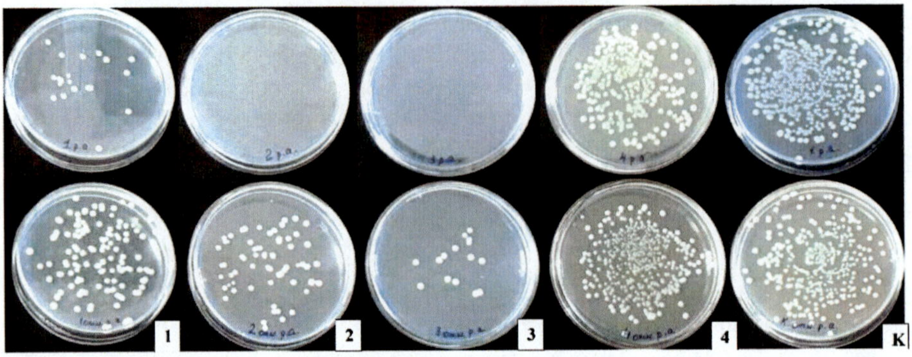

Fig. 7. Bactericidal activity of experimental and control samples against p. aeruginosa ATCC 27853; quantitative inoculation on solid media. Top row - intact samples, bottom row - samples after 24-h washing in distilled water (Concentration: 1 - per 0.20 mol Cu° per 1 mol SiO_2 in the final sol; 2 - by 0.30 mol Cu° per 1 mol SiO_2 in the final sol; 3 - by 0.40 mol Cu° per 1 mol SiO_2 in the final sol; 4 - per 0.4 mol of CuO per 1 mol of SiO_2 in the final sol; K - xerogel based on aqueous dispersion of aerosil A-380 (control sample).

Table 2. Quantitative assessment of the bactericidal activity of xerogels against *p. aeruginosa* ATCC 27853.

Xerogel Composition	Bactericidal activity			
	intact samples		washed samples	
	number of colonies	calculated concentration of viable microbial cells after incubation with xerogel, CFU/ml	number of colonies	calculated concentration of viable microbial cells after incubation with xerogel, CFU/ml
SiO_2:Cu° 1:0.20	17	340	88	1760
SiO_2:Cu° 1:0.30	0	0	53	1060
SiO_2:Cu° 1:0.40	0	0	14	280
SiO_2:CuO 1:0.40	>300	>6000	>300	>6000
SiO_2	>300	>6000	>300	>6000

5 Conclusion

In this way,

- By the method of measuring the specific surface area (Brunauer–Emmett–Teller), the dimensional characteristics of the obtained micropowders of the compositions SiO_2, SiO_2:CuO and SiO_2:Cu°, pore volumes, the pore diameter was determined based on the results of nitrogen adsorption and desorption. The pore diameter has a wide range and ranges from 5 to 85 nm.
- The study of the surface of micropowders that underwent structuring heat treatment by scanning electron microscopy made it possible to establish that copper oxide is formed in the SiO_2 matrix in the form of a separate microdisperse phase, "enveloping" the surface of SiO_2 globules. At high concentrations, reduced copper "coapsulates" into individual spherical particles with a diameter of ~30–70 nm.
- Transmission electron microscopy (TEM) was used to study xerogels of SiO_2:CuO and SiO_2:Cu° composition. Processing of micrographs for synthesized materials made it possible to determine the features of localization in the formed carrier matrix of agglomerates of reduced Cu° sorbed on the surface of SiO_2 globules.
- The study of the antibacterial activity of the resulting xerogels makes it possible to propose them as biologically active substances to create antimicrobial agents. The presence of reduced agglomerates of Cu° particles in the structure of SiO_2:Cu° compounds can effectively stabilize their antibacterial properties by reducing the mechanisms of their natural oxidation when localized in the system of the xerogel closed pores.

Acknowledgments. The work was carried out within the framework of task 2.2.6 of the state research program "Convergence-2025" (subprogram "Microcosm, plasma and the Universe"), funded from the republican budget for state research programs for 2021–2025 in the Republic of Belarus. "All experiments were prepared in the Laboratory of Technical Ceramics and Nanomaterial's, Sukhoi State Technical University of Gomel, Gomel, Belarus."

References

1. Segal, D.L.: Sol-gel processing: routes to oxide ceramics using colloidal dispersions of hydrous oxides and alkoxide intermediates. J. Non-Cryst. Solids **63**(1–2), 183–191 (1984)
2. Sheppard, L.M.: Low-temperature synthesis of ceramics. Adv. Vat. Process. inc. Metal Progr. **130**(5), 47–51 (1986)
3. Ogihara, T.: Hydrolysis of metal alkoxides in homogeneous solutions. In: Sugimoto, T. (ed.) Fine Particles Sythesis, Characterization and Mechanism of Grouth, Basel, pp. 35–57 (2000)
4. Mirzaei, M., et al.: Silica mesoporous structures: effective nanocarriers in drug delivery and nanocatalysts. Appl. Sci. **10**, 7533 (2020). https://doi.org/10.3390/app10217533
5. Satyvaldiev, A.S., et al.: Copper nanoparticles: synthesis and biological activity. In: IOP Conference Series: Materials Science and Engineering, Volume 302, 2017 International Conference on Aerospace Technology, Communications and Energy Systems (ATCES 2017), Samara, Russian Federation, 28–30 September 2017 (2017). https://doi.org/10.1088/1757-899X/302/1/012075

6. Babushkina, I.V., Borodulin, V.B., Korshunov, G.V., Puchinjan D.M.: Comparative study of antibacterial action of iron and copper nanoparticles on clinical Staphylococcus aureus strains. Saratov J. Med. Sci. Res. **6**(1), 11–14 (2010). (in Russian)
7. Alekseenko, A.A., Boika, A.A., Submonetary, E.N.: Functional materials based on silicon dioxide obtained by the sol-gel method. Bull. Sukhoi State Tech. Univ. Gomel **183** (2008). ISBN 978-985-420-725-4. (in Russian)
8. JEM-2100 Electron Microscope – [Electronic resource]. https://www.jeol.com/products/scientific/tem/JEM-2100.php. Accessed 10 Sept 2023
9. PDF SEM-Hitachi-S-4800 Database. – [Electronic resource]. – chrome-extension://efaidnbmnnnibpcajpcglclefindmkaj/ https://www.scu.edu/media/school-of-engineering/photos/cns/SEM-Hitachi-S-4800-Instruction-Manual.pdf. Accessed 25 Sept 2023
10. Alekseenko, A.A., Al-Kamali, M.F.S.H., Titenkov, O.A.: Synthesis and properties of SiO_2:Cu° xerogels intended for use in biomedical research. Bull. Sukhoi State Tech. Univ. Gomel (3–4), 40–47 (2020). (in Russian)
11. Kadlec, O., Dubinin, M.M.: Comments on the limits of applicability of the mechanism of capillary condensation. J. Colloid Interface Sci. **31**(4), 479–489 (1969). https://doi.org/10.1016/0021-9797(69)90049-6. ISSN 0021-9797
12. Boika, A.A., Podnezhny, E.N., Melnichenko, I.M., Baglai, A.K., Ostrovsky, L.K.: Physico-chemical study of gels formed in the process of obtaining doped quartz glass by the sol-gel method. J. Proc. Natl. Acad. Sci. Belarus Chem. Ser. **4**, 101–104 (1994)
13. Al'-Kamali, M.F.S.H., Alekseenko, A.A., Titenkov, O.A.: Structure formation of SiO_2-xerogels containing copper compounds of different phase composition. PFMT **3**(44), 7–12 (2020). (in Russian)
14. Al-Kamali, M.F.S.H., Boika, A.A., Al-Shamiri, H.A.S.: SiO_2:CuO(Cu°) targets for depositing thin films of Ion-Beam spraying obtained by sol-gel method. Doklady Natl. Acad. Sci. Belarus **66**(3), 348–355 (2022). https://doi.org/10.29235/1561-8323-2022-66-3-348-355. (in Russian)
15. Olenin, A.Y., Nizamov, T.R., Lisichkin, G.V.: Chemical modification of the surfaces of silver nanoparticles: synthesis of Janus particles. Nanotechnol Russia **9**, 467–473 (2014). https://doi.org/10.1134/S1995078014050103
16. Pshenova, A.S., Klyukin, D.A., Sidorov, A.I., Andreeva, O.V.: Porous glasses with silver nanoparticles as the sensitive material for sensors to measure the index of refraction of analytes. J. Opt. Technol. Opt. Technol. **83**, 438–440 (2016)

Variability of Soil Health Status Using X-ray Fluorescence Spectroscopy (XRF) and Fourier Transformed Infrared Spectroscopy (FTIR)

Mariana Rusu(✉), Iuliana Motrescu, Irina Gabriela Cara, Denis Topa, and Gerard Jitareanu

Ion Ionescu de la Brad University of Life Sciences, 3 Sadoveanu Alley, 700490 Iasi, Romania
mrusu@uaiasi.ro

Abstract. Soil health is a complex concept that includes the biological, physical, chemical properties of soil on long term such as to ensure sustainable productivity with reduced impact on the environment. Detail analysis depends on the targeted information and can be performed by a wide range of methods from simple observational status of the soil up to specifically measure some properties. The acquired information is used to establish the present status of the soil for predicting the behaviour in sustainable agricultural systems.

In this work we focused on studying the properties of soil (e.g., organic compounds, mineral composition, trace elements) based on using two non-invasive fast obtaining results methods: X-ray fluorescence spectroscopy (XRF) and Fourier transform infrared spectroscopy (FTIR). Based on these analyses we were able to monitor some properties of the soil at different horizons in the studied area, and make recommendations for soil management such as to increase the sustainable management of the soil.

Keywords: soil health · XRF · FTIR · heavy metals · molecular bonds

1 Introduction

Assessment of soil quality status helps in understanding soil fertility dynamics and can therefore perform optimal management practices [1]. Soil quality status relies on the chemical composition of soil organic matter and the elemental composition of macro and micronutrients [2, 3]. In recent years, innovative techniques and technologies have emerged to assess soil health with unprecedented precision and efficiency [4]. Among these, X-ray Fluorescence Spectroscopy (XRF) and Fourier Transformed Infrared Spectroscopy (FTIR) have gained prominence as powerful analytical tools [5]. These advanced spectroscopic techniques have been revolutionising the assessment of soil health, providing rapid and comprehensive insight into the chemical composition and structural characteristics of soils, ultimately contributing to sustainable land management practices.

X-ray Fluorescence Spectroscopy is a non-destructive analytical technique used to determine the elemental composition of soils [6]. It operates on the principle that when

Y. Ono and J. Kondoh (Eds.): Inter-Academia 2023, LNNS 939, pp. 151–163, 2024.
https://doi.org/10.1007/978-3-031-54450-7_17

X-rays are directed onto a sample, they excite the electrons in the atoms of the sample material. As these excited electrons return to their ground state, they emit X-ray photons with characteristic energies that correspond to specific elements. By measuring the energy and intensity of these emitted X-rays, XRF can provide quantitative information about the elemental composition of the soil sample [7]. This technique can detect Si, Pb, Ca, S, Cl, Al, Fe, Cr, Ti, V, Mn, Ni, Zn, Br, and Cu in the analysed samples [8–11]. Several working conditions can be used for the detection of light elements (such as Si and Al) or heavier elements (such as Pb) [12]. The limit of detection for most elements is in the mg/kg order. A lower sensitivity was reported in lighter elements, such as those with an atomic number lower than 12 [13].

Fourier Transformed Infrared Spectroscopy is another powerful tool in the realm of soil health assessment [14]. Unlike XRF, FTIR focuses on the molecular composition of soils. It relies on the principle that different chemical bonds in soil molecules absorb infrared radiation at characteristic frequencies [15]. By analysing the absorption patterns in the infrared spectrum, FTIR can identify and quantify various organic and inorganic compounds present in the soil [16]. FTIR is particularly valuable for assessing soil organic matter, which is a key determinant of soil health [17]. This information is used to identify the functional groups present in the sample and the types of bonds that are formed between atoms [18].

The applications of XRF and FTIR in soil health assessment are vast and impactful. Here are some key benefits and applications:

- Precision Agriculture: XRF and FTIR enable precision agriculture by providing real-time information about soil nutrient status. This allows farmers to adjust fertilizer applications according to the specific needs of their fields, reducing costs and minimizing nutrient runoff [19].
- Contaminant Detection: XRF can identify contaminants such as heavy metals, which are often of concern in polluted soils. It helps in locating contamination sources and planning remediation strategies [20].
- Carbon Sequestration: FTIR aids in understanding soil organic matter dynamics, contributing to strategies for carbon sequestration and climate change mitigation [21].
- Land Use Planning: Both techniques provide data on soil variability within a landscape, helping land managers make informed decisions about land use, conservation, and restoration [22].
- Soil Health Monitoring: Continuous monitoring using XRF and FTIR can track changes in soil health over time, providing valuable data for long-term land management [23].

While XRF and FTIR offer substantial advantages in soil health assessment, they are not without challenges:

- Cost and Accessibility: The instruments required for XRF and FTIR analysis can be expensive, limiting access for smaller agricultural operations or resource-constrained regions [24].
- Complex Data Interpretation: Analysing XRF and FTIR data requires expertise in spectroscopy and chemistry, making data interpretation and calibration critical [4].

- Sample Preparation: Proper sample preparation is essential for accurate results, which can be time-consuming and labour-intensive [25].
- Depth Limitations: These techniques primarily analyse surface soil, which may not capture the full profile of soil health [26].

Based on the above-mentioned facts we chosen, XRF and FTIR as techniques to provide a rapid, suitable and non-destructive alternative to classical soil analysis methods. Namely, FTIR was used to identify the presence of molecular bonds in both organic and inorganic soil horizons while portable XRF was employed to determine the elemental content of major soils essential macro and microelements as well as traces contaminants, when present. To resume, in recent years, there has been a growing interest in sustainable and alternative approaches to chemical synthesis and processes. Chemical protocols have several disadvantages, including being time-consuming, costly due to the need to use numerous reagents, and potentially hazardous to both human health and the environment. Consequently, the importance of the transition to alternative methods is stressed, which should pursue and address the shortcomings of traditional protocols by promoting efficiency, cost-effectiveness, human health safety and environmental sustainability.

2 Materials and Methods

2.1 Study Area

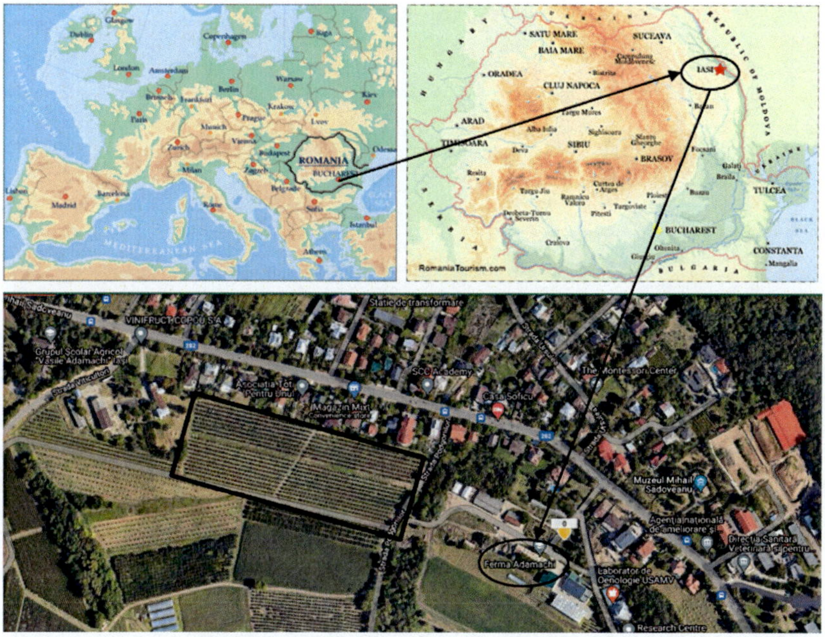

Fig. 1. Location of the research area

The study was carried out at Adamachi Farm - Iasi University of Life Sciences (47°15′N-27°30′E), located in Moldavian Plateau, northwest Romania, Europe (Fig. 1).

The climate belongs to the temperate continental climate zone, the annual average temperature is about 10 °C and the annual rainfall is about 518 mm (according to the weather station from Iasi Institute for Agriculture and Environment). The growing conditions for Prunus domestica L. in 2022 (April–August) presented the following average values: air temperature (19,1 °C), dew point (10,4 °C), solar irradiance (208.8 W/m^2), vapor pressure deficit (1.0 kPa) (Fig. 2), relative humidity (61,8%), precipitation (50.1 mm), humidity at leaf surface (7 min) and Delta T (4.8 °C) (Fig. 3).

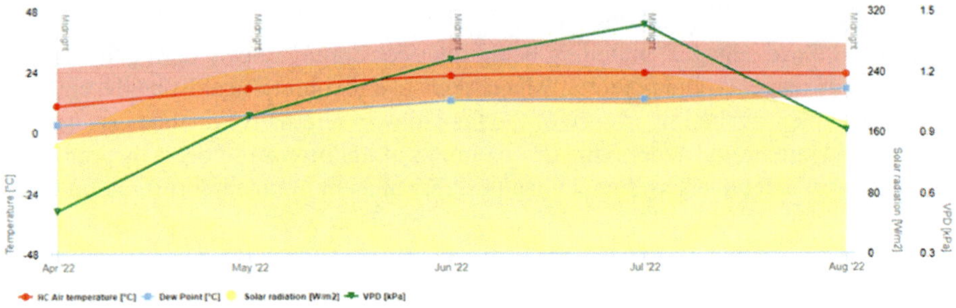

Fig. 2. Dates of the growth period (months)

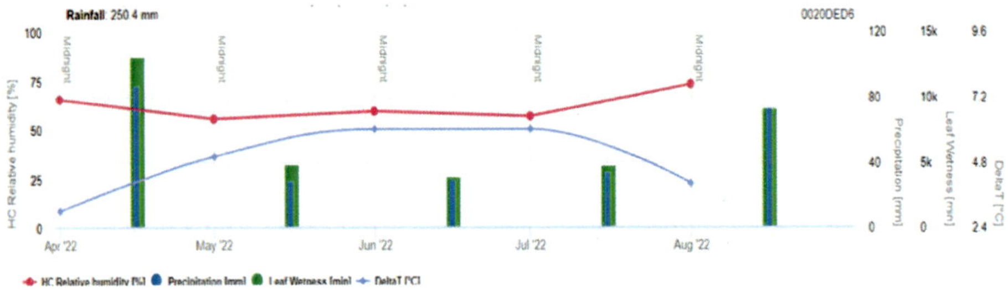

Fig. 3. Dates of the growth period (months)

2.2 Soil Characterization

The study was conducted in a plum orchard (Prunus domestica L.), which was planted in autumn 2014, on aric - cambic chernozem type soil, with loamy-clay texture, medium to good fertility. The plum orchard fields were subdivided into three (3) sampling areas with five (5) sampling points in each area.

Soil samples from different depths of 0–20 and 20–40 to 80–100 cm were collected from Adamachi Farm – Iasi University of Life Sciences. The procedure consists of three stages suggested for rapidly identifying the optimum quality status of the soil. Soil description was made first by classic methods to determine the pH, organic carbon and humus content, total nitrogen, phosphorus, and potassium (Table 1).

Table 1. Soil analysis description.

Depth (cm)	pH	Organic C %	Humus content (%)	N_t (%)	P (mg/kg)	K (mg/kg)
0–20	6.9 ± 0.12a	2.81 ± 0.05a	4.9 ± 0.08a	0.245 ± 0a	94.2 ± 1.3a	403.1 ± 5.04a
20–40	6.3 ± 0.07b	1.67 ± 0.03b	2.90 ± 0.06b	0.145 ± 0b	52.1 ± 0.99b	325.6 ± 4.92b
40–60	6.4 ± 0.02b	0.76 ± 0.02c	1.80 ± 0.01c	0.09 ± 0c	10.1 ± 0.18c	145.8 ± 2.78d
60–80	6.4 ± 0.05b	0.65 ± 0.01c	1.07 ± 0.01d	0.054 ± 0d	3.15 ± 0.04d	108.5 ± 0.66e
80–100	6.5 ± 0.03b	0.30 ± 0d	0.52 ± 0.01e	0.026 ± 0e	1.91 ± 0.01d	178.8 ± 1.29c

Values associated with different letters are significantly different according to Tukey's test at $p < 0.05$; * - WHO/FAO, 2022.

The identification and analysis of aliphatic, phenolic, aromatic, and carbonyl functional groups present in infrared spectroscopy allows the differentiation of the chemical composition of soil organic matter [27]. For these analyses, an Interspec 200-X FTIR spectrometer was used. The soil samples were grinded before the measurements. For XRF analysis, a portable XRF Vanta was employed. The XRF measurements helped quantify the chemical elements presented in the samples, analysed as they were with very good accuracy, with no prior preparation [28, 29]. The integration of FTIR and X-ray spectroscopy techniques is a rapid and cost-effective method to predict soil quality status.

2.3 Statistical Analysis

Statistical data was processed using the IBM SPSS v20 soft (IBM Corp, Armonk, NY, USA) and results were presented as means ± standard errors.

2.4 Bivariate Correlation Analysis

Bivariate correlation coefficients demonstrate how much two variables are dependent on each other. The impact of the other independent variables is not considered in the case of the multiple linear regression model. It is marked by ρyxj at the population level and by $r$yxj at the sample level, with values in the range $[-1, 1]$.

If there is a perfect link between variables at $[-1, 1]$, a direct link at $\rho \in (0, 1)$, a reverse link at $\rho \in [-1, 0)$ and no link at all at $\rho \in \{0\}$ then. When $|\rho| \geq 0.7$, the correlation between the variables is strong, for a smaller value $|\rho| \in [0.4; 0.7)$ there is a moderate link, and for $|\rho| \in (0; 0.4)$ the link between variables is low.

3 Results and Discussions

Understanding soil nutrient composition and dynamics is essential for sustainable agriculture and environmental management. The study of soil properties and their relationships is complex and interrelated. They play a key role in understanding soil health and fertility. Maintaining adequate levels of soil pH, organic matter and humus content is essential for optimizing nutrient availability to plants, reducing nutrient leaching and reducing risks associated with heavy metal contamination.

Among essential soil components, organic carbon (C) and total nitrogen (Nt) play a key role in soil fertility, microbial activity and plant growth. A strong and direct correlation was observed between the organic carbon, **C**, and total nitrogen, $\mathbf{N_t}$. This bivariate correlation is statistically significant as $p = 0.002 \leq$ significance threshold (α) $= 0.05$ (Table 2).

Table 2. Correlations between organic C and N_t.

		Organic C	N_t
Organic C	Pearson Correlation	1	0,986**
	Sig. (2-tailed)		0,002
	N	5	5
N_t	Pearson Correlation	0,986**	1
	Sig. (2-tailed)	0,002	
	N	5	5

** Correlation is significant at the 0.01 level (2-tailed).

There is a strong and direct correlation between organic **C** and **P**. This bivariate correlation is statistically significant as $p = 0.012 \leq$ significance threshold (α) $= 0.05$ (Table 3).

Table 3. Correlations between organic C and P.

		Organic C	P
Organic C	Pearson Correlation	1	0,953*
	Sig. (2-tailed)		0,012
	N	5	5
P	Pearson Correlation	0,953*	1
	Sig. (2-tailed)	0,012	
	N	5	5

* Correlation is significant at the 0.05 level (2-tailed).

From the table below (Table 4), a strong and direct correlation between organic **C** and **K** is observed. This bivariate correlation is statistically significant as $p = 0.024 \leq$ significance threshold (α) $= 0.05$.

A strong and direct correlation between $\mathbf{N_t}$ and **K** was evidenced, as it can be seen in the data presented below. This bivariate correlation is statistically significant as $p = 0.006 \leq$ significance threshold (α) $= 0.05$ (Table 5).

Table 4. Correlations between organic C and K.

		Organic C	K
Organic C	Pearson Correlation	1	0,926*
	Sig. (2-tailed)		0,024
	N	5	5
K	Pearson Correlation	0,926*	1
	Sig. (2-tailed)	0,024	
	N	5	5

* Correlation is significant at the 0.05 level (2-tailed).

Table 5. Correlations between N_t and K.

		N_t	K
N_t	Pearson Correlation	1	0,972**
	Sig. (2-tailed)		0,006
	N	5	5
K	Pearson Correlation	0,972**	1
	Sig. (2-tailed)	0,006	
	N	5	5

** Correlation is significant at the 0.01 level (2-tailed).

A strong and direct correlation was observed between **P** and **N_t**. This bivariate correlation is statistically significant as $p = 0.009 \leq$ significance threshold $(\alpha) = 0.05$ (Table 6).

Table 6. Correlations between P and N_t.

		N_t	P
N_t	Pearson Correlation	1	0,962**
	Sig. (2-tailed)		0,009
	N	5	5
P	Pearson Correlation	0,962**	1
	Sig. (2-tailed)	0,009	
	N	5	5

** Correlation is significant at the 0.01 level (2-tailed).

There is a strong and direct relationship between **K** and **P**. This bivariate correlation is statistically significant as $p = 0.039 \leq$ significance threshold $(\alpha) = 0.05$ (Table 7).

Table 7. Correlations between K and P.

		K	P
K	Pearson Correlation	1	0,898*
	Sig. (2-tailed)		0,039
	N	5	5
P	Pearson Correlation	0,898*	1
	Sig. (2-tailed)	0,039	
	N	5	5

* Correlation is significant at the 0.05 level (2-tailed).

A strong and direct correlation was observed between the threshold value (mg/kg) and the lower guideline value (mg/kg), the threshold value (mg/kg) and the higher guideline value (mg/kg), the higher guideline value (mg/kg) and the lower guideline value (mg/kg). These bivariate correlations are statistically significant as $p_1 = 0.010$, $p_2 = 0.009$, $p_3 = 0.001 \leq$ significance threshold $(\alpha) = 0.05$ (Table 8).

Table 8. Correlations between Threshold value (mg/kg), Lower guideline value (mg/kg) and Higher guideline value (mg/kg).

		Threshold	Lower	Higher
Threshold value (mg/kg)	Pearson Correlation	1	0,959*	0,962**
	Sig. (2-tailed)		0,010	0,009
	N	5	5	5
Lower guideline value (mg/kg)	Pearson Correlation	0,959*	1	0,992**
	Sig. (2-tailed)	0,010		0,001
	N	5	5	5
Higher guideline value (mg/kg)	Pearson Correlation	0,962**	0,992**	1
	Sig. (2-tailed)	0,009	0,001	
	N	5	5	5

* Correlation is significant at the 0.05 level (2-tailed).
** Correlation is significant at the 0.01 level (2-tailed).

XRF analysis provided information related to the main chemical elements contained in soils, as well as important data about the trace elements, such as heavy metals. Soil contamination with such elements might present risks as pollutant metals ca be incorporated in the crops, reaching and endangering the people when they enter the food chain (Table 9).

Understanding the origin of heavy metals in soil is a vital aspect of soil science and environmental research. It is important to understand the sources and mechanisms

Table 9. Correlations between chemical properties of soil.

Depth (cm)	pH	Organic C (%)	Humus content (%)	N_t (%)	P (mg/kg)	K (mg/kg)
0–20	0,852(*)	0,993(*)	0,673	0,878(*)	0,752	0,647
20–40	0,976(*)	0,516	0,989(*)	0,988(*)	0,958(*)	0,898(*)
40–60	0,772	0,755(*)	0,346	0,971(*)	0,212	0,764
60–80	0,876(*)	0,426	0,799(*)	0,958(*)	0,818(*)	0,997(*)
80–100	0,975(*)	0,758(*)	0,802(*)	0,442	0,564	0,883(*)

* Correlation is significant at the 0.05 level (2-tailed).

that lead to the accumulation of heavy metals in soil to effectively manage soil, control pollution, and protect the environment. The origin of heavy metals in soil is a complex interplay of geological, anthropogenic, agricultural, atmospheric, and biological factors. Researchers are continuously working on developing strategies to reduce the adverse effects of heavy metal contamination to safeguard the health of both ecosystems and human populations.

Table 10 presents some of the results on Zn, Cu, Cd, Ni, Co and Pb. The data shows that the upper soil horizon has relatively higher concentration of heavy metals, all within the recommended legal limits.

The concentration and distribution of heavy metals in soil is a serious concern for environmental health, soil quality and human well-being. The correlation of heavy metals content was performed from 20 to 20 cm to 1 m soil dept. For values <0.5 the correlation is weak, between 0.5–0.7 the correlation is medium and >0.7 the correlation is strong. All correlations in the tables are positive. Those marked with (*) are significant correlations (Table 11).

Soil organic matter (SOM) consists of microorganisms and non-decomposed plant materials, and it influences almost all the properties of the soil. Intensive agriculture has negative consequences on the SOM status. Diffuse reflectance FTIR of the analysed soils under agricultural use indicated absorbance bands corresponding to the SOM content such as amide (N-H) and alcohol (O-H), aliphatic (C-H), aromatic (C=C), amide (C=O), and carboxyl (C-O). A clear quantification is hard to be done due to the overlapping of these bands with mineral absorbances [30].

In the FTIR spectra we identified the following peaks: 995 cm^{-1} corresponding to Si–O bond, 1350–1445 cm^{-1} from C–H, 1700 cm^{-1} due to C=O, and 3000–3600 cm^{-1} due to O–H and N–H. There are not major changes with depth regarding the presence of these bonds, except that for higher depths there is a higher transmittance which translates in a decreased concentration of organic material. In the same time, the broad peak corresponding to O–H, N–H is the larger for the most upper layer of the soil, indicating the presence of increased and more complex organic material near the surface (Fig. 4).

Table 10. The influence of depth sampling on soil heavy metals concentrations.

Depth (cm) / Heavy metals	Zn (mg/kg)	Cu (mg/kg)	Cd (mg/kg)	Ni (mg/kg)	Co (mg/kg)	Pb (mg/kg)
0–20	111 ± 3.2a	218 ± 1.15a	0.8 ± 0.01a	53 ± 0.2a	13 ± 0.19a	20 ± 0.17a
20–40	88 ± 1.47b	157 ± 1.96b	0.4 ± 0b	38 ± 0.32b	10 ± 0.29b	19 ± 0.1b
40–60	55 ± 0.76c	98 ± 1.48c	0.1 ± 0c	31 ± 0.16c	8 ± 0.13c	15 ± 0.19c
60–80	56 ± 1.06c	82 ± 1.56d	0 ± 0e	30d ± 0.38c	5 ± 0.07d	14 ± 0.21d
80–100	39 ± 0.69d	52 ± 0.32e	0 ± 0d	27 ± 0.41d	1 ± 0.02e	10 ± 0.19e
Threshold value (mg/kg)	**200**	**100**	**1**	**50**	**20**	**60**
Lower guideline value (mg/kg)	250 (e)	150 (e)	10(e)	100(e)	100(e)	200(e)
Higher guideline value mg/kg	400 (e)	200 (e)	20(e)	150(e)	150(e)	750(t)

Values associated with different letters are significantly different according to Tukey's test at $p < 0.05$; * - WHO/FAO, 2022; The guideline values have been defined on the basis of ecological risks (e) or health risks (t).

Table 11. The correlation on heavy metals in the soil.

Depths (cm) / Heavy metals	Zn (mg/kg)	Cu (mg/kg)	Cd (mg/kg)	Ni (mg/kg)	Co (mg/kg)	Pb (mg/kg)
0–20	0,102	0,785(*)	0,07	0,788(*)	0,231	0,998(*)
20–40	0,806(*)	0,426	0,897(*)	0,958(*)	0,818(*)	0,912(*)
40–60	0,822(*)	0,983(*)	0,505	0,812(*)	0,752	0,584
60–80	0,916(*)	0,516	0,761(*)	0,897(*)	0,902(*)	0,800(*)
80–100	0,832(*)	0,804(*)	0,044	0,191	0,422	0,192

* Correlation is significant at the 0.05 level (2-tailed).

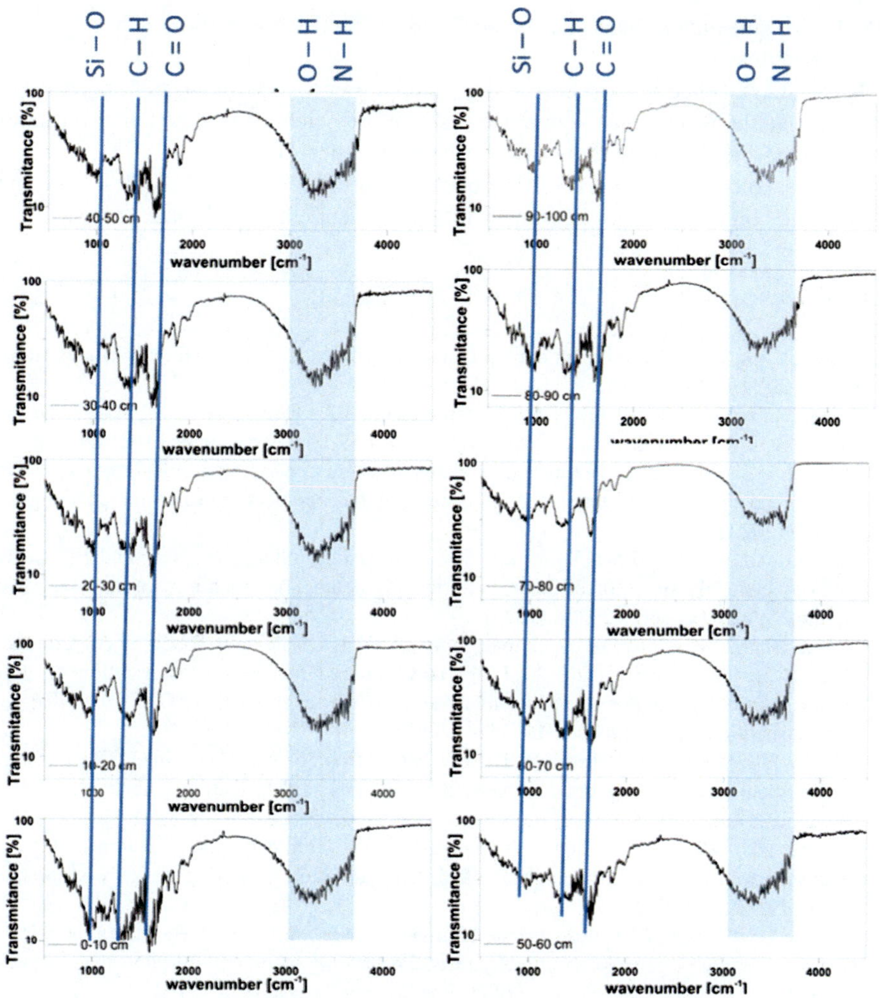

Fig. 4. FTIR spectra of soil samples as function of depth.

4 Conclusions

In this work we focused on diagnosing the soil health using non-invasive physico-chemical methods. We found that based on XRF results we can detect trace amounts of heavy metals in the soil from a orchard that could pose danger if passing into the food chain. FTIR also gives important information related mostly to the organic compounds in soil, and we found out that the differences are mostly near the surface probably due to the higher consumption of organic matter here and/or increase due to the presence of more organic sources.

Soil health status is crucial for a wide range of ecological, agricultural, and societal reasons. It affects food production, water quality, climate change mitigation, biodiversity, and the overall health of our planet. Recognizing and addressing soil health issues is essential for sustainable land management and the well-being of both ecosystems and human societies.

X-ray fluorescence spectroscopy (XRF) and Fourier transform infrared spectroscopy (FTIR) are valuable analytical techniques used in soil science by providing information on soil composition and characteristics to assess soil health.

Healthy soils reduce the need for chemical fertilizers and pesticides, minimize erosion, and contribute to carbon sequestration, helping combat climate change.

References

1. Jităreanu, G., et al.: Treatise on agrotechnics. "Ion Ionescu de la Brad" Publishing House, Iași (2020). (in Romanian)
2. Arias, M.E., Gonzalez-Perez, J.A., Gonzalez-Vila, F.J., Ball, A.S.: Soil health – a new challenge for microbiologists and chemists. Int. Microbiol.Microbiol. **8**, 13–21 (2005)
3. Nath, D., et al.: Strategies to admittance soil quality using mid-infrared (mid-IR) spectroscopy an alternate tool for convenient lab analysis: a global perspective. Environ. Challenges **7**, 100469 (2022)
4. Baddini, A.L., et al.: PLS-DA and data fusion of visible reflectance, XRF and FTIR spectroscopy in the classification of mixed historical pigments. Spectrochimica Acta Part A: Mol. Biomol. Spectrosc. **265**, 120384 (2022)
5. Correa-Ochoa, M.A., Rojas, J., Gómez, L.M., Aguiar, D., Palacio-Tobón, C.A., Colorado, H.A.: Systematic search using the Proknow-C method for the characterization of atmospheric particulate matter using the materials science techniques XRD, FTIR, XRF, and Raman spectroscopy. J. Sustain. **15**, 8504 (2023)
6. Galvão, E.S., Santos, J.M., Lima, A.T., Reis, N.C., Orlando, M.T.D.A., Stuetz, R.M.: Trends in analytical techniques applied to particulate matter characterization: a critical review of fundaments and applications. Chemosphere **199**, 546–568 (2018)
7. Motrescu, I., Calistru, A.E., Jitareanu, G., Miron, L.D.: Monitoring the environmental quality of marine waters through the analysis of biomineralization in bivalve shells. Eng. Sustain. Future **101**, 269–277 (2020)
8. Lima, F.D.M., Pérez-Martínez, P.J., de Fatima Andrade, M., Kumar, P., de Miranda, R.M.: Characterization of particles emitted by pizzerias burning wood and briquettes: a case study at Sao Paulo, Brazil. Environ. Sci. Pollut. Res. **27**, 35875–35888 (2020)
9. Senlin, L., Longyi, S., Minghong, W., Zheng, J., Xiaohui, C.: Chemical elements and their source apportionment of PM_{10} in Beijing urban atmosphere. Environ. Monit. Assess.Monit. Assess. **133**, 79–85 (2007)
10. Popovicheva, O., et al.: Microscopic characterization of individual particles from multicomponent ship exhaust. J. Environ. Monit.Monit. **14**, 3101–3110 (2012)
11. Tshehla, C., Wright, C.Y.: Spatial variability of PM10, PM2.5 and PM chemical components in an industrialised rural area within a mountainous terrain. S. Afr. J. Sci. **115** (2019)
12. Marcazzan, G.M., Vaccaro, S., Valli, G., Vecchi, R.: Characterisation of PM10 and PM2.5 particulate matter in the ambient air of Milan (Italy). Atmos. Environ. **35**, 4639–4650 (2001)
13. Marguí, E., Queralt, I., de Almeida, E.: X-ray fluorescence spectrometry for environmental analysis: basic principles, instrumentation, applications and recent trends. Chemosphere **303**, Part 1, 135006 (2022)
14. Rosi, F., Burnstock, A., Van den Berg, K.J., Miliani, C., Brunetti, B.G., Sgamellotti, A.: A non-invasive XRF study supported by multivariate statistical analysis and reflectance FTIR to assess the composition of modern painting materials. Spectrochim. Acta Part A Mol. Biomol. Spectrosc.. Acta Part A Mol. Biomol. Spectrosc. **71**(5), 1655–1662 (2009)

15. Cloete, K.J., Šmit, Ž, Gianoncelli, A.: Multidimensional profiling of human body hairs using qualitative and semi-quantitative approaches with SR-XRF, ATR-FTIR, DSC, and SEM-EDX. J. IJMS **24**, 4 (2023)
16. Swiatly-Blaszkiewicz, A., Pietkiewicz, D., Matysiak, J., Czech-Szczapa, B., Cichocka, K., Kupcewicz, B.: Rapid and accurate approach for honeybee pollen analysis using ED-XRF and FTIR spectroscopy. J. Mol. **26**, 19 (2022)
17. Țopa, D., Cara, I.G., Jităreanu, G.: Long term impact of different tillage systems on carbon pools and stocks, soil bulk density, aggregation and nutrients: a field meta-analysis. J. Catena **199** (2021)
18. Kwan, Y.C.G., Ng, G.M., Huan, C.H.A.: Identification of functional groups and determination of carboxyl formation temperature in graphene oxide using the XPS O 1s spectrum. Thin Solid Films **590**, 40–48 (2015)
19. Sharma, V., Chauhan, R., Kumar, R.: Spectral characteristics of organic soil matter: a comprehensive review. Microchem. J.. J. **171**, 106836 (2021)
20. Leite, F.D., et al.: Analysis of Zebrafish contamination with heavy metals using a FF-XRF imaging system based on a MPGD. J. Elsevier **198**, 106545 (2022)
21. Anandakumar, S., et al.: Impact of long-term nutrient management on sequestration and dynamics of soil organic carbon in a semi-arid tropical Alfisol of India. Appl. Soil Ecol. **177**, 104549 (2022)
22. Gianoncelli, A., Žižić, M., Bonanni, V., Birarda, G., Bedolla, D.E., Mikuš, V.: Soft X-rays radiation damage on plunge-frozen and freeze-dried maize roots evaluated by FTIR spectromicroscopy. J. Electron Spectrosc. Relat. Phenom.Spectrosc. Relat. Phenom. **267**, 147384 (2023)
23. Aitkenhead, M., Gwatkin, R., Coull, M., Donnelly, D.: Assessment of soil health and fertility indicators with mobile phone imagery. Digit. Soil Morphometrics **17**, 89–110 (2015)
24. Liu, Y., et al.: Clinoptilolite based zeolite-geopolymer hybrid foams: potential application as low-cost sorbents for heavy metals. J. Environ. Manage. **330**, 117167 (2023)
25. Surowka, A.D., Czyzycki, M., Ziomber-Lisiak, A., Migliori, A., Szczerbowska-Boruchowska, M.: On 2D-FTIR-XRF microscopy – a step forward correlative tissue studies by infrared and hard X-ray radiation. J. Elsivier/Ultramicroscopy **232**, 113408 (2022)
26. Trueman, C.N., Behrensmeyer, A.K., Tuross, N., Weiner, S.: Mineralogical and compositional changes in bones exposed on soil surfaces in Amboseli National Park, Kenya: diagenetic mechanisms and the role of sediment pore fluids. J. Archaeol. Sci.Archaeol. Sci. **31**(6), 721–739 (2004)
27. O'Rourke, S.M., Minasny, B., Holden, N.M., McBartney, A.B.: Synergistic use of Vis-NIR, MIR, and XRF spectroscopy for the determination of soil geochemistry. Soil Sci. Soc. Am. J. **80**, 888–899 (2016)
28. Radu, T., Diamond, D.: Comparison of soil pollution concentrations determined using AAS and portable XRF techniques. J. Hazard. Mater. **171**, 1168–1171 (2009)
29. Kaniu, M.I., Angeyo, K.H., Mangala, M.J., Mwala, A.K., Bartilol, S.K.: Fesability for chemometric energy dispersive X-ray fluorescence and scattering (EDXRFS) spectroscopy method for rapid soil quality assessment. X-Ray Spectrom.Spectrom. **40**, 432–440 (2011)
30. Margenot, A.J., Calderon, F.J., Goyne, K.W., Goyne, K.W., Mukome, F.N.D., Parikh, S.J.: IR spectroscopy, soil analysis applications. In: Encyclopedia of Spectroscopy and Spectrometry, 3rd edn., vol. 2, pp. 448–454 (2017)

Spiral Wire Microplasma Inducing Growth and Viability of Nasal Cell

Abubakar Hamza Sadiq[1(✉)], Jaroslav Kristof[2], Alam Md Jahangir[3], Sadia Afrin Rimi[1], Yoshinori Mizuno[1,4], and Kazuo Shimizu[1,2,3]

[1] Graduate School of Science and Technology, Shizuoka University, Hamamatsu 432-8011, Japan
{h.s.abubakar.22,shimizu.kazuo}@shizuoka.ac.jp
[2] Organization for Innovative and Social Collaboration, Shizuoka University, Hamamatsu 432-8011, Japan
[3] Graduate School of Medical Photonics, Shizuoka University, Hamamatsu 432-8011, Japan
[4] Research and Development Center, Asahikogyosha Co., Ltd., Narashino 275-0001, Japan

Abstract. In medicine, the promising potential of plasma treatment requires the measurement of possible risks and the prevention of cell damage by direct or indirect application of plasma devices. This study aimed to investigate the impact of plasma discharge generated from spiral wire microplasma on cell viability and the delivery of fluorescein isothiocyanate-dextran 150 (FD-150). At discharge voltages of 2, 3, 4 $kV_{p\text{-}p}$, the growth and viability of the cell was investigated. The influence of the optimized discharge at 3 $kV_{p\text{-}p}$ was further studied in three different volumes (1, 2, 3 ml) of cell culture medium, as well as the delivery of FD-150 with a molecular weight of 150 kDa into cells. The olfactory dark basal cell (DBC1.2) from mouse nasal septum, similar to the nasal epithelium, was used. Cell viability was evaluated using a cell counter and a hemocytometer, while drug delivery was measured by the fluorescence intensity of a microplate reader. The results revealed an increased cell number and viability for plasma-irradiated samples compared with control samples at an incubation time of 30 min, whereas at 24 h incubation, the results were approximately the same. Similarly, plasma discharge in 1, 2, and 3 ml volumes of Dulbecco's Modified Eagle Medium (DMEM) revealed no cell damage by plasma but rather a 3%, 4%, and 5% increase compared with control cells. The results observed for delivery of the drug sample also indicate that the microplasma not only prevent epithelial cell damage, but also increase drug absorption in cells.

Keywords: spiral wire microplasma · cell viability · drug absorption · Nasal cell · DBC1.2

1 Introduction

Plasma medicine is an emerging interdisciplinary field covering plasma physics, life science, and clinical medicine [1]. These plasma devices use low-temperature atmospheric-pressure plasmas to generate controlled doses of electrically and chemically reactive

Y. Ono and J. Kondoh (Eds.): Inter-Academia 2023, LNNS 939, pp. 164–173, 2024.
https://doi.org/10.1007/978-3-031-54450-7_18

species that can react with biological targets [2–4]. The discharges generated from the surfaces of the plasma devices, disperse energies that are transferred to the air-liquid interface [5], and when in contact with the liquid, additional species are created within. These additional species interact with the cells and tissue [6] in the culture medium. It was recently demonstrated that plasma has anticancer, sterilizing, blood-coagulation, tissue-regenerating, and anti-inflammatory effects [7–11]. As a new therapeutic option in this field, microplasma treatment has also shown promising potential in the delivery of low and high molecular weight drugs into cells and tissues [12]. The interactions of the cell membrane with these reactive species and electric fields help in the delivery of drugs to cells [13]. However, a reduction in cell viability has been reported in most studies of epithelial cells [12, 14]. Improvement has been described for wound healing in human S9 bronchial epithelial cells with plasma treatment at certain intensities by a comprehensive study of the variations in cellular response [15].

In respiratory medicine and drug delivery, direct and future cell damage due to interactions from plasma discharge should be prevented. Therefore, the use of microplasma that is well tolerated by epithelial cells is important.

In this study, we designed a dielectric barrier discharge (DBD) microplasma by winding a ground electrode over the dielectric barrier coating covering the high-voltage electrode. The microplasma is a thin flexible electrode coiled into a spiral form. This study contributes to the existing novel designs of DBD microplasma. We investigated the effect of the discharge generated from the spiral wire microplasma on the viability of olfactory dark basal cells (DBC1.2), and the delivery of fluorescein isothiocyanate-dextran 150 (FD-150). These DBC1.2 cell lines are similar to cells of the nasal cavity and express keratin [16]. The results of cell viability at three different discharge voltages were observed. Furthermore, the viability of the DBC1.2 cell line as well as the delivery of FD-150 into the cells cultured in three different volumes of Dulbecco's Modified Eagle Medium (DMEM) was also evaluated.

2 Experimental Method

2.1 Spiral Wire Microplasma Electrode Design

Figure 1a shows the wire microplasma electrode in a spiral form before plasma discharge and in Fig. 1b during plasma discharge when activated under atmospheric conditions to generate numerous streams of discharge that are transferred to the dielectric surface. The geometry of the spiral wire microplasma electrode consists of a ground electrode and a high-voltage (HV) electrode covered by a dielectric barrier coating. The high-voltage electrode with a diameter of 1 mm is a copper wire, while the ground electrode is a copper wire with a diameter of 0.4 mm. The design structure was such that the ground electrode was coiled round the surface of the high-voltage dielectric coating to a length of 20 cm. The discharge occurs only in this region when energized. At the top, 1 cm dielectric coating was peeled off to allow for energy transfer to the electrode from a power source. The winding distance from one turn to the next was approximately 4 mm (Fig. 1).

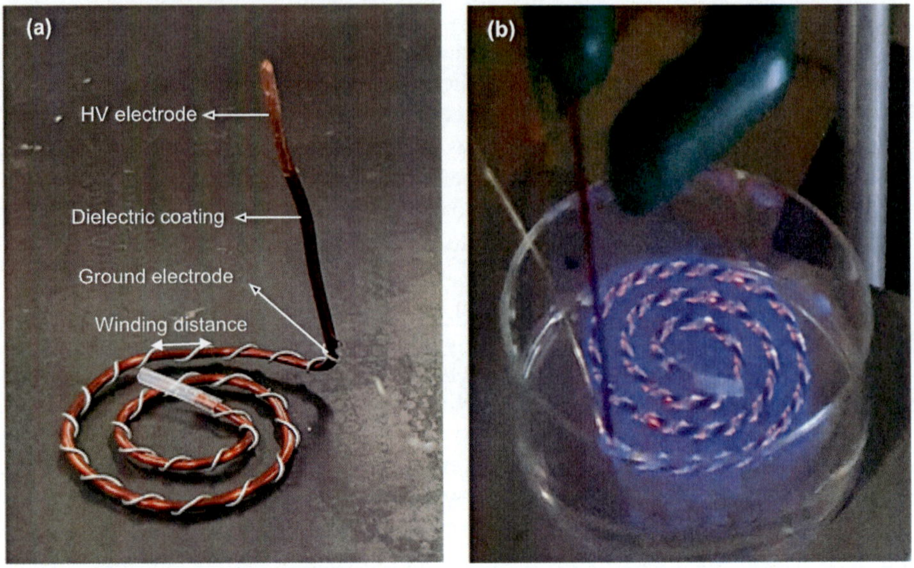

Fig. 1. (a) Spiral wire microplasma (b) Spiral wire microplasma discharge ignited in air

2.2 Electrical Characteristics Setup

Figure 2a shows the experimental setup for generating the plasma discharge. The peak-to-peak sine waveforms was set by a function generator (Tektronix, AFG3102) and amplified by a high-voltage amplifier (Trek, model 5/80) (Fig. 2). The discharge voltage was measured using a high-voltage probe (Tektronix, P6105A) connected to an oscilloscope (Tektronix, TDS 2014B). The discharge current is measured by a Pearson current monitor connected to the oscilloscope.

2.3 Cell Culture

The DBC1.2 cell line used in this study was purchased from Nibiohn JCRB cell bank. The cells were thawed from -80 °C following the manufacturer's culture protocol and seeded into 25 cm^2 culture flask [17] until confluency of approximately 95% under a microscope. The cells in 25 cm^2 culture flask were maintained in 3 ml Dulbecco's Modified Eagle Medium (DMEM). The DMEM powder (4.75 g) was dissolved in 500 mL distilled water and supplemented with 5 mL L-glutamine solution (Fujifilm Wako Pure Chemical), 5 mL penicillin-streptomycin solution (Fujifilm Wako Pure Chemical), 8 mL of 10% $NaHCO_3$ (Fujifilm Wako Pure Chemical), 25 mL fetal bovine serum (FBS, Serana), 200 μL insulin solution, human recombinant (Fujifilm Wako Pure Chemical) [12]. The culture environment was set to 37°C in 5% CO_2 incubator (E-22, As One) until confluency. The confluent flasks were washed twice with 2 ml phosphate buffer saline (PBS) and detached with 0.5 ml trypsin (0.25 w/v% Trypsin–1 mmol/l EDTA·4Na solution with Phenol Red). Once cells are detached from the plate, 3 ml of DMEM solution was immediately added to neutralize the effect of trypsin and homogenously mixed with a

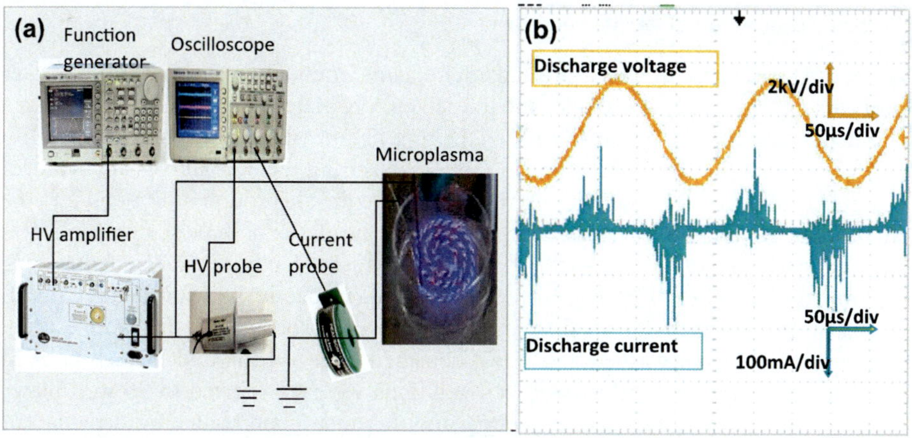

Fig. 2. (a) Experimental setup for generating plasma discharge on the surface of the cell culture medium. (b) Schematic of the oscilloscope voltage and current waveform at a discharge voltage of 4 kV peak-peak sine wave.

pipette until a single cell suspension was observed, before splitting into the culture plates. At 24 h after incubation, cells adhere to the culture plate and old DMEM is aspirated and replaced with fresh DMEM. To sustain cell growth, DMEM was replaced every 2 days until a confluency of 75% or above was observed under a microscope. For three sample replicates, cells were treated with wire microplasma for 2 min at an irradiation distance of ~ 2–3 mm for a discharge voltage of 2, 3, 4 kV_{P-P} at 5 kHz frequency. After plasma irradiation, cells were immediately returned to a 37 °C, 5% CO_2 incubator for an appropriate time before cell counting with a countess automatic cell counter and evaluation of FD-150 absorption into cells using TECAN microplate reader.

2.4 Cell Viability

The viability of cells was evaluated using a countess automatic cell counter. After all experiments and incubation periods, cells were washed twice with 2 ml PBS. 500 μl trypsin are further added and incubated for 3 min to detach the cells. Afterwards, 3 ml DMEM was added to neutralize the effect of trypsin and homogenously mixed with a pipette until a single cell suspension was approximately obtained. These cells were stained with trypan blue dye. Upon dilution, 10 μl was transferred to a hemocytometer with a cover slip and inserted into a countess automatic cell counter for total cell count, live cell count, and dead cell count. The percentage cell viability was calculated by dividing the number of live cells by the total number of cells and multiplied by 100. Trypan blue from sigma life science has a molecular weight of 960.81 Da. It works on the principle that healthy cells have a strong membrane to prevent the penetration of trypan blue dye, whereas dead and weak cell membranes do not.

2.5 Physiocochemical Measurement

The pH of DMEM was tested before and after plasma irradiation using a pen-type water quality tester. 3 ml DMEM was added to the cell culture plate and irradiated with plasma at discharge voltages of 2, 3, and 4 kV_{p-p}. During pH measurements, the water quality tester was held at 90° and submerged in DMEM, ensuring that air bubbles are avoided.

TECAN microplate reader was used to determine the absorption of FD-150 into the cell, based on the amount of fluorescence intensity measured. For three sample replicates and in each treatment well, FD-150 of 3 μl, 6 μl, 9 μl (15 mg/ml) was added to 1, 2, and 3 ml volumes of DMEM and treated with plasma respectively. After treatment, cell samples were incubated for 15 h to allow for proper absorption of the sample drug. Afterwards, cells were washed with PBS, detached with trypsin, added fresh DMEM and homogenously mix. 250 μl/well cell solutions were transferred to 96-well plates (Violamo; F-bottom with lid) and inserted into the microplate reader for detection of fluorescence in cells at excitation and emission wavelength of 485 nm and 535 nm.

The emission spectrum from the microplasma was measured using a spectrometer (ASEQ instruments, LR1-B). The spectrometer was connected to a computer software to visualize the line spectra when the microplasma electrode was energized under room air at a frequency of 5 kHz and discharge voltages of 2, 3, and 4 kV_{p-p}.

3 Results and Discussion

Prior to cell culture, the pH value of the DMEM culture medium was measured with a water quality tester, and the same values of 6.86 was recorded in both control and plasma-irradiated samples at a discharge voltages of 2, 3, and 4 kV_{p-p}. The neutrality of the pH result in both control and the discharge voltages showed that the plasma discharge does not increase or decrease the acidity or alkalinity of the DMEM culture medium.

The impact of the microplasma discharge on cells based on percentages of viable cells was investigated. The electrode distance to the surface of the DMEM medium is approximately 2–3 mm. Plasma discharge due to electrical breakdown of air in the gap between high-voltage electrode and the surface of DMEM releases heat, UV light, electromagnetic waves, reactive oxygen and nitrogen species (RONS), charged and uncharged species into the DMEM culture medium, which further interact with the cell within [6]. The treatment time is 2 min at a frequency of 5 kHz. This interaction is experimentally evaluated in Fig. 3 to measure cell survival at plasma discharge voltages of 2, 3, and 4 kV_{p-p}. In Fig. 3, at approximately 75% confluency, the effect of plasma discharge on the viability of cells after 30 min and 24 h of incubation time was evaluated using a countess automatic cell counter. At 30 min of incubation, the percentage viability of control cells (75%) and plasma-irradiated cells (81%, 92%, and 91%), calculated as the ratio of live cells to total (live and dead) cell counts showed high viability in plasma-irradiated samples. This increase indicates that cellular responses are induced by various reactive species generated from the microplasma discharge. However, plasma irradiation at 3 kV appears to have the highest positive impact. Further incubation of the same culture dishes for an additional 24 h and evaluated for viability shows that the percentage viability of control cells (92%) and plasma-irradiated cells (93%, 94% and 89%) remained almost intact. The positive effect from discharge was sustained at 2 kV_{p-p} and 3 kV_{p-p}, while a decrease was observed at 4 kV_{p-p} discharge voltage. This decrease is an indication of negative impact on cell viability for a prolonged incubation, thereby inducing cellular stress or damage leading to reduced cell survival.

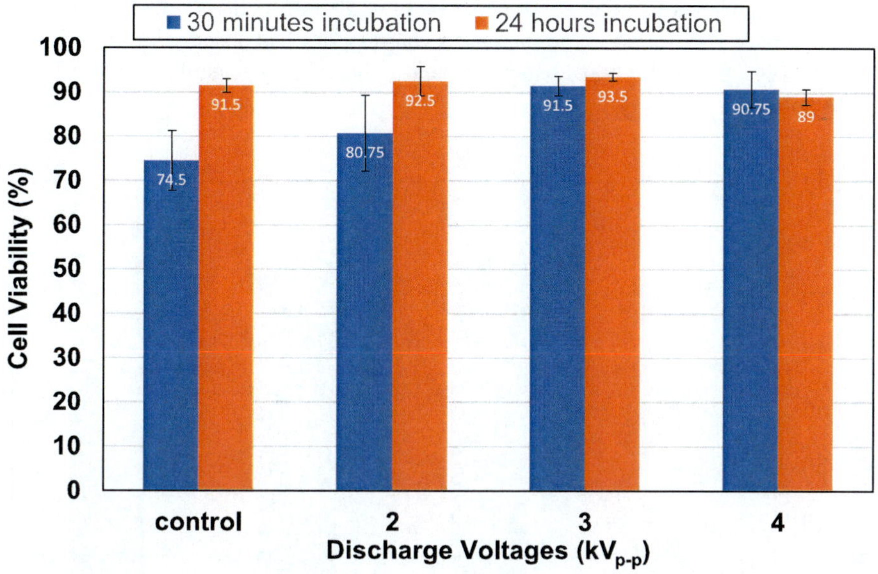

Fig. 3. Measurement of cell viability at 30 min incubation after plasma irradiation at discharge voltages of 2 kV, 3 kV, 4 kV and an additional 24 h incubation of the same cell samples.

Furthermore, cells were cultured in three different volumes (1, 2, 3 ml) of DMEM and subjected to plasma discharge at a discharge voltage of 3 kV_{p-p} to evaluate for cell viability and drug absorption into cells. Figure 4 shows the result of cell viability over DMEM volumes after plasma irradiation and incubation for 15 h. The cell count indicates an increase in cell viability with plasma irradiation compared with control samples. The cell cultured in 3 ml volume of DMEM showed the highest viability at 87.2%, followed by 2 ml of DMEM with viability of 86.2% and 1 ml of DMEM with viability of 85.7%. These results indicate no cell damage caused by plasma discharge and have a modest positive impact on cell survival over the long term when compared with the control viability at 82.3%.

Figure 5 shows the results of drug absorption into cells. The sample drug used was FD-150, with molecular weight of 150 kDa, purchased from Cosmo Bio Co., LTD (TdB Labs). According to Ikeda et al. [18], molecular weight determines the amount of molecules that are absorbed into the cells by plasma discharge. We investigated the intracellular absorbance of FD-150 by fluorescence intensity from a microplate reader with excitation and emission wavelengths of 485 nm and 535 nm in three replicate samples. Figure 5 shows the result of FD-150 absorption in the cell. Where the negative control is without FD-150 in 3 ml DMEM, the control has 9 μl FD-150 in 3 ml DMEM, and the plasma treated samples has 3 μl, 6 μl, 9 μl (15 mg/ml) of FD-150 added to 1 ml, 2 ml, and 3 ml volumes of DMEM, respectively. After treatment, the absorption of

FD-150 into the cellular membrane was observed to have higher fluorescence in plasma-treated samples, as shown in Fig. 5. With 3 ml DMEM volume being the highest and the least being the control. The increased absorption in plasma-treated samples could be due to the distortion of the cell membrane, created by RONS and the physical activities of the wire microplasma. The absorbance of FD-150 into cells is a dose-dependence response, which also supports the increased viability of cells observed in 3 ml DMEM volume in Fig. 4. The highest absorbance and cell viability was observed in 3 ml of DMEM volume.

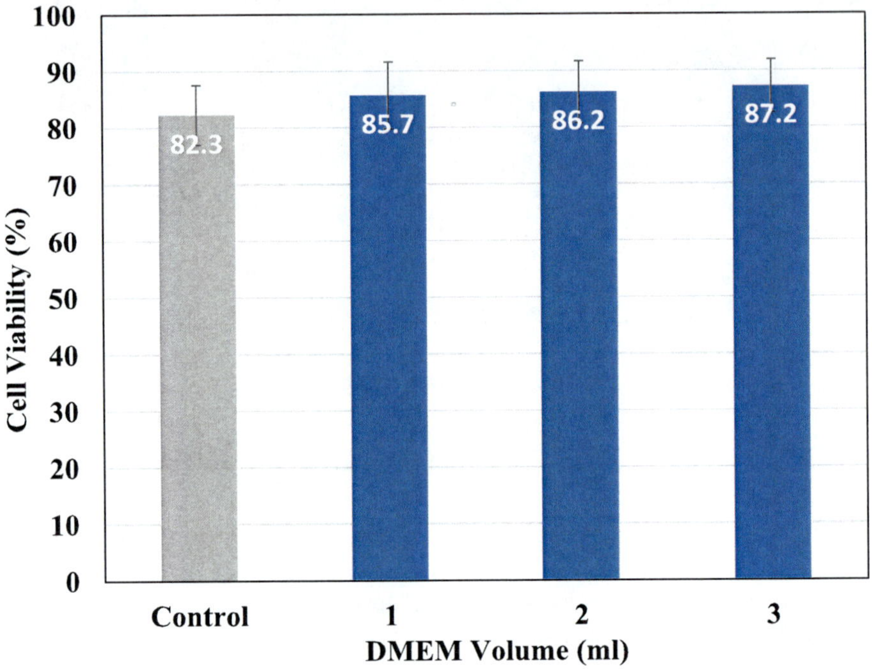

Fig. 4. Measurement of cell viability in different volumes of DMEM culture medium after plasma discharge and incubation at 37 °C and 5% CO_2 for 15 h.

To confirm the result of FD-150 absorption in cells, emission spectra of different plasma conditions were observed. Figure 6 shows the emission spectra of air microplasma at discharge voltages of 2, 3, and 4 $kV_{p\text{-}p}$. At each discharge voltage, the emitted light was distributed across various intensities and wavelengths, showing the excited states of atoms and molecules in the plasma. At 2 $kV_{p\text{-}p}$, the distributed lines showed a small intensity of the excited state of the plasma. At a higher energy input of 3 $kV_{p\text{-}p}$, there was an increase in the intensity of the spectra lines, indicating an increase in the excited state of the plasma. For more energy input of 4 kV, there is a more pronounced excitation. These results show that increasing the voltages is an indication of the energy dynamics in plasma discharge. As discharge voltages increase, collision and excitation of atoms and molecules also increased.

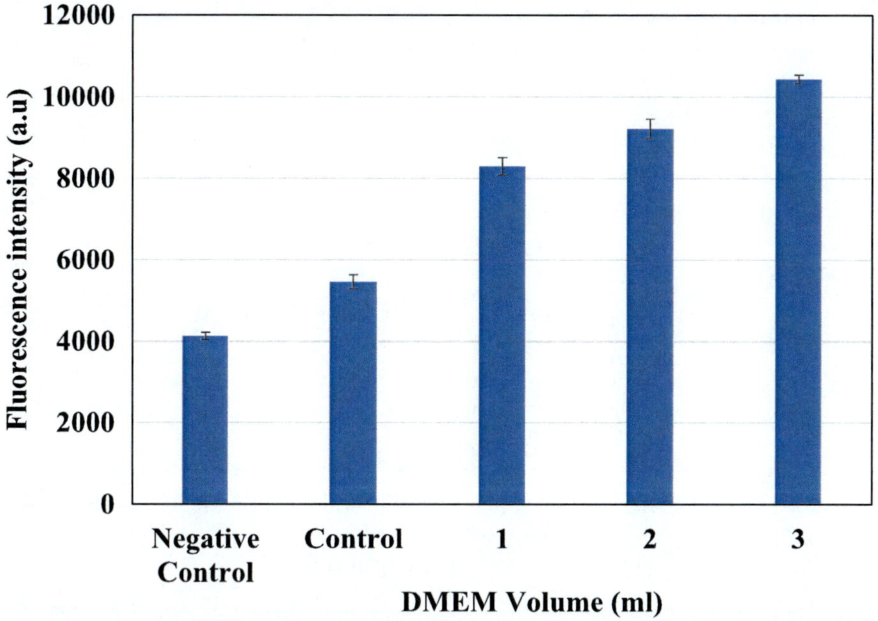

Fig. 5. Measurement of fluorescence intensity from a microplate reader for FD-150 drug samples absorbed into the intracellular membrane after plasma discharge at 3 kV_{p-p}, 5 kHz, 2 min, and incubated for 15 h.

It was further observed that the intensity of the spectra lines increased with increasing voltage, but the wavelength remained constant. While the increased intensities indicate a more energetic collisions and excitation of the plasma particles, the constant wavelength indicates that the microplasma generates a consistent and similar excited species at all discharge voltages. Similarly, N_2 second positive band with peak positions of 296; 315; 337; 358; 376; 381; 400 nm and N_2 first negative band with peak positions of 391; 428 nm were identified [19, 20] from the emission spectra.

4 Conclusions

A new DBD microplasma electrode is designed and tested on the DBC1.2 cell line to observe the impact of plasma discharge on the cell viability as well as the absorption of FD-150 into the cell. Prior to plasma discharge on cell, the pH of culture medium was evaluated before and after plasma irradiation, and results showed no changes in the pH of medium both immediately after 24 h of incubation. Furthermore, cells were irradiated with plasma at discharge voltages of 2, 3, 4 kV_{p-p} and incubated for 30 min and 24 h. Upon comparison with control samples, no cell damage was observed in the viability of plasma-irradiated samples. An increased viability is observed in 2 kV_{p-p} but a more positive increase and continuous long-term effect on cell was observed at discharge voltage of 3 kV_{p-p}, whereas 4 kV_{p-p} discharge voltage expresses a potential negative impact over time.

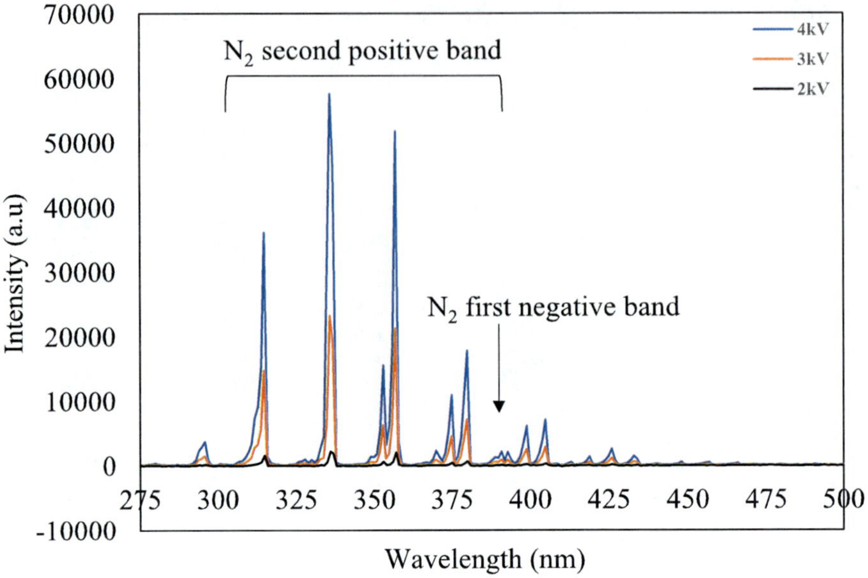

Fig. 6. The combined emission spectra of spiral wire microplasma energized in room air at discharge voltages of 2 kV, 3 kV, and 4 kV and frequency of 5 kHz.

In addition, the plasma discharge at 3 $kV_{p\text{-}p}$ was studied on the viability of cells cultured in three different volumes (1, 2, 3 ml) of DMEM, as well as the absorption of FD-150 with molecular weight of 150 kDa into the cells. The highest viability of cells and delivery of FD-150 was observed at 3 ml DMEM. This result indicates a dose-dependence response to plasma discharge.

The emission spectra of plasma condition at 2, 3, 4 $kV_{p\text{-}p}$ was further evaluated to identify the intensities of excited species generated by the air microplasma. The results showed that while the intensities of the emitted spectra lines increased with increasing discharge voltages, the wavelength was the same for all excited species generated by the discharge parameters.

Further studies are required to observe the molecular and cellular mechanisms triggering these observed viabilities. This will provide insight into the specific pathways influenced by plasma discharges.

Acknowledgement. This work was supported by the Japan Society for the Promotion of Science and Technology Research Grant (JP23K17473). In addition, the authors would like to express their gratitude to Prof. Kosuke Shimizu, Prof. Mitsutoshi Seto, Dr. Shuhei Aramaki, Dr. Al Mamun from the Graduate School of Photonics and Medical Engineering, Hamamatsu University School of Medicine, who have greatly cooperated in this research.

References

1. Mounir, L.: Plasma medicine: a brief introduction. Plasma **1**(1), 47–60 (2018)

2. Shimizu, K.: Biological effects and enhancement of percutaneous absorption on skin by atmospheric microplasma irradiation. Plasma Med. **5**(2–4), 205–221 (2016)
3. Haralambiev, L., et al.: Effects of cold atmospheric plasma on the expression of chemokines, growth factors, TNF superfamily members, interleukins, and cytokines in human osteosarcoma cells. Anticancer Res. **39**, 151–157 (2019)
4. Shimizu, K., Fukunaga, H., Blajan, M.: Biomedical applications of atmospheric microplasma. Curr. Appl. Phys. **14**(2), S154–S161 (2013)
5. Lu, X., Laroussi, M., Puech, V.: On atmospheric-pressure non-equilibrium plasma jets and plasma bullets. Plasma Sources Sci. Technol. **21**(3), 034005–034022 (2012)
6. Bao, P., Lu, X., He, M., Liu, D.: Kinetic analysis of delivery of plasma reactive species into cells immersed in culture media. IEEE Trans. Plasma Sci. **44**, 2673–2681 (2016)
7. Kaushik, N.K., et al.: Nitric-oxide enriched plasma-activated water inactivates 229E coronavirus and alters antiviral response genes in human lung host cells. Bioactive Mater. **19**, 569–580 (2023)
8. Yahaya, A.G., Okuyama, T., Kristof, J., Blajan, M.G., Shimizu, K.: Direct and indirect bactericidal effects of cold atmospheric-pressure microplasma and plasma jet. Molecules **26**(9), 2523 (2021)
9. Wu, A.S., et al.: Porcine intact and wounded skin responses to atmospheric nonthermal plasma. J. Surg. Res. **179**, e1–e12 (2013)
10. Choi, J.W., et al.: Novel therapeutic effects of non-thermal atmospheric pressure plasma for muscle regeneration and differentiation. Sci. Rep. **6**, 28829 (2016)
11. Kang, S.U., et al.: N2 non-thermal atmospheric pressure plasma promotes wound healing in vitro and in vivo: potential modulation of adhesion molecules and matrix metalloproteinase-9. Exp. Dermatol. **26**, 163–170 (2017)
12. Kristof, J., Yokoyama, R., Yahaya, A.G., Rimi, S.A., Blajan, M., Shimizu, K.: Absorption of FD-150 into intestinal cells by microplasma. Plasma Med. **12**(4), 11–28 (2023)
13. Vijayarangan, V., Delalande, A., Dozias, S., Pouvesle, J., Pichon, C., Robert, É.: Cold atmospheric plasma parameters investigation for efficient drug delivery in hela cells. IEEE Trans. Radiat. Plasma Med. Sci. **2**, 109–115 (2018)
14. Hirst, A.M., Simms, M.S., Mann, V.M., Maitland, N.J., O'Connell, D., Frame, F.M.: Low-temperature plasma treatment induces DNA damage leading to necrotic cell death in primary prostate epithelial cells. Br. J. Cancer **112**(9), 1536–1545 (2015)
15. Scharf, C., Eymann, P., Emicke, J., Bernhardt, M., Wilhelm, et. al.: Improved Wound Healing of Airway Epithelial Cells Is Mediated by Cold Atmospheric Plasma: A Time Course-Related Proteome Analysis. Oxidative Medicine and Cellular Longevity (2019)
16. Satoh, M., Yoshida, T.: Expression of neural properties in olfactory cytokeratin-positive basal cell line. Dev. Brain Res. **121**, 219–222 (2000)
17. https://cellbank.nibiohn.go.jp/wordpress/wp-content/uploads/jcrb/How-to-start-the-culture-20180601ver.pdf
18. Ikeda, Y., Motomura, H., Kido, Y., Satoh, S., Jinno, M.: Effects of molecular size and chemical factor on plasma gene transfection. J. Appl. Phys. **55**, 07LG06 (2016)
19. Kossyi, I.A., Kostinsky, A.Y., Matveyev, A.A., Silakov, V.P.: Kinetic scheme of the nonequilibrium discharge in nitrogen–oxygen mixtures. Plasma Sources Sci. Technol. **1**(3), 207–220 (1992)
20. Shimizu, K., Ishii, T., Blajan, M.: Emission spectroscopy of pulsed power microplasma for atmospheric pollution control. IEEE Trans. Ind. Appl. **46**(3) (2010)

Measurement, Monitoring, and Modeling

Integration of Localized Surface Plasmon Resonance Sensor on Surface Acoustic Wave Device

Atsuya Kida and Jun Kondoh(✉)

Shizuoka University, Hamamatsu-shi 432-8561, Japan
kondoh.jun@shizuoka.ac.jp

Abstract. When a localized surface plasmon resonance (LSPR) sensor is fabricated on a surface acoustic wave (SAW) propagating surface, transportation of a droplet by the SAW and measurement of it by the LSPR are possible. It is important to know the effect of a SAW on the LSPR sensor. Especially, it is necessary to clarify the effect of acoustic streaming caused by the SAW in a liquid on the LSPR sensor. In this paper, two sample liquids, distilled or ultrapure waters, are placed on the SAW device, and the LSPR sensor response is monitored during SAW excitation. Whereas the LSPR sensor responses are stable for ultrapure water, those for distilled water are variable. Comparison with the calculation results reveals that the reason for this is the cavitation effect caused by acoustic streaming. Moreover, to avoid the cavitation effect, it is found that the applied voltage to the interdigital transducer for exciting SAW must be less than 20 V.

Keywords: Surface acoustic wave · localized surface plasmon resonance · acoustic streaming · bubble · cavitation

1 Introduction

Currently, liquid testing in medical and biotechnological experiments requires mixing liquids in one device to make a new liquid, moving the liquid to another device, and testing the liquid. This process runs a risk of contaminations being mixed with the liquid when moving it from one device to another, or that the liquid may be spilled. To solve this problem, this study integrates a surface acoustic wave (SAW) device, which can mix liquids, and a localized surface plasmon resonance (LSPR) sensor, which can measure the refractive index or permittivity of liquids. The integration of the LSPR sensor on the SAW device will allow for the transportation (see Fig. 1) [1] and mixing of droplets, and measuring of droplets on a single device, eliminating the need to move the liquid and thus solving the problems of impurities in the liquid and spills.

The SAW is generated by applying a high-frequency signal to an interdigital transducer (IDT) which is fabricated on a piezoelectric substrate. The SAW radiates the longitudinal wave into a liquid when the SAW propagates at the interface of the liquid and piezoelectric substrate [2]. In 1989, Shiokawa et al. reported that a droplet can be

Y. Ono and J. Kondoh (Eds.): Inter-Academia 2023, LNNS 939, pp. 177–184, 2024.
https://doi.org/10.1007/978-3-031-54450-7_19

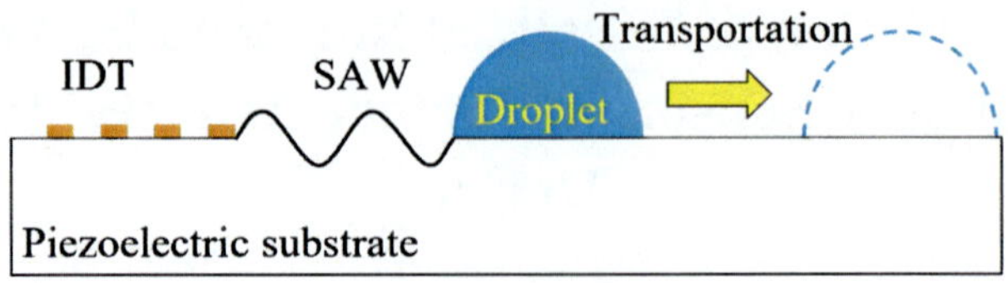

Fig. 1. Schematic illustration of a droplet transportation by SAW.

transported by the SAW [1]. In droplets, nonlinear phenomena, such as acoustic streaming, are caused in a liquid by the radiated longitudinal wave, when the SAW amplitude is increased [3–6].

An LSPR is a localized and emphasized electric field around metal particles [7] when metal nanoparticles are illuminated by an incident light. Silver and gold nanoparticles are usually used. In this study, the permittivity of a liquid is measured using the LSPR sensor, so gold nanoparticles (AuNPs) are chosen. The AuNPs were fabricated using the thermal annealing method [8]. The reflected light intensity from AuNPs was measured using a spectrometer. The spectra depend on the refractive index around AuNPs and the wavelength at which the maximum intensity is obtained. This maximum wavelength is called the peak wavelength.

In our previous research [8], the IDT was damaged during the AuNPs fabrication. To solve this problem, we proposed a fabrication method for AuNPs that combines short-time annealing at high temperature and rapid quenching [9]. In this study, we adopted the proposed method and integrated the LSPR sensor on the SAW device. When the LSPR sensor is integrated on the SAW device, the effect of SAW on the LSPR sensor needs to be investigated (see Fig. 2). Reference [10] concluded that the peak wavelength is affected by the temperature rise of the liquid due to the acoustic streaming. However, there was no reproducibility. We considered that there were other causes besides temperature. Therefore, in this paper, we conducted experiments using distilled water and ultrapure water as samples and investigated the peak wavelength change when SAW was excited. As a result, the reproducibility of the results was low for distilled water, whereas the results were highly reproducible for ultrapure water. It was found that bubbles were generated in distilled water. This bubble generation is the cause of the unstable response.

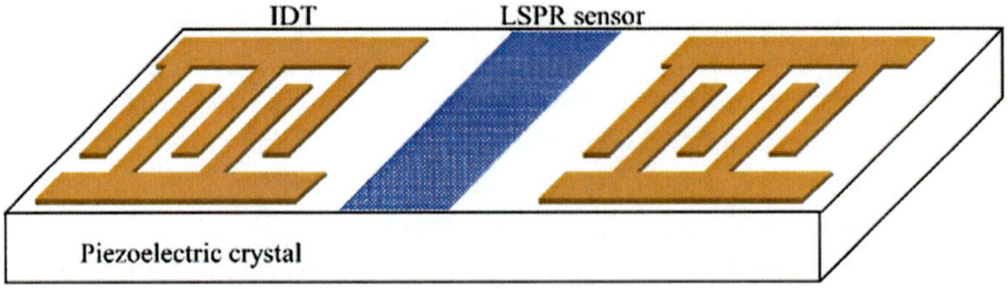

Fig. 2. Illustration of integrated LSPR sensor on SAW device.

2 Fabrication of SAW Device and LSPR Sensor

The transverse type SAW device (see Fig. 2) was fabricated on 128YX-$LiNbO_3$. The center frequency of the SAW device was 50 MHz. Then thin gold film was thermally deposited on the propagation surface between IDTs. During gold evaporation, IDTs were protected by a metal mask. The SAW device with gold thin film was kept in an electric furnace at 400 ℃ for 5 min and then quenched to room temperature. Figure 3 shows the atomic force microscope (AFM) image of the AuNPs fabricated.

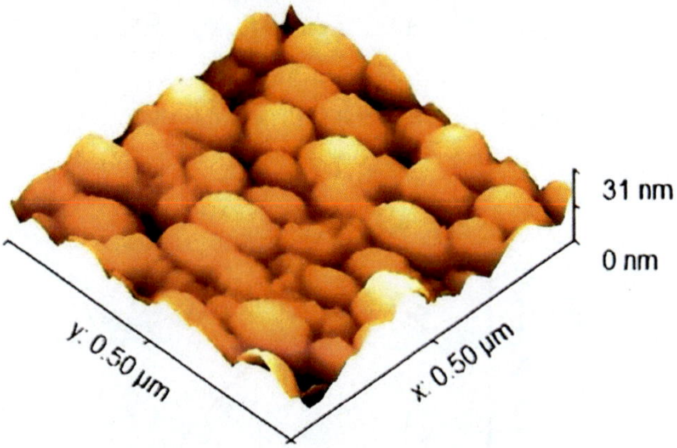

Fig. 3. The AFM image of the AuNPs.

3 Experimental Method

The experimental system is shown in Fig. 4. Silicone rubber was placed on the AuNPs to keep liquid samples. AuNPs were irradiated with white light from a halogen lamp and the reflected light was measured with a spectrometer. Figure 5 shows the illustration of the measured results. The peak wavelength shift was measured in this study. The SAW excitation method is also shown in Fig. 4. The burst signal was created by multiplying the sinusoidal signal from a signal generator and the pulse signal from a function generator by a high-frequency power amplifier and applied to the IDT.

4 Results and Discussion

Figures 6 and 7 show the relationship between excitation time and peak wavelength when SAW was excited. The voltage of the applied signal to the IDT was 70 V. Sample liquid for Fig. 6 is distilled water and that for Fig. 7 is ultrapure water. The measurement was repeated. In the case of Fig. 6 using distilled water, the measured results are not reproducible. The peak wavelengths at time 0 are different. The peak wavelength depends on the particle size and particle spacing of AuNPs. In AuNPs prepared by thermal

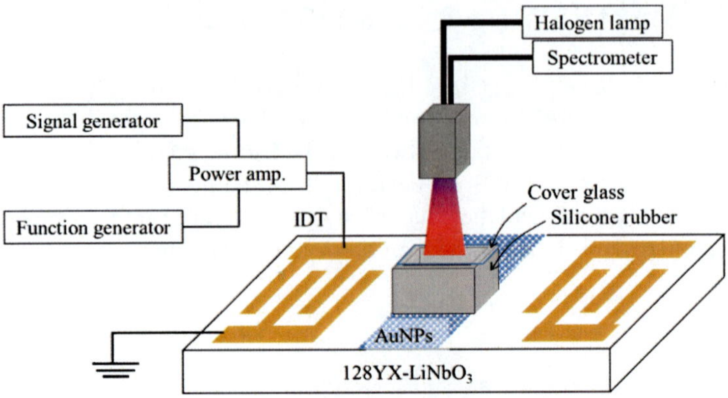

Fig. 4. Experimental system in this study.

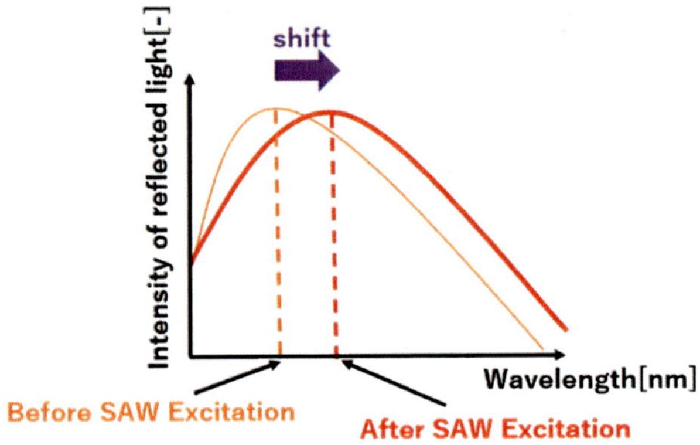

Fig. 5. Schematic illustration of measured spectrum and peak wavelength. The peak wavelength shift due to SAW excitation.

annealing and quenching, the particle size varies and the particle spacing is not constant. Therefore, we consider the difference in the peak wavelength at time 0 in Fig. 6 to be due to the difference in the irradiation position of the incident light. When ultrapure water is used, the measurement results are reproducible. At the same time, there was no difference in the liquid temperature measured with a non-contact thermometer. Observation of the surface before and after the measurement revealed that bubbles were generated in distilled water (Fig. 8).

Here the influence of bubbles is discussed. Microbubbles, which are bubble nuclei, exist in distilled water. The bubble nuclei grew due to the cavitation effect [11] caused by acoustic streaming. The dielectric constants of water and air are 1.77 and 1.0, respectively [12]. Figure 9 shows the peak wavelengths for the calculated dielectric constants. When bubbles exist around the AuNPs, the dielectric constant is smaller than that of water. As the bubbles are moved away from the AuNPs' surroundings due to acoustic streaming (see Fig. 10), the dielectric constant increases and the peak wavelength also increases, as shown in the calculation results. Therefore, it is reasonable to assume that the increase

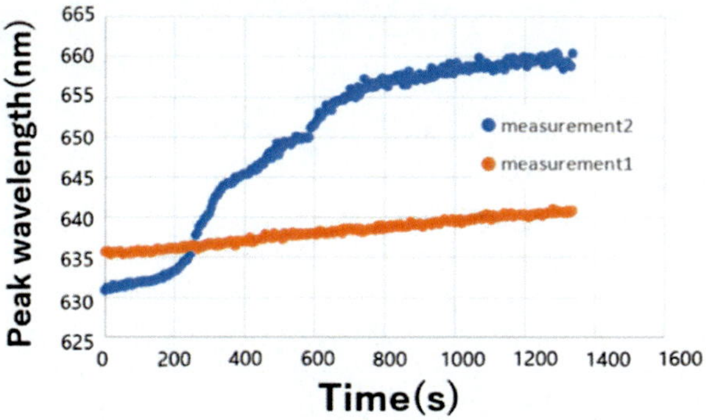

Fig. 6. Peak wavelength change during SAW excitation for distilled water.

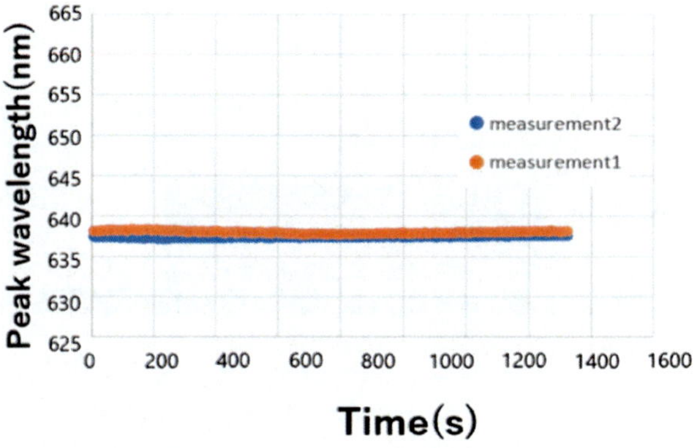

Fig. 7. Peak wavelength change during SAW excitation for ultrapure water.

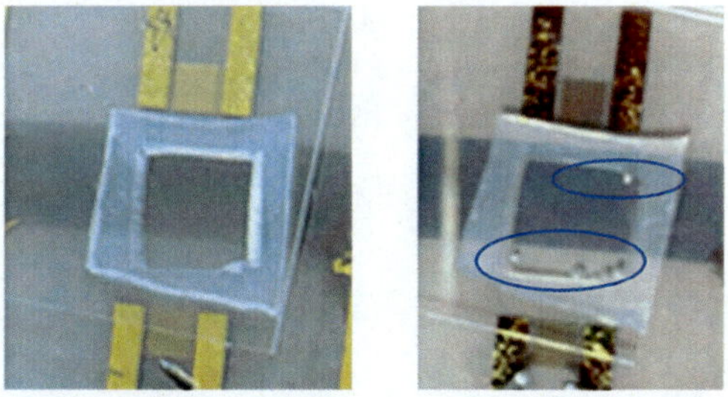

(a) (b)

Fig. 8. Observation results (a) before and (b) after SAW excited for distilled water. There is a bubble generated in the blue ellipse.

in the peak wavelength in Fig. 6 with increasing SAW excitation time is due to the movement of the bubbles from the surroundings of the AuNPs. It is also reasonable that the reproducibility of the measurement results is low because bubble generation is difficult to control.

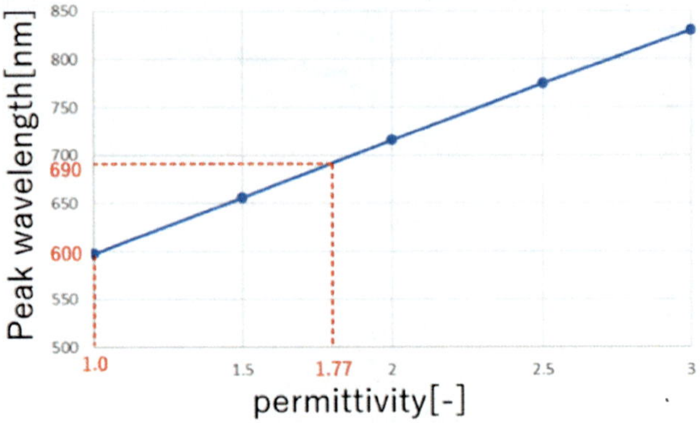

Fig. 9. Calculated peak wavelength as a function of permittivity,

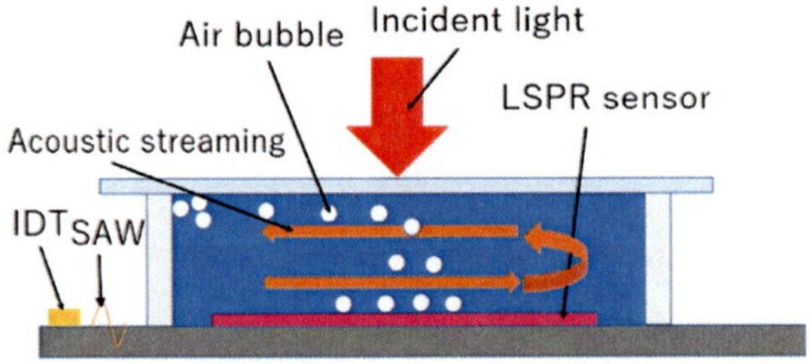

Fig. 10. Image of bubbles being moved by acoustic streaming.

The same measurements were performed with decreasing applied voltage to the IDT. Figure 11 shows the results for distilled water at the applied voltage of 20 V. In Fig. 11(a), the peak wavelength is initially affected by bubbles and then remains constant. On the other hand, in Fig. 11(b), the peak wavelength is constant. In other words, as the applied voltage is reduced to 20 V, the frequency of bubble growth due to acoustic streaming is reduced. Therefore, when measuring with distilled water, the applied voltage to the IDT must be less than 20 V.

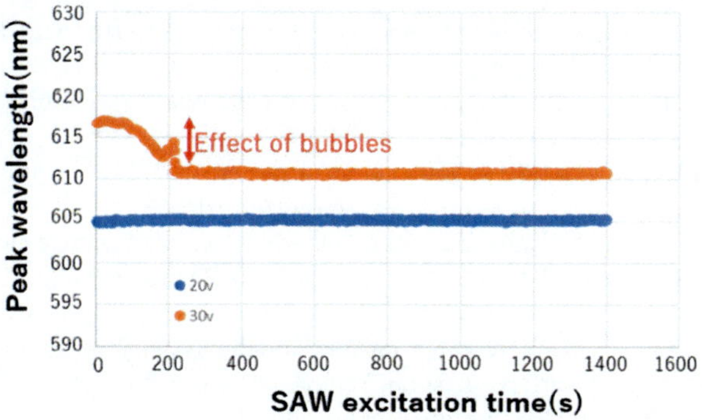

Fig. 11. Measured results peak wavelength change during SAW excitation, when the applied voltage to IDT is 20 V and 30 V.

5 Conclusion

In this study, the effect of SAW on the LSPR sensor was experimentally investigated. Comparison was made for distilled water and ultrapure water as the measurement samples. In the case of ultrapure water, the peak wavelength was not affected by SAW excitation even when the applied voltage to the IDT was 70 V. However, in the case of distilled water, the peak wavelength was not reproducible and increased. In the case of distilled water, it was also found that bubbles were generated after the measurement. Therefore, we calculated the effect of bubbles and found that the measurement results could be explained by the movement of bubbles created by acoustic streaming around the AuNPs. Next, when the applied voltage to the IDT was lowered to 20 V, it was found that the effect of bubbles appeared in the peak wavelength change in some cases and not in others. These results suggest that when distilled water is used as a sample, the voltage applied to the IDT must be less than 20 V or degassed water should be used. In this study, droplets were held in silicone rubber. We will conduct further experiments using droplets without silicone rubber in the future.

References

1. Shiokawa, S., Matsui, Y., Moriizumi, T.: Experimental study on liquid streaming by SAW. Jpn. J. Appl. Phys. **28**(S), 126–128 (1989)
2. Bertoni, H.L., Tamir, T.: Unified theory of Rayleigh-angle phenomena for acoustic beams at liquid-solid interfaces. Appl. Phys. **2**, 157–172 (1973)
3. Chono, K., Shimizu, N., Matsui, Y., Kondoh, J., Shiokawa, S.: Development of novel atomization system based on SAW streaming. Jpn. J. Appl. Phys. **43**, 2987–2991 (2004)
4. Ito, S., Sugimoto, M., Matsui, Y., Kondoh, J.: Study of surface acoustic wave streaming phenomenon based on temperature measurement and observation of streaming in liquids. Jpn. J. Appl. Phys. **46**, 4718–4722 (2007)
5. Kondoh, J., Shimizu, N., Matsui, Y., Sugimoto, M., Shiokawa, S.: Development of temperature-control system for liquid droplet using surface Acoustic wave devices. Sens. Act. A **149**, 292–297 (2009)

6. Kondoh, J.: Nonlinear acoustic phenomena caused by surface acoustic wave and its application to digital microfluidic system. Jpn. J. Appl. Phys. **57**, 07LA01 (2018)
7. Mayergoyz, D.: Plasmon Resonances in Nanoparticles. World Scientific Publishing, Singapore (2013)
8. Sano, H., Kondoh, J.: A study of interaction surface acoustic wave with localized surface plasmon resonance. Jpn. J. Appl. Phys. **58**, SGGA02 (2019)
9. Handoyo, T., Kondoh, J.: Enhancement of ultrathin localized surface plasmon resonance sensitivity using sequential temperature treatment for liquids sensing. Sens. Act. A **307**, 112006 (2020)
10. Kasai, K., Kondoh, J.: Basic research on microfluidic systems integrating surface acoustic wave and localized surface plasmon resonance sensors. In: Proceedings of the 43rd Symposium on Ultrasonic Electronics, 1Pb3-4 (2021)
11. Choi, P.-K. (ed.): Onkyo-Baburu to sonokemisutori (Acoustic-bubble and sonochemistry). Corona Publishing, Tokyo (2012). (in Japanese)
12. Ito, A., Goto, A.: Measurements of refractive index for several liquids and its dependence on temperature. Trans. Jpn. Soc. Mech. Eng. B **60**, 2875–2881 (1994). (in Japanese)

Vacuum Gripper-Based Practical Method of Gentle Deposition of Living Cells and Its Filter Substrates onto SiN Films for Electron Beam Irradiation Experiments

Maciej Kretkowski[1(✉)], Junya Katai[2], Hiroyuki Futamata[1], Wataru Inami[3], and Yoshimasa Kawata[3]

[1] Research Institute of Green Science and Technology, Shizuoka University, Shizuoka 432-8011, Japan
kretkowski.maciej@shizuoka.co.jp
[2] Department of Applied Chemistry and Biochemical Engineering, Graduate School of Engineering, Shizuoka University, Shizuoka 432-8011, Japan
[3] Research Institute of Electronics, Shizuoka University, Shizuoka 432-8011, Japan

Abstract. Electron beam irradiation of living cells requires very careful preparation of samples which are typically placed onto SiN thin films that form the barrier between the vacuum of the electron gun and the atmosphere. This paper presents the development of a specialized vacuum gripper and a method of gentle deposition of living cells onto a silicon nitride film. The presented method allows for safe and repetitive preparation of samples mitigating the difficulties resulting from the fragile nature of thin SiN films which are prone to destruction if in direct contact with mechanical tools such as tweezers.

Keywords: silicon nitride film · electron beam irradiation · vacuum gripper

1 Introduction

Silicon nitride (SiN) films are widely used in sample imaging using atmospheric scanning electron microscopy (ASEM) [1]. The films are strong enough to form a barrier between the vacuum of the electron gun column and the atmospheric pressure. Moreover, the film is transmissible to an electron beam EB [2], hence the SiN films can be used for high resolution imaging, irradiation and testing of the effects of direct electron beam irradiation of living specimens [3]. The films (also called "windows" or "membranes") are manufactured in different sizes and thicknesses to accommodate variety of uses and sizes of the target samples. They are very fragile however, and any contact with another hard object such as tweezers can shatter it, therefore a careful handling is a necessity [4]. In the case of wet samples, the drying process may also cause the film to break. Our research involves precise preparation of cell samples on 5 × 5 mm 200 nm thick SiN film. For electron beam irradiation purposes, it is important that a uniform layer of cells (preferably thickness of a single cell) is spread over the film's area. This is

Y. Ono and J. Kondoh (Eds.): Inter-Academia 2023, LNNS 939, pp. 185–193, 2024.
https://doi.org/10.1007/978-3-031-54450-7_20

accomplished by using a 4 × 4 mm PTFE (polytetrafluoroethylene) based filter strip on top of the sample. It is very important to geometrically align the filter with the SiN film to prepare comparable batches of samples. Figure 1a) shows the principle of a prepared specimen. To place the cells on the film, typically, the cells are deposited on the 4 × 4 filter (4 μL sample) first and then transferred onto the SiN film using tweezers and placed gently on the surface of the film with the cells facing the film as shown on Fig. 1b. This process is very difficult to achieve manually and requires significant skill. It also has to be performed in a timely manner as the drying process can cause deformation to the SiN film [5]. Once the filter is deposed on the surface of the film, it has to be aligned with the film (all of it has to be within the area of the film). It is required for the EB to irradiate the whole sample uniformly. The alignment is particularly difficult when using tweezers, as they can very easily [6] damage the film resulting in a wasted sample.

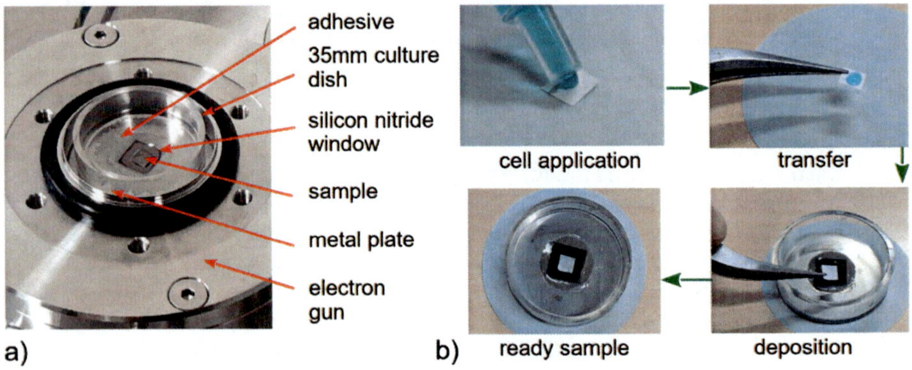

Fig. 1. a) Principle of a sample prepared for electron beam irradiation. b) Typical transfer method onto the SiN film.

Prepared samples are irradiated in a custom made EB irradiation system supplied by APCO Ltd. The system accepts standard 35 mm culture dishes and is composed of a miniature inverted electron gun with dedicated controller and vacuum pump system as shown on Fig. 2. The system operates by placing the sample in a receptacle, automatically evacuating the electron gun column to high vacuum and the irradiation with an EB of approx. 5 mm in diameter is performed automatically using preset time, voltage and current.

Samples that have been irradiated are incubated and the outcome of the irradiation procedure can be tested, compared and verified further in the research process. Difficulties and skills required for sample preparation can create a bottleneck in the research process. As an alternative to using tweezers as the main tool of holding the wet sample, a vacuum operated device similar to those used in semiconductor industry could be used [7], however there are many challenges related with bio-contamination and efficient sample release. Non-rigid objects such as the presented PTFE filters are difficult to hold and release reliably. Some research has been proposed regarding usage of capillary forces for holding small components [8], however in case of bio-active samples, safety

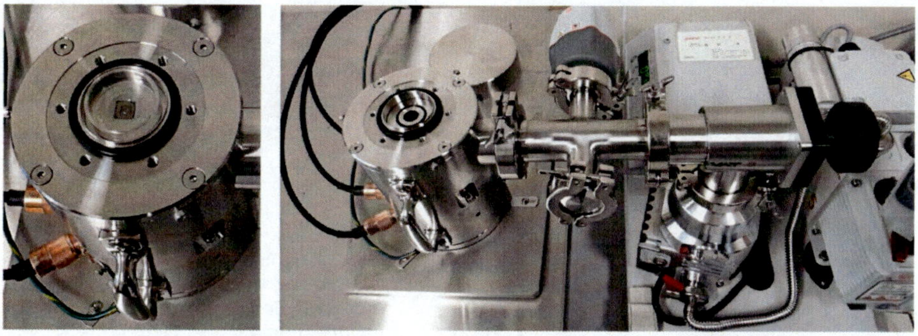

Fig. 2. Experimental electron beam irradiation system (APCO Ltd.)

measures have to be very strict, rendering the common gripping solutions [9] inadequate in our research. Therefore, the procedure for the cell deposition on a SiN film has been redesigned and made more efficient and user-friendly by introducing a specially developed system and transfer method presented in this paper.

2 Proposed Method

We propose a sample preparation procedure, where the bacteria are placed directly onto the center of the SiN film in a 4 μL liquid solution using a micro-pipette. The pre-cut 4×4 mm filter is then placed in a specially designed vacuum gripper nozzle and is lowered towards the deposited liquid using a guide jig. When the filter is close to the liquid, centered and its edges oriented parallel to the SiN film, the filter is released from the nozzle by a valve system that switches between vacuum and a small positive air pressure. The positive pressure creates a gentle soft release mechanism that eliminates a need for any mechanical manipulation of the filter substrate as shown in Fig. 3.

The nature of research with biological specimens dictates serious approach and safety measures for systems used in handling and bio-contamination prevention. Our method and experimental system address the safety concerns by introducing multi-stage contamination prevention design and material choice that takes advantage of antimicrobial and sterilizing properties of copper alloys [10].

3 Design of the Experimental System

3.1 Outline of the Vacuum/Release Pressure System

In principle, the proposed method is based on a vacuum suction cup in which the air flow can be reversed, effectively replacing the vacuum with positive pressure with regulated flow. The vacuum gripper system for the 4×4 mm PTFE-based filter is composed of a gripper nozzle, diaphragm electric pump and pneumatic valve system. Air flow reversal is accomplished using two solenoid 3-way valves controlled via electric foot switch. The system is composed of standard pneumatic components thus ensuring interchangeability and serviceability as well as low cost.

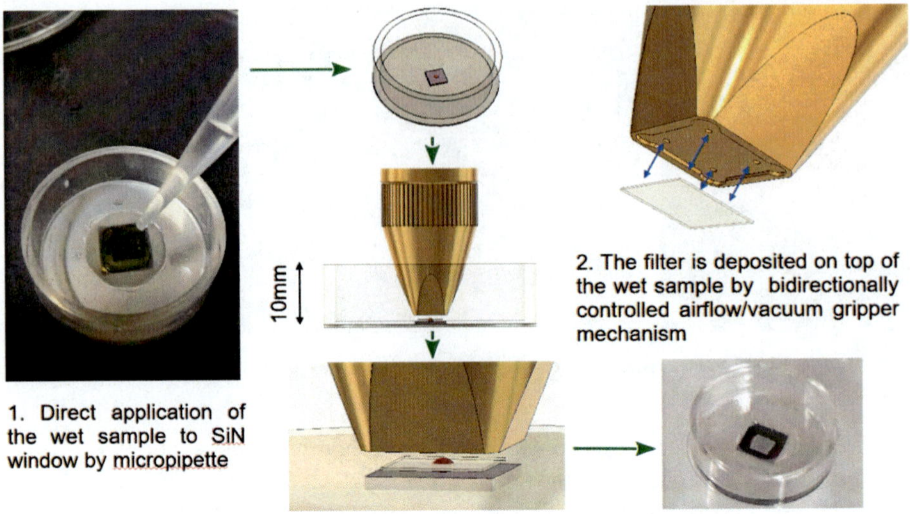

Fig. 3. Proposed method. At first the sample is transferred directly onto the SiN film using a micropipette. The filter is subsequently deposited on top of the liquid by using the vacuum gripper tool

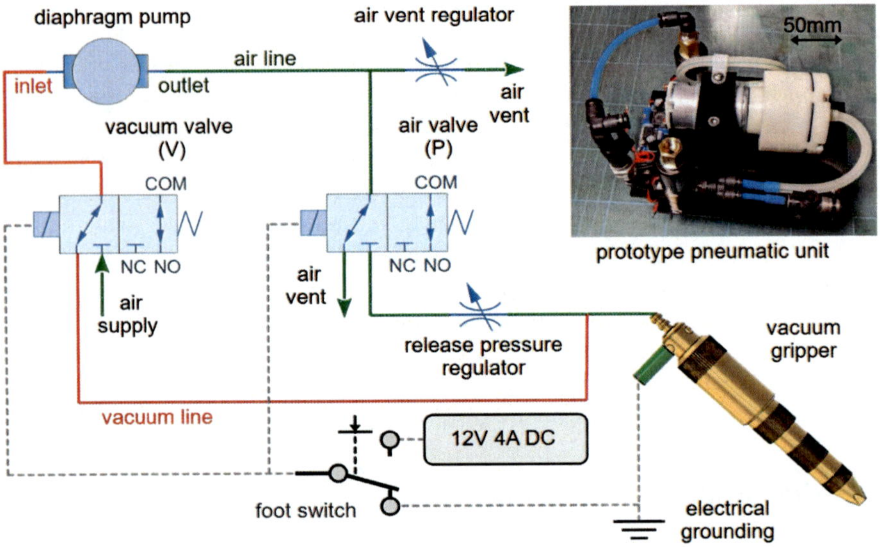

Fig. 4. Outline of the air flow reversal and pressure control pneumatic system

Figure 4 shows the pneumatic outline of the system. During gripping phase (foot switch is un-pressed), the valve V is directing the airflow from the vacuum gripper directly to the vacuum side of the pump. Valve P during this phase is directing the outlet of the pump directly to the atmosphere. This creates vacuum inside the gripper which allows to hold on to the manipulated object.

Release of the object is done by pressing the foot switch which reverses the polarization of the valves' solenoid coils. In the release phase, valve V is allowing air to flow into the vacuum side of the pump, while the output of the pump is directed to the gripper via valve P. In the release configuration, the pump pressure is divided into two regulated streams creating a simple yet effective pressure regulator that allows only a fraction of the pressurized air to enter the vacuum gripper. By adjusting the flow valves in both streams, an optimal setting can be achieved for gentle release of the gripped object. The gripper connects to the pump assembly by a silicone tube.

3.2 Design of the Vacuum Gripper

Primary function of the gripper is to hold on to a 4 × 4 mm PTFE-based filter composed of 0.45 μm pores and partially permeable to air and moisture. In principle, the filter is deposited onto the wet sample by releasing it on top of the liquid via an air cushion, soft-release mechanism induced by the pneumatic system. However, due to the porous structure of the filter and possibility of misuse error, there is a potential risk of some of the liquid being ingested inside the vacuum gripper and contaminating the pneumatic system. Therefore a multistage protection system is incorporated in the design to prevent biohazard to the operator and the environment.

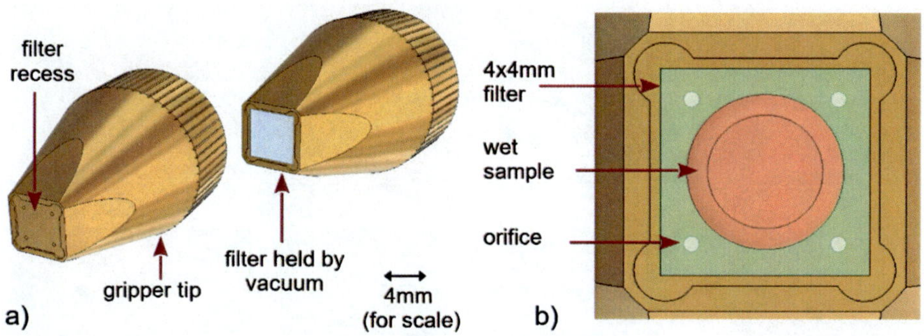

Fig. 5. a) Design of the tip of the gripper. b) Size relation of the filter, liquid sample and the tip.

Figure 5a) shows the tip of the vacuum nozzle. The design is based on a recess (4.4 × 4.4 mm, 0.2 mm deep) with four orifices 0.3 mm in diameter placed in the corners of it. The filter fits inside the recess and is vacuum held in the corners by the orifices acting as rigid suction cups. The diameter of the sample deposited onto the SiN film by a micropipette is typically around 2–3 mm in diameter, therefore the diagonal distance between the orifices has been set to 4 mm. This is the first stage of protection against accidental liquid ingestion as there is no direct path for a standard 4 μL liquid sample to enter an orifice as shown in Fig. 5b.

The vacuum gripper is constructed from several cylindrical modules held together by threaded joints. It is easy to disassemble, sterilize and service. The cross section of the device is shown on Fig. 6. As described above, the tip is the first stage of protection from bio-contamination, however if a liquid would pass through the tip, it would be contained

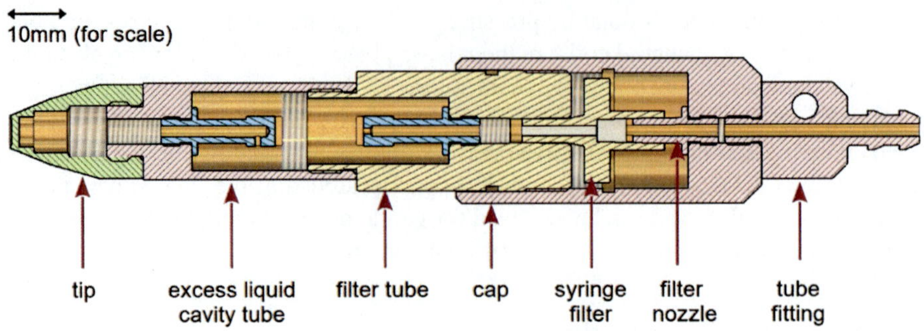

Fig. 6. Cross section of the vacuum gripper.

by the excess liquid cavity that is designed as integral part of the gripper. The lower nozzle has a port at 90 degrees to the central axis of the gripper. In conjunction with the upper nozzle which has its port in line with the axis, it forms an effective liquid trap. The idea is shown on Fig. 7a. The excess liquid volume possible to be safely contained is 0.4 mL which is 100× amount of a standard sample (4 μL). The cavity also protects the reverse-contamination of a sample when it is being released by a positive air flow.

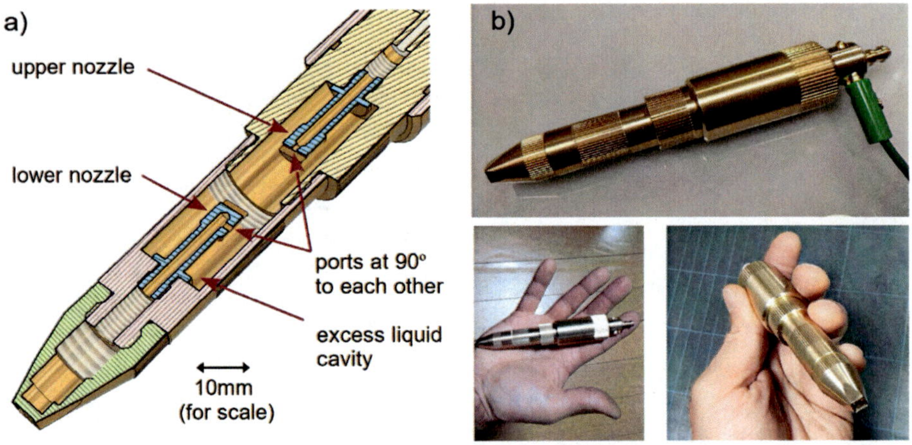

Fig. 7. a) Excess liquid cavity protection. b) Prototype of the vacuum gripper.

To increase the safety of operation even further, the gripper also contains a standard syringe filter. The syringe filter's pore size of 0.2 μm provides complete filtration against most known bacteria. The all-metal construction from C3604 brass ensures robust design and provides additional protection due to antimicrobial properties of copper alloys. The gripper's diameter is φ24 mm (max), φ16 (min) and 149 mm in length. Figure 7b shows the prototype of the device.

3.3 Design of the Alignment System

Key feature of the presented vacuum gripper system is the gentle release of the filter on top of the wet sample. In order to provide repeatable results and precise deposition of the filter with regard to height and alignment with the silicon nitride film, an alignment jig has been introduced as shown on Fig. 8. The system is composed of a grounding plate, jig body and an adjuster ring. The 35 mm dish is securely placed in a recess in the grounding plate and is consequently covered with the jig body. The body has viewing ports on both sides to allow for visual alignment of the tip of the gripper with the SiN film. The vacuum gripper is inserted into the adjuster ring. It has a step designed to limit the insertion depth into the adjuster ring and by rotating the adjuster, the gripper's tip can be carefully lowered towards the sample maintaining the alignment between the tip and the SiN film. To facilitate as good visual aid as possible, the outer shape of the tip is rectangular in shape.

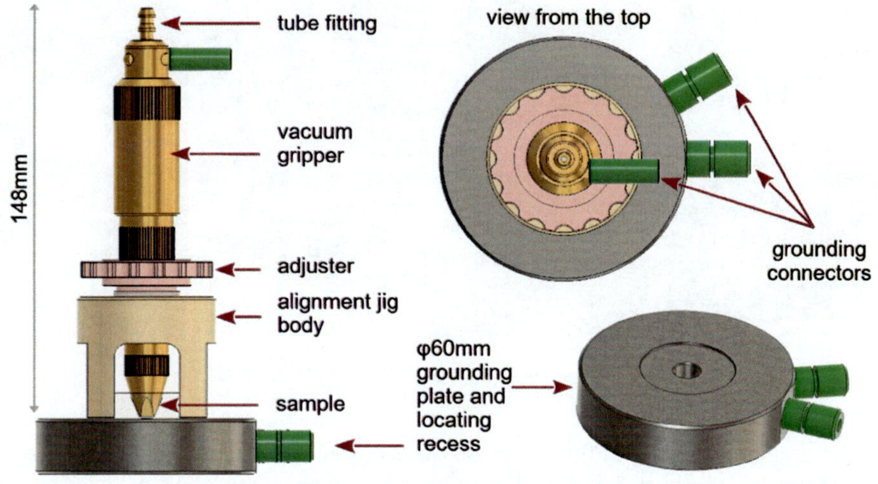

Fig. 8. Alignment system

The grounding plate is connected to the gripper as well as to the control box and consequently to the grounding of the electrical system's earth. It is very important to avoid static electricity buildup on the deposited filter, especially in dry environment. Failing to do so affects the release of the filter from the vacuum cleaner because static electricity charge will cause adhesion of the filter to any ungrounded surfaces.

4 Results and Conclusion

Prior to the development of the vacuum gripper-based method, the number of samples that were damaged during preparation was estimated at 10–20% (depending on the skill level of the researcher). The main cause for damage was mechanical contact of the SiN film with a tool used to deposit and align the samples onto the surface. The damage is always critical in such a way that the film is detached from its frame and the whole section of it, where the specimen would be placed is destroyed. Identification of damaged samples is therefore performed by a visual inspection. Prior to deposition attempts, the culture dish assembly is tested for vacuum leaks, therefore with 5 × 5 mm size of the film used in the presented research methodology, visual inspection is sufficient. The preparation of the whole sample consists of bonding the culture dish with a backing metal plate and the silicon nitride film using a low outgassing adhesive. After curing of the adhesive, the dish assembly is tested for vacuum leaks. This process is in itself time consuming, therefore a subsequent, failed attempt at deposition of a wet sample and breakage of the film is discouraging to the researchers and the costs can become significant.

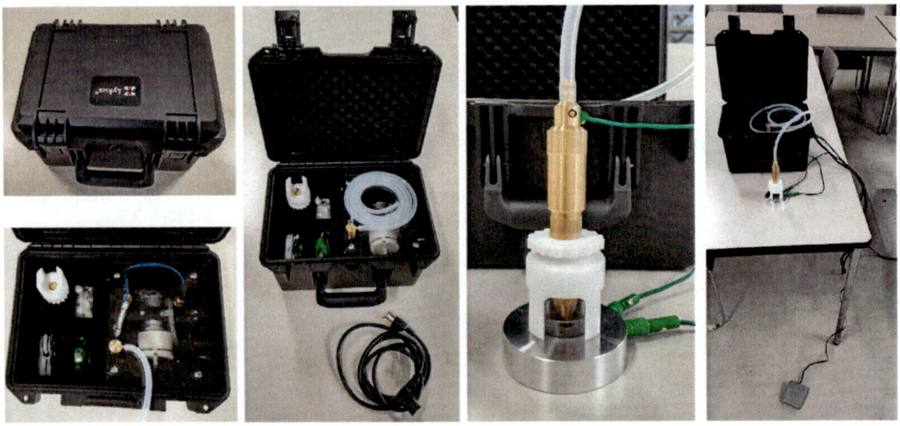

Fig. 9. Accomplished vacuum-gripper system

The system described in this paper (as shown on Fig. 9) has greatly improved the repeatability of sample preparation. The process has become less labor-intensive and the risk of damaging the SiN film has been significantly reduced. Effectively, the breakage of the film due to mechanical trauma has been eliminated. Over the time span of several months between accomplishing of the prototype system and the time of writing this paper, no damaged samples have been reported (with over 100 samples prepared). Moreover, our method requires less training for preparation of the specimens, therefore the research can be conducted more efficiently. The scalable and portable nature of the design of the vacuum gripper system can also be beneficial to other applications, where soft, flexible and very light objects have to be manipulated with precision and low risk of biochemical contamination. The multi stage protection provided in a compact form design can be applied to a variety of instruments. Further development of this technology is aimed at simplifying workflows and making the sample preparation process easy and trouble free for the researchers.

Acknowledgments. This work was supported by CORE RESEARCH FOR EVOLUTIONARY SCIENCE AND TECHNOLOGY (JPMJCR2003) of the Japan Science and Technology (JST) Agency.

References

1. Ring, E.A., Peckys, D.B., Dukes, M.J., Baudoin, J.P., de Jonge, N.: Silicon nitride windows for electron microscopy of whole cells. J. Microsc. **243**(3), 273–283 (2011)
2. Inami, W., Nakajima, K., Miyakawa, A., Kawata, Y.: Electron beam excitation assisted optical microscope with ultra-high resolution. Opt. Express **18**(12), 12897–12902 (2010)
3. Inami, W., Horiba, D., Kawata, Y.: Cell stimulation by focused electron beam of atmospheric SEM. Ultramicroscopy **206**, 112823 (2019)
4. Merle, B., Göken, M.: Fracture toughness of silicon nitride thin films of different thicknesses as measured by bulge tests. Acta Materialia **59**(4), 1772–1779 (2011). ISSN 1359-6454
5. Yang, J., Paul, O.: Fracture properties of LPCVD silicon nitride thin films from the load–deflection of long membranes. Sens. Actuators A: Phys. **97–98**, 520–526 (2002). ISSN 0924-4247
6. Shamray, S., Azarhoushang, B., Paknejad, M., Buechler, A.: Ductile-brittle transition mechanisms in micro-grinding of silicon nitride. Ceramics Int. **48**(23), Part A, 34987–34998 (2022). ISSN 0272-8842
7. Fontana, G., Iacono, N., Negri, S.P., Papadia, G.: Design of a multi-mode hybrid micro-gripper for surface mount technology component assembly. Micromachines **14**(7), 1464 (2023). https://doi.org/10.3390/mi14071464
8. Fantoni, G., Hansen, H.N., Santochi, M.: A new capillary gripper for mini and micro parts. CIRP Ann. **62**(1), 17–20 (2013). https://doi.org/10.1016/j.cirp.2013.03.005. ISSN 0007-8506
9. Cecil, J., Vasquez, D., Powell, D.: A review of gripping and manipulation techniques for micro-assembly applications. Int. J. Prod. Res. **43**(4), 819–828. https://doi.org/10.1080/00207540512331311813
10. Dick, R.J., Wray, J.A., Johnston, H.N.: A Literature and Technology Search on the Bacteriostatic and Sanitizing Properties of Copper and Copper Alloy Surfaces. Battelle Columbus Laboratories; Columbus, OH, USA: 1973. Phase 1 Final Report, INCRA Project No. 212

Check for updates

Effects of Magnetic Field Configuration on the Performance of 100 W-Class Hall Thruster Using Carbon Dioxide

Seiya Tsuchikawa[1](✉), Hiroki Watanabe[2], Shinatora Cho[2], Yasushi Ohkawa[2], and Makoto Matsui[3]

[1] Department of Engineering, Shizuoka University, Hamamatsu, Japan
tsuchikawa.seiya.19@shizuoka.ac.jp
[2] Research and Development Directorate, JAXA, Chofu, Japan
{watanabe.hiroki,choh.shinatora,okawa.yasushi}@jaxa.jp
[3] Department of Mechanical Engineering, Shizuoka University, Hamamatsu, Japan
matsui.makoto@shizuoka.ac.jp

Abstract. We focused on Carbon dioxide (CO_2) as an alternative propellant for Hall thrusters which is inexpensive and advantageous for Mars exploration missions (ISRU: In-Situ Resource Utilization). In our previous study, a significant reduction in thrust efficiency compared to Xe was confirmed. From the result, it is suggested that CO_2 is not sufficiently ionized, and ions could not be accelerated enough because Hall thruster used in the study was designed for Xe and had a smaller plasma generation region. Therefore, in this study, the yoke, which was originally installed to shield the magnetic field lines was removed. The plasma generation region was expanded by changing the distribution of the magnetic field to improve previous results. However, the thrust was about 30% lower than before the change, and the same was true for the thrust, current utilization, and mass utilization efficiencies. This was considered to be caused by a decrease in the number density of ions due to the expansion of the plasma generation region.

Keywords: Electric propulsion · Hall thruster · Alternative propellant · Carbon dioxide

1 Introduction

Mainly in Europe and the United States the use of electric propulsion for spacecrafts has been expanding for the purpose of commercial communication satellites and deep space exploration since the 1990s. Besides, all-electric propulsion satellites which do not equip any chemical propulsion system to generate thrust by chemical reaction energy have been focused on in recent years [1]. Particularly, Hall thrusters have been increasingly used due to their simpler structure, higher thrust density than ion engines and specific impulse. Not only thrusters of more than 1 kW but also Hall thrusters of several hundred watts are also focused on as the main thrust for nano-satellites [2, 3]. However, Hall thrusters with a nominal discharge power of less than 100 W have been found to have lower

Y. Ono and J. Kondoh (Eds.): Inter-Academia 2023, LNNS 939, pp. 194–201, 2024.
https://doi.org/10.1007/978-3-031-54450-7_21

anode efficiency [4–6]. Hall thrusters are miniaturized as the discharge power decreases to maintain the propellant flow density in the channel. This increases the surface-to-volume ratio and leads to high energy losses to the channel wall. Hence, it is necessary to evaluate the distribution of the magnetic field for low-power Hall thrusters that operate stably and efficiently.

Hall thrusters operate almost exclusively on the noble gas Xe. This is because it has the advantages of high relative density, low ionization cost, and non-toxicity. On the other hand, Xe exists in the earth's atmosphere only in small quantities, and its price has soared due to expanding its use in the space engineering, the medical industry, and semiconductor industry [7].

From this background, research, and development of alternative propellants to Xe and design optimization of Hall thrusters have been conducted. Although typical alternative propellants are noble gases, solid propellants such as bismuth and iodine are also being considered because of their high density in recent years [8–10]. However, these have not yet been put to practical use since these propellants require heaters for gas supply systems. Therefore, this study focused on CO_2 (dry ice) as an inexpensive propellant. There are several benefits of CO_2. First, its relative density is almost the same as that of Xe nevertheless, its price is about 1/5,000 of Xe. Second, by storing dry ice at the triple point, low-pressure storage at 0.52 MPa is possible, which allows for lighter tanks than in the case of Xe. Furthermore, CO_2 can be procured on the destination site for a Mars exploration mission (ISRU) [11]. However, the ionization energy of CO_2 is higher than that of Xe.

Only thrust measurements have been performed using CO_2 [12]. In our previous study, it was found that both thrust, and anode efficiency were significantly lower when the thruster was operated with CO_2 compared to Xe. It was suggested that the reason for this is that the ion beam current is low due to the low ionization of CO_2. Therefore, in this study, the plasma generation region was expanded by changing the distribution of the magnetic field to investigate the thrust performance.

2 Experimental Procedure

Figure 1 shows a schematic of the experimental setup. A 100 W class magnetic Hall thruster was used. The discharge channel and anode are composed of boron nitride and molybdenum, respectively. The mean diameter of the discharge channel was 20 mm, the channel width was 6 mm, and the distance from the anode downstream surface to the channel exit plane was 8 mm. A plasma bridge cathode with tungsten wire was externally mounted.

Figure 2 shows a magnetic field profile. The magnetic field was applied only near the channel exit because the thruster used in this study was designed for using Xe. Thus, the two points below have been changed so that a magnetic field can also be applied to the region around the anode. The experimental conditions are shown in Table 1.

1. The iron inner screen yoke was removed to allow the magnetic field lines to penetrate.
2. The material of the outer screen yoke was changed to Al with low specific permeability because the coil cannot be wrapped if it is removed.

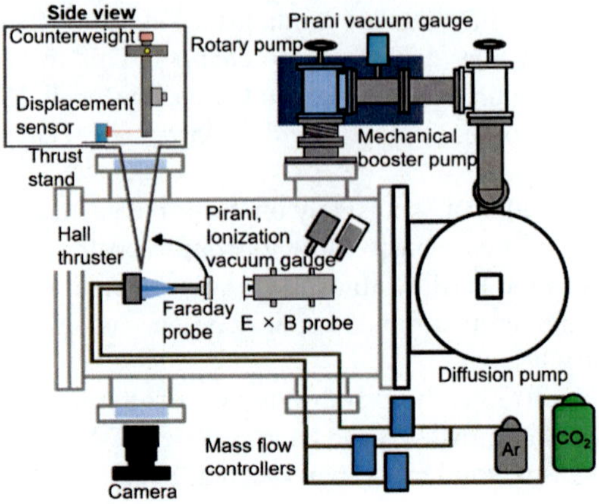

Fig. 1. Top view schematic of the experimental apparatus.

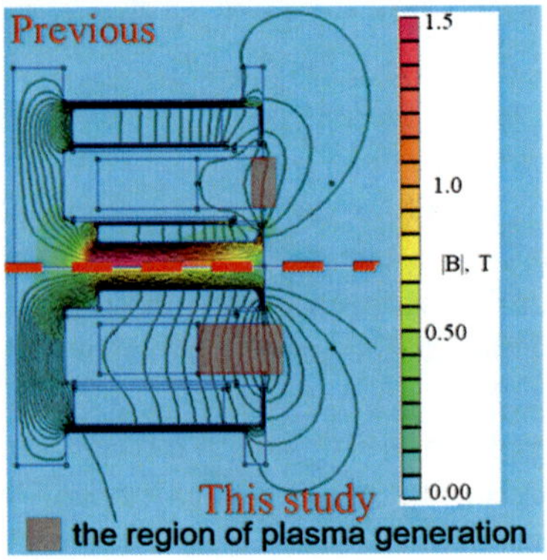

Fig. 2. Magnetic field profile.

Table 1. Experimental conditions.

Anode propellant	CO_2
Anode mass flow rate	15–50 sccm
Cathode propellant	Ar
Cathode mass flow rate	2.0 sccm

(*continued*)

Table 1. (*continued*)

Discharge voltage	150–200 V
Magnetic flux density	17 mT
Faraday probe voltage	−30 V
Radial measurement distance	0.3 m
E × B probe voltage	10–85 V
Distance from thruster exit to E × B probe	0.45 m
Background pressure (CO_2 equivalent)	1.35×10^{-2} Pa

3 Results and Discussion

3.1 Discharge Characteristics

First, we investigated the thrust by changing the distribution of the magnetic field using a pendulum-type thrust stand. Figures 3 and 4 show the experimental results and the anode efficiency calculated from them, respectively. The anode efficiency η_A was calculated using Eq. 1.

$$\eta_A = \frac{T^2}{2\dot{m}_A P_{in}} \tag{1}$$

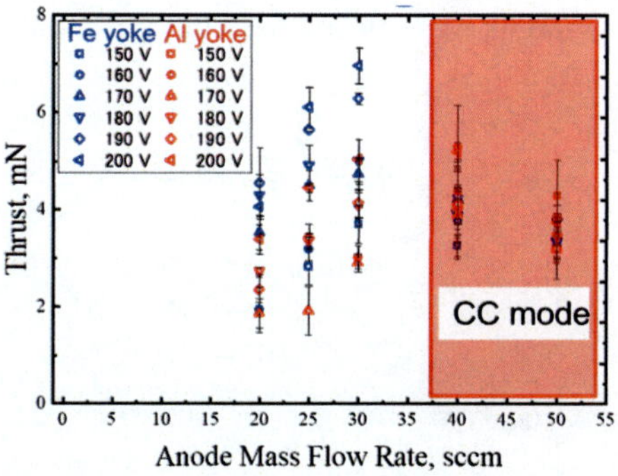

Fig. 3. Thrust as a function of $\dot{m}_A$.

Where, T is the thrust, $\dot{m}_A$ is the anode mass flow rate, and P_{in} is the input power. In this experiment, to avoid over-heating of the thruster, a current limit was applied to the regulated DC power supply to prevent operation above 2 A and set to shift from Constant Voltage (CV) mode to Constant Current (CC) mode. Figure 3 indicates that in both cases, the thrust increases with increasing discharge voltage and anode mass flow rate. The maximum thrust under these conditions was 7.0 mN for Fe and 5.2 mN for Al. This indicates that changing the distribution of the magnetic field reduces performance

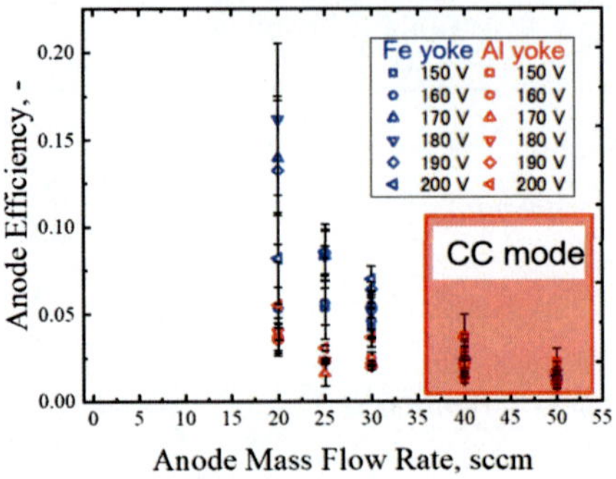

Fig. 4. Anode efficiency as a function of $\dot{m}_A$.

by about 30%. Figure 4 shows that the anode efficiency is maximum at an anode mass flow rate of 20 sccm in both cases, but the efficiency increases as the anode mass flow rate decreases. In this study, the point of best efficiency could not be found because no plasma was generated at an anode mass flow rate of 15 sccm in either case. The cause of the reduced thrust performance is discussed in the section on plume characteristics.

3.2 Plume Characteristics

Figure 5 shows the plume appearance operated with CO_2. Figure 5 shows that plasma is also generated downstream of the channel outlet in the case of using the Al yoke, indicating that the plasma generation region is expanded. Figure 6 shows the current utilization efficiency η_I and mass utilization efficiency η_u versus anode mass flow rate and discharge voltage. η_I and η_u were calculated from Eq. 2 and 3, respectively. The ion beam current is obtained by Eq. 4.

$$\eta_u = \frac{I_{beam}}{\overline{I_d}} \tag{2}$$

$$\eta_I = \frac{m_i I_{beam}}{e\dot{m}_A} \sum \frac{\Omega_i}{Z_i} \tag{3}$$

$$I_{beam} = 2\pi R^2 \int_0^{\frac{\pi}{2}} j(\theta)\sin(\theta)\mathrm{d}\theta \tag{4}$$

where, $\overline{I_d}$ is the average discharge current, m_i is the mass of an ion, e is the elementary charge, Ω_i is the current fractions of the ith ion species, Z_i is the charge-state of the ith ion species, R is the radial distance from thruster centerline, $j(\theta)$ is the angular distribution of the ion current density, and θ is the probe angle.

Figure 6 shows that Fe is superior to Al in almost all conditions. Both η_c and η_m are indicators of whether the propellant is ionized, and the poorer these indicators are, the smaller the discharge current is. The cause of this is considered to be related to the number density of the plasma. In this study, the region of plasma generation was expanded by changing the distribution of the magnetic field. However, the number density of plasma decreases because the discharge voltage is kept in the same condition. Moreover, the acceleration region of the ions is only at the position where the steep potential drop occurs, which depends on the position where the magnetic flux density is at its maximum. This is considered to have reduced the number of ions being accelerated.

Next, Fig. 7. Shows the typical E × B probe trace. The values shown in Fig. 7 represent the percentage of each ion species calculated from the area ratio by Gaussian fitting of the original plot. Figure 7 shows that the fraction of ions excluding O^+ increases as the distribution of the magnetic field is changed. It was found that changing to an Al yoke was superior because CO_2^+, which has the largest molecular weight, contributes most to the thrust has a large abundance ratio. However, the results of this study showed that the thrust performance was better in the case of the Fe yoke. As a result of these reasons, it was suggested that the weak ion beam is the most significant factor.

Therefore, thrust performance may be improved by increasing the discharge voltage and making the power input equivalent to that of Fe. It is also necessary to measure the number density of the plasma using a Langmuir probe.

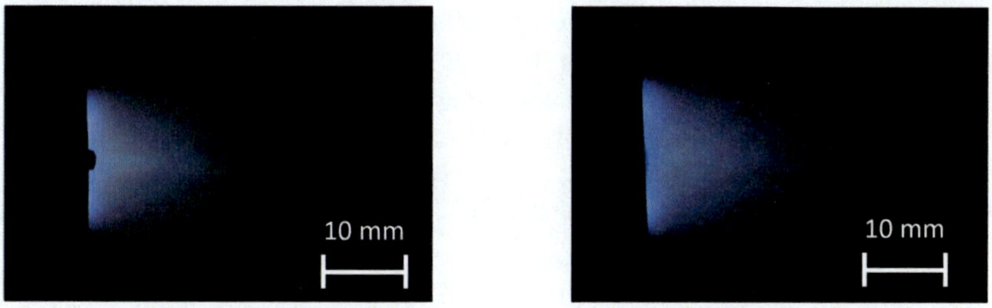

(a) Previous study (Fe yoke) (b) This study (Al yoke)

Fig. 5. Plume appearances operated with CO_2.

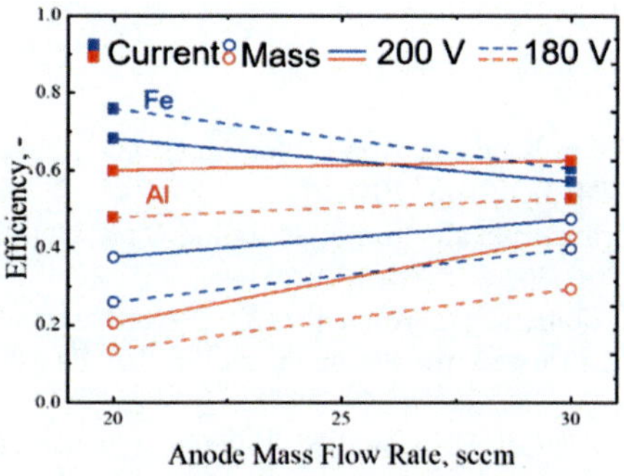

Fig. 6. Efficiency as a function of $\dot{m}_A$.

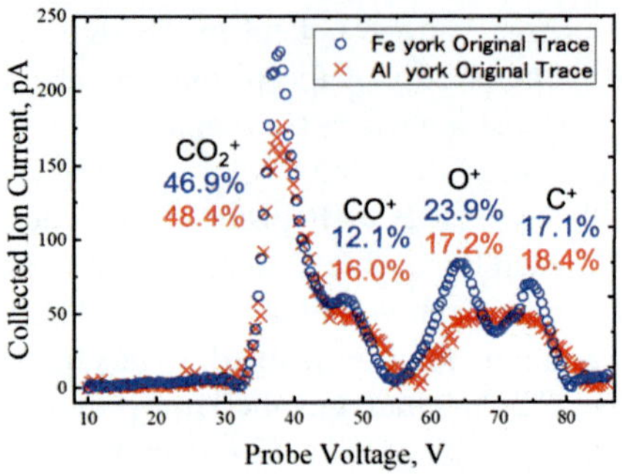

Fig. 7. Typical E × B probe trace: 200 V, 20 sccm.

4 Conclusions

The conclusions obtained from this study can be summarized as follows.

1. The maximum thrust of the Fe yoke was 7.0 mN at 30 sccm and 200 V. In the case of the Al yoke, it was 5.2 mN under the same conditions.
2. Both the current utilization efficiency and mass utilization efficiency of the Al yoke were inferior to those of the Fe yoke. The reason for this may be that the number density of ions decreased due to the expanded plasma generation region.
3. Ion species measurements showed that the Al yoke produced a larger fraction of heavier ions. This result confirmed that the cause of the performance degradation is the low discharge current, and the weak ion beam due to the constant discharge voltage condition set in the experiment.
4. It was suggested that the effect of the number density of plasma should be investigated by the Langmuir probe to find the cause of the performance degradation.

References

1. Nishiyama, K., Kinefuchi, K.: Current and future missions using electric rocket propulsion. J. Plasma Fusion Res. **94**(2), 60–65 (2018)
2. Szabo, J., et al.: Performance evaluation of an iodine-vapor Hall thruster. J. Propuls. Power **28**, 848–857 (2012)
3. Ikeda, T., Togawa, K., Tahara, H., Watanabe, Y.: Performance characteristics of very low power cylindrical Hall thrusters for the nano-satellite PROITERES-3. Vacuum **88**, 63–69 (2013)
4. Szabo, J., Tedrake, R., Metivier, E., Paintal, S., Taillefer, Z.: Characterization of a one hundred watt, long lifetime hall effect thruster for small spacecraft. In: 53rd AIAA/SAE/ASEE Joint Propulsion Conference (2017)
5. Mazouffre, S., Grimaud, L.: Characteristics and performances of a 100-W Hall thruster for microspacecarft. IEEE Trans. Plasma Sci. **46**, 330–337 (2018)
6. Watanabe, H., Cho, S., Kinefuchi, K.: Performance and plume characteristics of an 85 W class Hall thruster. Acta Astronaut. **166**, 227–237 (2020)

7. Cho, S., Watanabe, H., Kinefuchi, K.: Discharge characteristics of a small hall thruster using CO_2 propellant. In: Proceedings of the Space Sciences and Technology Conference, p. 2K08 (2022)
8. Yamasaki, J., Nonaka, M., Yokota, S., Shimamura, K.: Transportation cost by hall thruster using low cost propellnat. Space Solar Power Syst. Soc. **5**, 82–85 (2020)
9. Kuwabara, H., et al.: Performance estimation of a water Hall thruster by a Faraday probe and comparison with Xe. STEP, p. 008 ((2021))
10. Szabo, J., Robin, M., Paintal, S., Pote, B., Hruby, V.: High density hall thruster propellant investigations. In: 48th AIAA/ASME/SAE/ASEE Joint Propulsion Conference &Exhibit 30 July–01 August 2012, Atlanta, Georgia (2012)
11. Maki, T., Kinefuchi, K., Cho, S., Watanabe, H.: Dry ice propellant for electric propulsion with triple-point storage. Acta Astronaut. **202**, 283–291 (2023)
12. Hohman, K.: Atmospheric breathing electric thruster for planetary exploration. Busek Co (2012)

Test Rig for MEMS Run-out Sensor

Maciej Zams and Sergiusz Łuczak(✉)

Warsaw University of Technology, Warsaw, Poland
sergiusz.luczak@pw.edu.pl

Abstract. A custom test rig for experimental studies of a novel axial run-out sensor, based on a low-g MEMS accelerometer, is presented. It enables to manually apply precise angular position of the tested sensor and features computer-controlled data acquisition. A custom mechanical unit makes it possible to preset a desired angular misalignment of the sensitive axis of the tested sensor with respect to the rotation axis. The misalignment simulates a twist of a rotating element that is to be monitored by means of the sensor. The related geometrical relationships are presented. The main metrological features of the test rig are evaluated and briefly discussed; its scope of application as well as the related limitations are considered. A prototype of the axial run-out sensor is briefly presented. Exemplary results of experimental study of the prototype sensor realized by means of the test rig are reported.

Keywords: axial run-out · MEMS · accelerometer

1 Introduction

Development of technology and increasing miniaturization of devices, as well as the desire to create smart devices capable of communicating with the surroundings, is forcing equipment manufacturers to use more and more sensors. One type of sensor that allows devices to gather information about various physical quantities are accelerometers; their application makes it possible to determine quantities not only directly related to acceleration, e.g. tilt [1, 2]. A good example may be enhanced traction control/ABS systems developed recently for motorcycles [3], where MEMS accelerometers and gyroscopes are used to determine tilt of the motorcycle while taking a bend. In this paper, we focus on determining one of the position deviations [4]: the axial run-out of a rotating axle or shaft, employing low-g MEMS accelerometer, what is a novel concept we have developed and patented [5] recently. The axial run-out can be also expressed in a conventional way, using other tolerances of orientation specified e.g. in [6, 7].

Untypical measurement method, such as in our case, requires experimental verification proving its validity. Besides, when the accuracy of sensors used for research purposes is of particular importance, a series of tests must be carried out to determine the actual performance of a particular unit, what requires application of a precise test rig [8, 9]. Therefore, the aim of the presented study was to build an appropriate testing device and to test the novel axial run-out sensor based on MEMS accelerometer.

Y. Ono and J. Kondoh (Eds.): Inter-Academia 2023, LNNS 939, pp. 202–208, 2024.
https://doi.org/10.1007/978-3-031-54450-7_22

2 The Test Rig

The test rig presented in Fig. 1 is based on a rotary optical head that allows precise setting of the angular position of the rotated shaft [10]. Axial run-out is set using a linear stage with micrometer screw mounted directly on the spindle of the head, within the range of 0–15 mm. The tested sensor was fixed to a selected shaft mounted in two centres. Since the live centre can be positioned eccentrically, misalignment the sensitive axis of the tested sensor with respect to the rotation axis simulates a twist of the rotating shaft.

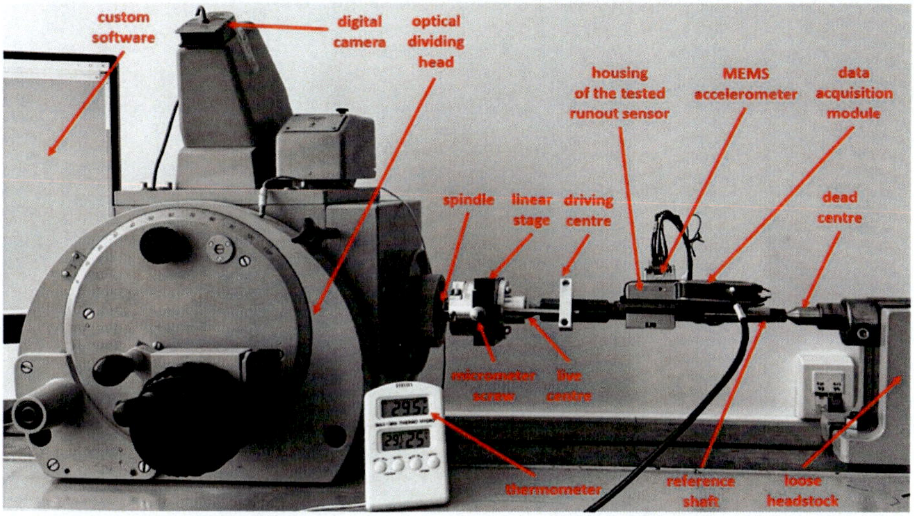

Fig. 1. Mechanical structure of the constructed test rig.

The reference shaft used for initial measurements (a chucking reamer) is made of HSS steel and its length is 235 mm. Such length allows to tilt the tested sensor within the range 0°–3°39' (what is illustrated in Fig. 2) with an error of $\varepsilon \leq 9''$. The test rig is suitable for both longer and shorter shafts, what makes it possible to test and calibrate sensors featuring different measurement range and accuracy. The tested sensor can be rotated about the horizontal axis by full angle.

In the middle of the tested shaft a custom-made axial run-out sensor (patent pending [5]) was mounted. The draft of the sensor is presented in Fig. 3. The sensor is based on a new-generation uniaxial MEMS accelerometer (4) fixed to an aligning unit (3). Its two-part casing (2) is mounted on the shaft (1) by means of two screws (8). The purpose of the aligning unit, which is secured in the casing by means of two screws (5), is to align the accelerometer sensitivity axis with respect to to the casing basis, as reported and minutely described in the case of the tilt sensor presented in [11], and thus to the axis of the tested shaft. Slight angular displacement of the accelerometer is achieved by using aligning screws (6), (7) which deform the aligning unit and enable precise alignment of the accelerometer sensitive axis. Alignment procedure minimizes misalignment errors and eliminates a need for their subsequent compensation, what significantly simplifies further calculations and increases the accuracy.

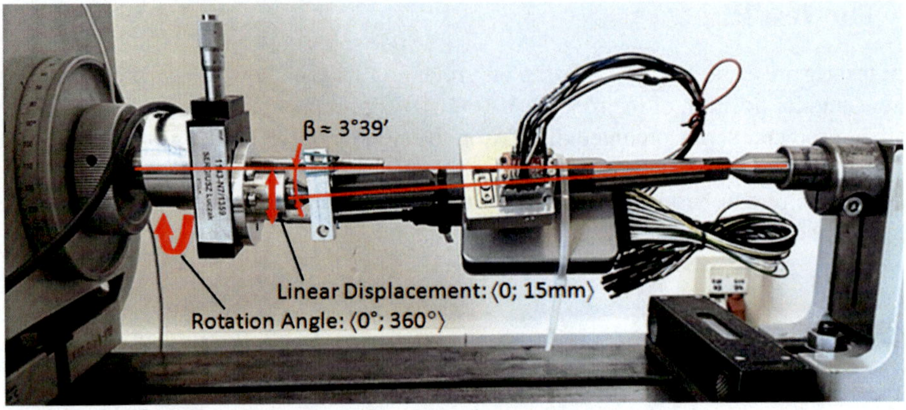

Fig. 2. Kinematics of the test rig.

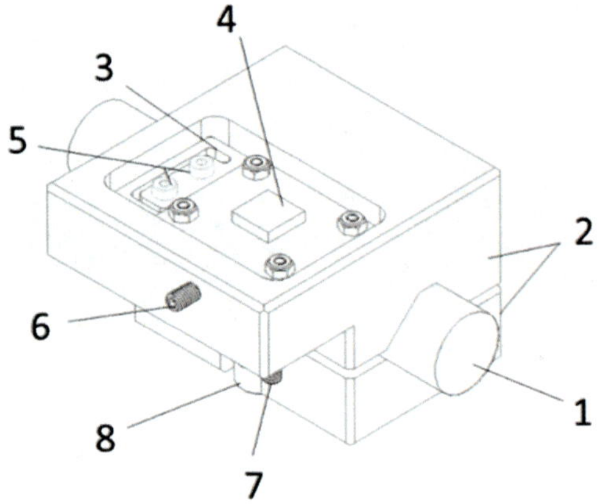

Fig. 3. The axial run-out sensor.

In the sensor, a single-axis low-g MEMS accelerometer was used. The choice was SCA830-D06-PCB fabricated by Murata due to its high precision (low noise, small cross-axis sensitivity). Owing to application of a USB-8452 data acquisition module by National Instruments [12] a seamless acquisition and transmission of the digital data generated by the accelerometer was achieved (employing SPI protocol).

3 Alignment Procedure

When installing a sensor in a device, there is never guaranteed that it is correctly aligned. Sometimes this is impossible, and in some cases it requires too much effort and cost.

Failure to align the sensor is a source of significant measurement error [13], since the measured acceleration is a vector quantity.

In order to adjust the sensor, we need to act on the compliant mechanism in such a way that the sensitive axis of the sensor is parallel to the axis of the rotating system, so the signal acquired from the sensor while rotating it by one full revolution should not fluctuate.

The alignment procedure should be repeated until the magnitude of the noise equals or exceeds the amplitude of the measured signal [14]. Usually, after several times, it is possible to achieve a state in which the amplitude of the signal is less than or equal to the amplitude of the noise of the measurement determined on the basis of 30 measurements at each point. Such situation is shown in Fig. 4 (the initial value of the rotation angle was accepted approximately at the horizontal orientation of the twist plane - as in Fig. 1).

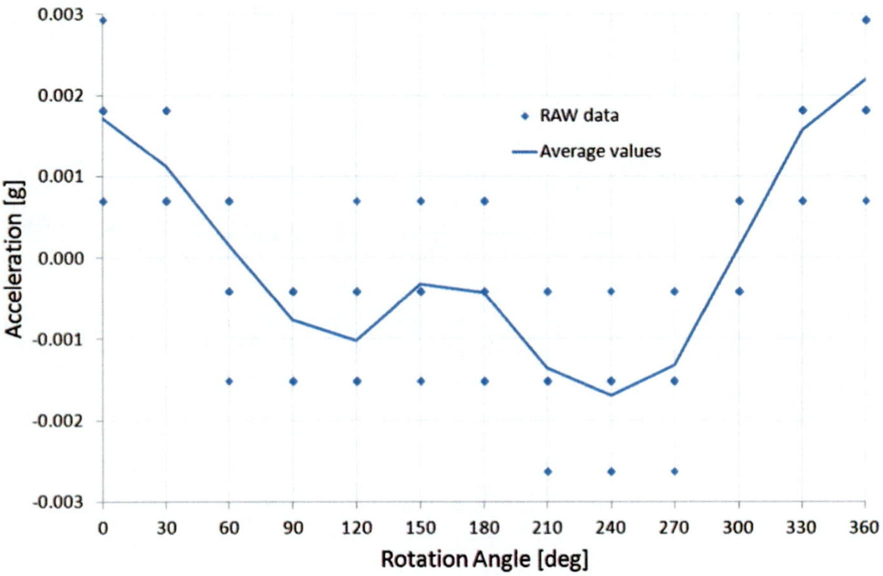

Fig. 4. Residual misalignment vs. measurement noise.

Once the alignment is completed, we can determine the angular position error due to the magnitude of the noise using a simple equation:

$$\varepsilon_n \leq \operatorname{asin}(\max(\varepsilon_\alpha)) \tag{1}$$

where ε_α is the magnitude of noise at rotation angle α.

Assuming $\alpha = 360°$:

$$\varepsilon_n \leq 7'38'' \tag{2}$$

In this case, the error of determining the angular position is about 0.1 degree arc; in order to decrease it, a more accurate accelerometer, with a smaller measurement range, higher resolution and lower noise should be applied.

4 Initial Tests

In order to evaluate the performance of the test rig and assess its accuracy, two runs were made at extreme positions of the linear stage. The first run was performed for a shaft aligned coaxially with both centres. At each angular position, 30 measurements were performed and averaged. Then the second course was obtained for the maximal displacement of the reference shaft, i.e. 15 mm eccentricity of its end with respect to the rotation axis determined by the centres. As for the first run, 30 signals were recorded at each angular position and the results were averaged. The results are presented in Fig. 5.

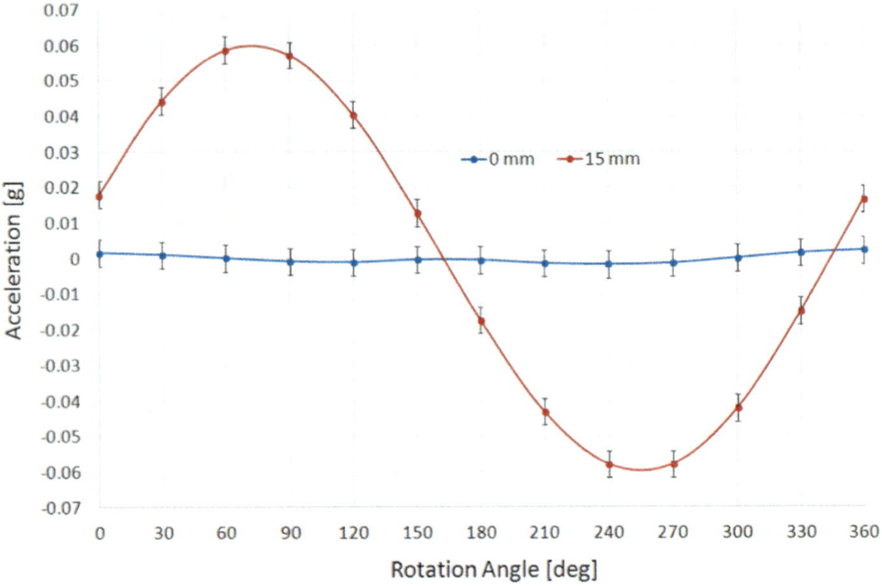

Fig. 5. Acceleration variations measured for coaxial (eccentricity of 0 mm) and twisted (eccentricity of 15 mm) shaft (error bars are determined as 3-sigma errors).

The course of the output signal representing acceleration a obtained for the twisted shaft can be described by the following formula:

$$\mathrm{a} = \mathrm{g}\sin(\alpha + \gamma) \cdot \sin\varphi \tag{3}$$

where g is the gravitational acceleration, φ is the actual reference twist angle and γ is the phase shift (in this case of approximately $-15°$).

Since we are looking for the axial run-out, we have to consider the extreme values of the signal: $\max(a)$ and $\min(a)$ – the components of the gravitational acceleration sensed by the accelerometer at the rotation angle α of approximately 75° and 255°. The axial run-out can be determined as follows:

$$\beta = \operatorname{asin}\left(\frac{\max(a) - \min(a)}{2}\right) = 3°21' \tag{4}$$

On the other hand, the actual reference twist angle of the shaft is:

$$\varphi = \mathrm{atan}\left(\frac{15\,\mathrm{mm}}{235\,\mathrm{mm}}\right) = 3^{\circ}39' \tag{5}$$

So, we can calculate the relative error as:

$$\varepsilon_{\beta} = \frac{|\varphi - \beta|}{\varphi} \cdot 100\% = 8.2\% \tag{6}$$

For some applications such a value may be too large. However, after the preliminary tests, many issues that can be improved have been identified and will be addressed in the future.

On the other hand, in the case of many applications, such relative error is acceptable, especially when no precise value of the axial run-out is to be determined, but only exceeding a certain threshold is to be signaled instead, e.g. in order to trigger an alarm or other safety system.

5 Conclusion

A test rig for testing axial run-out sensors was designed and integrated. The mechanical structure of the test rig allows the reference shaft to be replaced for either a shorter or a longer element. In this way, range of the axial run-out to be applied can be changed, however the bigger the range the larger the error of the twist created between the symmetry axis of the reference shaft and its rotation axis, and vice-versa.

In order to maintain high precision of applying the angular position of the tested run-out sensor, application of any coupling was purposely avoided [15].

The test rig is equipped with the NI USB-8452 data acquisition module for digital communication with new-generation MEMS accelerometers. Owing to digital data transmission via USB, it is also possible to perform data acquisition using any PC, without need to install a dedicated card in an internal computer socket.

The structure of the application controlling the data acquisition, and the LabView environment in which it was developed, enables easy code modifications, what may be required while testing new sensors or adding new software functions such as error compensation algorithms, what will be the subject of future works on the test rig.

During the tests we also monitor the ambient temperature and humidity; in the future, this information can be used while proposing new algorithms for compensation of thermal drifts, if the obtained sensor accuracy turns out to be insufficient.

Performance of the run-out sensor was tested under static condition only. Despite the fact that it is impossible to determine the run-out in the proposed manner under dynamic conditions (unless appropriate parameters are known), it will be still possible to monitor then variations of the output signal in qualitative terms: the centripetal acceleration will only increase the amplitude or the offset of the signal, and the sensed acceleration components due to vibration can be filtered out.

Acknowledgments. This research was funded by the Scientific Board of the Discipline of Automatics, Electronics, Electrical Engineering and Space Technologies at the Warsaw University of Technology, grant No. 504/04873/1143/43.022301.

References

1. Wilson, J.S.: Sensor Technology Handbook. Newnes, Burlington (2005)
2. Łuczak, S.: Guidelines for tilt measurements realized by MEMS accelerometers. Int. J. Precis. Eng. Manuf. **15**(9), 2012–2019 (2022)
3. Łuczak, S.: Tilt measurements in BMW motorcycles. In: Jabłoński, R., Szewczyk, R. (eds.) Recent Global Research and Education: Technological Challenges. AISC, vol. 519, pp. 287–293. Springer, Cham (2017). https://doi.org/10.1007/978-3-319-46490-9_39
4. Henzold, G.: Geometrical Dimensioning and Tolerancing for Design, Manufacturing and Inspection, 2nd edn. Elsevier, Oxford (2006)
5. Łuczak, S.: System for determining a measure of the axial runout of a shaft. Polish Patent Office, patent No. P.434917 (2020). (in Polish: Układ do wyznaczania miary bicia osiowego wałka)
6. Humienny, Z., Berta, M.: Using animations to support the understanding of geometrical tolerancing concepts. Tech. Mess. **82**(9), 422–431 (2015)
7. Geometrical product specifications (GPS) - Geometrical tolerancing - Tolerances of form, orientation, location and run-out, 4th edn. ISO 1101 international standard (2017)
8. Zams, M.: Laboratory Test Rig for experimental studies of tilt sensors. In: Proceedings of the Konferencja Młodych Naukowców nt.: Dokonania Naukowe Doktorantów VI. Zagadnienia aktualnie poruszane przez młodych naukowców, Kraków, Poland, pp. 174–175 (2018). (in Polish)
9. Łuczak, S.: Dual-axis test rig for MEMS tilt sensors. Metrol. Meas. Syst. **21**(2), 351–362 (2014)
10. Łuczak, S.: Experimental studies of hysteresis in MEMS accelerometers: a commentary. IEEE Sens. J. **15**(6), 3492–3499 (2015)
11. Łuczak, S., Zams, M., Dąbrowski, B., Kusznierewicz, Z.: Tilt sensor with recalibration feature based on MEMS accelerometer. Sensors **22**(4), 1504 (2022)
12. National Instruments USB-8452 data acquisition module specifications. Accessed 26 July 2023
13. Syed, Z., Aggarwal, P., Goodall, C., Niu, X., El-Sheimy, N.: A new multi-position calibration method for MEMS inertial navigation systems. Meas. Sci. Technol. **18**(7), 1897–1907 (2007)
14. Łuczak, S.: Fast alignment procedure for MEMS accelerometers. In: Jabłoński, R., Brezina, T. (eds.) Advanced Mechatronics Solutions. Advances in Intelligent Systems and Computing, pp. 481–487. Springer, Cham (2016). https://doi.org/10.1007/978-3-319-23923-1_70
15. Bodnicki, M., Łuczak, S.: Comments on "delay compensation of tilt sensors based on MEMS accelerometer using data fusion technique". IEEE Sens. J. **18**(3), 1333–1335 (2018)

Capacitive Tilt Micro-Sensor

Sergiusz Łuczak(✉), Maciej Zams, Michał Władziński, and Paweł Pieńczuk

Warsaw University of Technology, Warsaw, Poland
sergiusz.luczak@pw.edu.pl

Abstract. Division of the steregon (4π sr) into uniform sub-ranges is briefly discussed as a way of discrete determination of a tilt of an object. Mechanical designs of relevant patented sensors and their electronic circuits are considered. Then, a capacitive tilt micro-sensor is minutely presented. It features high reliability of operation and small dimensions. Its discrete principle of operation makes it possible to significantly simplify the signal post-processing. Results of experimental tests, where balls made of bearing steel with diameter of 2 to 5 mm were used are reported. A concept of a capacitive tilt sensor with analog output is proposed; such sensor determines the tilt not only in a discrete way but generates also a continuous output signal within sub-ranges of the steregon. Ways of further miniaturization of the micro-sensor are discussed.

Keywords: Tilt · Inclinometer · Capacitive Sensor · Bearing Ball

1 Introduction

At present, tilt measurements are usually performed employing low-g accelerometers, since it is considered one of the basic applications of these sensors [1]. A good solution is to use a MEMS-type accelerometer, since it ensures small size, low-cost, robustness and easy integration with electronics [2]. However, MEMS accelerometers feature also some considerable shortcomings like thermal errors, long-term drifts or drift of over lifetime (aging) [3] or necessity of their alignment [4] and calibration to be performed by the user [5, 6].

If the aforementioned shortcomings are too significant in certain applications, another measuring technique should be searched for, e.g. application of a liquid or mercury sensor. However, if tilt can be determined with low resolution and at the same time a simple and fast signal processing must be observed, simple electric sensors can be a good solution [7]. Since such sensors have a mechanical housing, their alignment is much easier comparing to alignment procedures for MEMS accelerometers [8]. However, to keep their overall dimensions as miniature as in the case of MEMS devices remains a challenge. Therefore, we developed a capacitive detection technique that makes it possible to employ free balls having diameter of 2 mm or less.

Tilt measurements have been successfully employed in numerous devices, with a significantly intensified tendency over the last years. There are three kinds of these measurements: 1. Single-axis tilt - usually referred to as inclination [9], 2. Dual-axis

Y. Ono and J. Kondoh (Eds.): Inter-Academia 2023, LNNS 939, pp. 209–217, 2024.
https://doi.org/10.1007/978-3-031-54450-7_23

tilt - usually represented by pitch and roll [10], 3. Axial tilt - where pitch of an axis is considered [11].

Except for typical application fields, such as leveling and navigation [12] or directional drilling [13], many examples of untypical use of tilt measurements can be suggested, e.g.: in diving equipment [14], motorcycle traction control systems [15], electronic gaming dice [16], monitoring of objects being elevated [17], wheel odometer [18], to name only a few.

2 Electric Tilt Sensors

A large group o such sensors are some liquid inclinometers or liquid tilt sensors. They feature high sensitivity, however often are quite expensive and their measurement range is usually limited, even in the case of inclinometers, which can sense only pitch. A full measurement range, equal to the steregon (4π sr), can be easily obtained using solid-state electric sensors. For instance, the steregon can be divided into 2 uniform sub-ranges by many switches commercially available [7]. Further division into 4, 8, 20 and 12 uniform sub-ranges can be obtained by means of some patented sensors [19–22]. Operation of the first three - presented in Fig. 1 - is based on short-circuiting three adjacent electrodes at a time by means of a free member: e.g. a ball. That makes it possible to generate the output signal in a very easy way: using TTL logics [20], a resistive technique [21] or a capacitive technique [19].

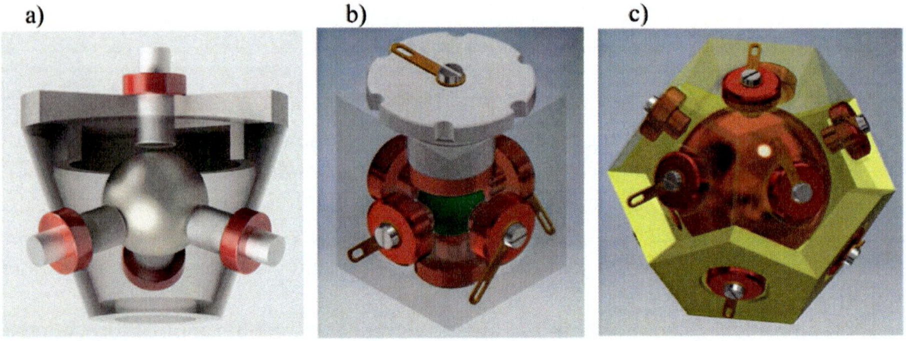

Fig. 1. Discrete tilt sensors based on regular polyhedrons providing 3-point contact: a) tetrahedron (4 vertexes); b) cube (8 vertexes); c) dodecahedron (20 vertexes).

2.1 Detection Based on TTL Logics

Detection of the ball using TTL logics is convenient to be used not only in the case of the cuboidal structure of the sensor, as in Fig. 1b, or tetrahedral structure, as in Fig. 1a, but also for any prismatic structure, as suggested in [20] and presented in Fig. 2. In the case of prismatic sensors the principle is that one base of the polyhedron has electric potential representing logical "0" and the other base has electric potential representing logical "1".

At any attitude of the sensor only two adjacent walls indicate logical “0” or “1”, as they are short-circuited by the ball with one of the two bases. In the case of the tetrahedral structure, one electrode has electric potential representing logical “1” (e.g. the upper one in Fig. 1a). The other three electrodes have electric potential representing logical “0”, what corresponds to orientation like in Fig. 1a, or only two of them have electric potential representing logical “1”, what corresponds to other three possible orientations.

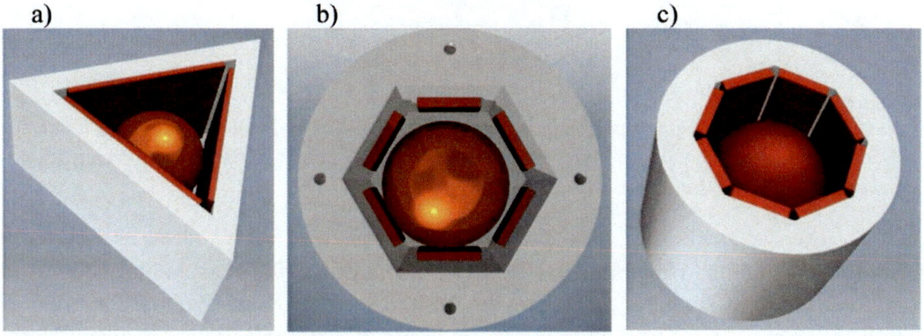

Fig. 2. Tilt sensors with prismatic shape (view without the upper electrode): a) triangular prism; b) hexagonal prism; c) octagonal prism.

2.2 Resistive Detection

Since in the case of the dodecahedronal structure (see Fig. 1c) there are too many electrodes to apply the detection based on TTL logics, a method based on resistance measurement was proposed in [21]. All the electrodes are electrically connected by resistors of unique resistance (in the order of kΩ). If the ball short-circuits certain group of adjacent electrodes, it results in a change of the resultant resistance of the whole circuit. For the ball located at each of the 20 vertexes, the resultant resistance is different. Thus it is possible to distinguish all the possible orientations of the sensor.

2.3 Sequential Method of Detection

In the case of the icosahedronal structure - presented in Fig. 3 - the operation is more complicated, because at each vertex the ball short-circuits three of five adjacent electrodes at a time. In order to detect the three short-circuited electrodes a dedicated microprocessor unit must sequentially apply electric potential representing logical “1” to a subsequent electrode and detect logical “1” on the other electrodes, until three adjacent electrodes altogether indicate logical “1”. In the worst case, it takes a large number of sequences to find the orientation of the ball.

As far as the movable member is concerned, its shape should be spherical most preferably. However, the geometrical shape deviations as well as the surface roughness should be possibly low in order to ensure low mechanical hysteresis of the sensor. Taking into account that the diameter of the movable member is to be less than 3 mm,

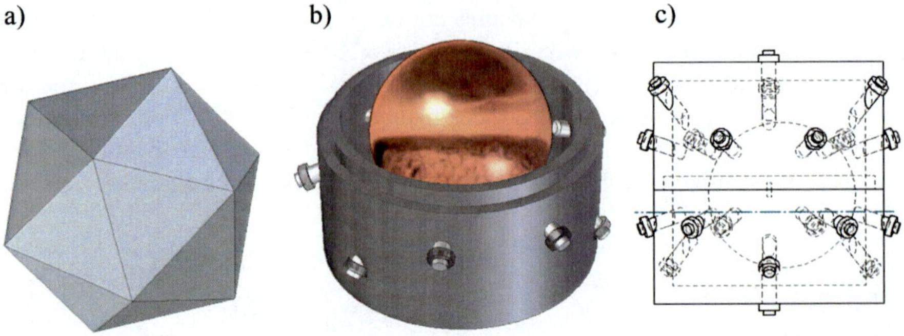

Fig. 3. Tilt sensors with icosahedronal shape: a) geometric model (12 vertexes); b) technical realization - view without the upper part of the cylindrical housing; c) technical realization - spatial arrangement of the 20 electrodes.

it would be difficult to manufacture such ball. Therefore, the reasonable solution is to apply a ball that is characterized by the desired features. Such balls are applied in rolling bearings and in optical engineering for fabrication of micro-lenses made of glass. However, whereas in the case of sliding micro-bearings a lot of different materials is available, e.g. various polymers [23], in the case of rolling micro-bearings the balls are made of only bearing steel or ceramic materials. Moreover, application of ceramic or glass balls would require having them coated with a metallic layer in order to feature appropriate electric properties (low resistance). The balls used in small bearings have typically diameters 3/64″ (1.19 mm), 5/64″ (1.98 mm) or 1/8″ (3.175 mm).

2.4 Capacitive Method of Detection

The basic principles of operation of the dedicated electronic circuit were successfully applied in proximity measurements presented in [24]. In this case, finding location of the ball at a given vertex of a polyhedron is carried out on the basis of detection of a large capacitance, or a short-circuit between two pairs of adjacent electrodes. The detection system uses a Schmitt gate along with additional elements and sensor electrodes, which constitute a rectangular wave generator circuit with an adjustable frequency of the generated signal; the frequency depends on the capacitance formed between the sensor electrodes. As the ball approaches the electrodes, the capacitance gets increased, what causes a decrease in the frequency of the generated waveform. If the electrodes are short-circuited, the generator ceases to operate. The rectangular waveform at the output of the generator is fed into the input of a monostable flip-flop, so that its output will generate a pulse signal with a fixed pulse duration and the same frequency as the input signal. As the result, the output will produce a digital signal with a variable pulse length, which depends on the frequency (thus on the capacitance between the electrodes). Then the digital signal with variable pulse length is integrated and at the output of the integrating circuit we get a voltage linearly dependent on the generated frequency. When this voltage falls below a predetermined threshold, a logical “1” (~5 V) will be created at the output of the circuit, which means a short-circuit of a given pair of electrodes, i.e., location of the ball in the the immediate vicinity of the electrodes. Electrical diagram of the capacitive electronic circuit is presented in [19].

3 Experimental Study

Experimental studies that have been carried out so far, proved that while decreasing the overall dimensions of such sensors it became more and more difficult to use TTL logics circuits or resistive technique for detection of the location of the ball. Therefore, the aforementioned capacitive technique has been proposed to overcome the problem related to low reliability of such electric sensors.

In order to verify performance of a tilt sensor employing capacitive detection of the ball, a special test rig was built. It is presented in Fig. 4.

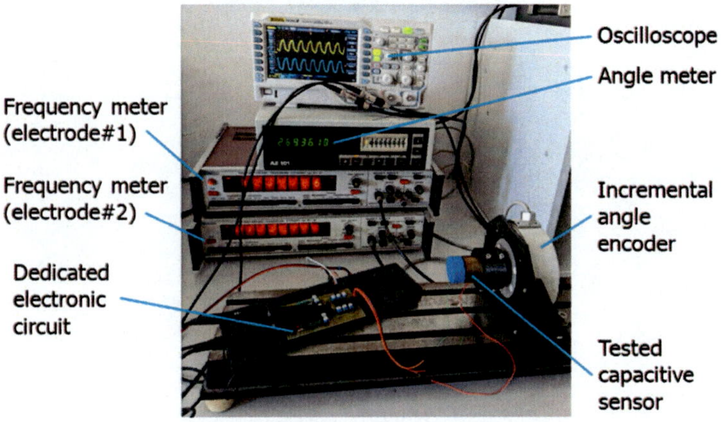

Fig. 4. The test rig.

The housing of the tested sensor containing 5 mm steel ball is presented in Fig. 5. The sensor was fixed to the incremental angle encoder used for applying tilt angles. The dedicated capacitive detection circuit was connected to the two electrodes of the sensor. Two meters and the oscilloscope indicated frequency of the signal generated on each electrode. The ball placed between the electrodes could freely move among them. The tests were repeated for three different balls with diameters of 2, 3 and 5 mm, respectively.

First, discrete detection of the ball was studied. All of the balls were tilted to be found in the vicinity of an electrode. Then, the respective generator stopped to operate - according to the assumption. So, reliability of discrete detection was proved.

However, the principle of operation of the electronic circuit enables not only detection of the location of the ball in the discrete way, but in a continuous way as well. Yet, due to increase of the rolling friction, the character of motion of the steel ball having diameter of 3 mm was so sudden that the recorded variations of the signal frequency where very irregular, as can be seen on the graph in Fig. 6. So, further tests will be carried out using a modified design of the inner part of the housing of the sensor. Much better results were obtained for a larger steel ball (diameter of 5 mm), as can be observed in Fig. 7.

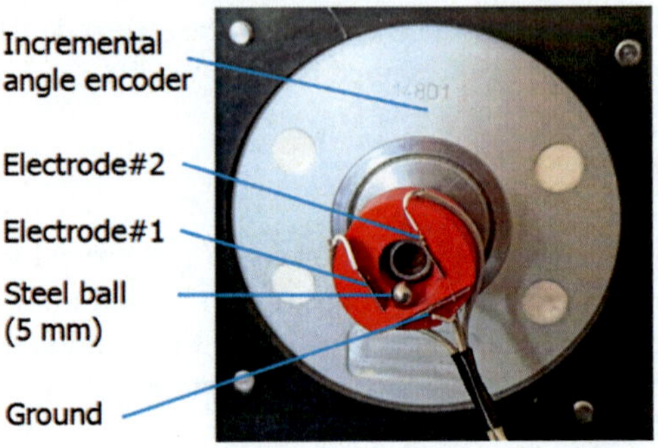

Fig. 5. The housing of the tested sensor.

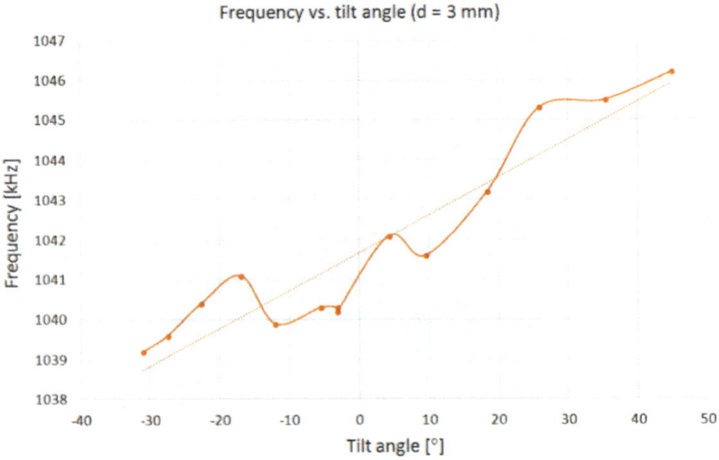

Fig. 6. Frequency of the output signal at electrode #1 vs. tilt angle of the sensor (steel ball with diameter of 3 mm).

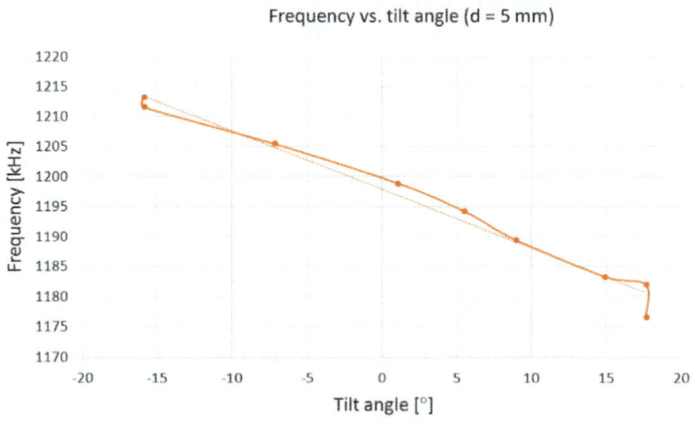

Fig. 7. Frequency of the output signal at electrode #1 vs. tilt angle of the sensor (steel ball with diameter of 5 mm).

4 Conclusion

When miniature dimensions of a simple and discrete electric tilt sensor are searched for, it is convenient to apply a capacitive method of tilt detection. Steel balls for rolling bearings should be employed as the free member, provided their diameter (being of at least 0.397 mm) is acceptable.

If larger scale of miniaturization is expected, balls made of soda-lime glass [25], manufactured for optical micro-lenses, should be applied, since their diameters commercially available are: 100, 50, 40, 30 μm. However, due to the capacitive detection method, they have to be coated with a metallic layer, which has to feature both appropriate electric and mechanical properties (high hardness, low roughness, constant thickness) in order to ensure small shape deviation, otherwise mechanical hysteresis of the tilt sensor will be significantly larger. Nevertheless, the smaller the ball diameter the bigger the mechanical hysteresis of the tilt sensor - it cannot be avoided or reduced by any detection technique.

A sensor featuring new mechanical structure (concave surface of the electrodes) with an analog characteristic will be developed in the future.

Acknowledgments. This research was funded by the Scientific Board of the Discipline of Automatics, Electronics, Electrical Engineering and Space Technologies at the Warsaw University of Technology, grant No. 504/04873/1143/43.022301.

References

1. Wilson, J.S.: Sensor Technology Handbook. Newnes, Burlington (2005)
2. Łuczak, S., Wierciak, J., Credo, W.: Effects of natural aging in biaxial MEMS accelerometers. IEEE Sens. J. **21**(2), 1305–1314 (2021)
3. Kaajakari, V.: Practical MEMS. Small Gear Publishing, Las Vegas (2009)

4. Syed, Z., Aggarwal, P., Goodall, C., Niu, X., El-Sheimy, N.: A new multi-position calibration method for MEMS inertial navigation systems. Meas. Sci. Technol. **18**(7), 1897–1907 (2007)
5. Łuczak, S., Zams, M., Dąbrowski, B., Kusznierewicz, Z.: Tilt sensor with recalibration feature based on MEMS accelerometer. Sensors **22**(4), 1504 (2022)
6. Won, S.P., Golnaraghi, F.A.: Triaxial accelerometer calibration method using a mathematical model. IEEE Trans. Instrum. Meas. **59**, 2144–2153 (2010)
7. Łuczak, S., Ekwińska, M.: Electric-contact tilt sensors: a review. Sensors **21**(4), 1097 (2021)
8. Łuczak, S.: Fast alignment procedure for MEMS accelerometers. In: Jabłoński, R., Brezina, T. (eds.) Advanced Mechatronics Solutions. Advances in Intelligent Systems and Computing, vol. 393, pp. 481–487. Springer, Cham (2016). https://doi.org/10.1007/978-3-319-23923-1_70
9. Luinge, H.J., Veltink, P.H.: Inclination measurement of human movement using a 3-D accelerometer with autocalibration. IEEE Trans. Neural Syst. Rehabil. Eng. **12**(1), 112–121 (2004)
10. Łuczak, S.: Advanced algorithm for measuring tilt with MEMS accelerometers. In: Jabłoński, R., Turkowski, M., Szewczyk, R. (eds.) Recent Advances in Mechatronics, pp. 511–515. Springer, Heidelberg (2007). https://doi.org/10.1007/978-3-540-73956-2_100
11. Łuczak, S.: Specific measurements of tilt with MEMS accelerometers. In: Jabloński, R., Brezina, T. (eds.) Mechatronics. Recent Technological and Scientific Advances, pp. 705–711. Springer, Heidelberg (2012). https://doi.org/10.1007/978-3-642-23244-2_85
12. Popowski, S.: Determining pitch and roll in inexpensive land navigation systems. J. Aeronaut. Integra **1**(3), 93–97 (2008). (in Polish)
13. Bortolami, S.B., Pierobon, A., et al.: Localization of the subjective vertical during roll, pitch, and recumbent yaw body tilt. Exp. Brain Res. **173**, 364–373 (2006)
14. Radeta, M., et al.: Lost in the deep?: performance evaluation of dead reckoning techniques in underwater environments. In: Proceedings of the ACM on Interactive, Mobile, Wearable and Ubiquitous Technologies, vol. 7, no. 2, pp. 1–27. Association for Computing Machinery, New York (2023)
15. Łuczak, S.: Tilt measurements in BMW motorcycles. In: Jabłoński, R., Szewczyk, R. (eds.) Recent Global Research and Education: Technological Challenges. AISC, vol. 519, pp. 287–293. Springer, Cham (2017). https://doi.org/10.1007/978-3-319-46490-9_39
16. Macdonald, E., et al.: 3D printing for the rapid prototyping of structural electronics. IEEE Access **2**, 234–242 (2014)
17. Łuczak, S., Zams, M.: Monitoring tilt of elevated loads using MEMS accelerometers. In: Szewczyk, R., Krejsa, J., Nowicki, M., Ostaszewska-Liżewska, A. (eds.) MECHATRONICS 2019. AISC, vol. 1044, pp. 274–284. Springer, Cham (2020). https://doi.org/10.1007/978-3-030-29993-4_34
18. Youssef, A.A., Al-Subaie, N., et al.: Accelerometer-based wheel odometer for kinematics determination. Sensors **21**(4), 1327 (2021)
19. Łuczak, S., Władziński, M., Zams, M.: Capacitive tetrahedral tilt sensor and method of determining tilt. Polish Patent Office, patent pending No. P.445643 (2023). (in Polish: Pojemnościowy czworościenny czujnik odchylenia od pionu oraz sposób określania odchylenia od pionu)
20. Łuczak, S.: Tilt sensor. Polish Patent Office, patent no. 239205 (2021). (in Polish: Czujnik odchylenia od pionu)
21. Łuczak, S., Władziński, M., Jodko-Władzińska, A.: Discrete tilt sensor and the way of determining the tilt. Polish Patent Office, patent no. 239207 (2021). (in Polish: Dyskretny czujnik odchylenia od pionu oraz sposób określania odchylenia od pionu)
22. Władziński, M., Jodko-Władzińska, A., Łuczak, S., Zams, M.: Discrete icosahedron tilt sensor and the way of determining the tilt. Polish Patent Office, patent no. P.438954 (2023). (in Polish:

Dyskretny dwudziestościenny czujnik odchylenia od pionu oraz sposób określania odchylenia od pionu)
23. Kusznierewicz, Z., Michałowski, M., Łuczak, S., Bagiński, K., Kabziński, B., Zams, M.: Selection of polymer materials for micro sliding bearings with respect to minimization of resistance to motion. IEEE Access **8**, 78622–78629 (2020)
24. Klimaszewski, J., Władziński, M.: Human body parts proximity measurement using distributed tactile robotic skin. Sensors **21**(6), 2138 (2021)
25. Milczarek, M., Jarząbek, D.M., Jenczyk, P., Bochenek, K., Filipiak, M.: Novel paradigm in AFM probe fabrication: broadened range of stiffness, materials, and tip shapes. Tribol. Int. **180**, 108308 (2023)

Evaluation of Ionization Efficiency in a Carbon Dioxide Hall Thruster by 0-D Discharge Simulation

Shu Tokuda[1(✉)], Shinatora Cho[2], Hiroki Watanabe[2], Yasushi Ohkawa[2], and Makoto Matsui[3]

[1] Department of Engineering, Shizuoka University, Hamamatsu, Shizuoka, Japan
tokuda.shu.19@shizuoka.ac.jp
[2] Research of Development Directorate JAXA, Sagamihara, Kanagawa, Japan
{choh.shinatora,watanabe.hiroki,okawa.yasushi}@jaxa.jp
[3] Department of Mechanical Engineering, Shizuoka University, Hamamatsu, Shizuoka, Japan
matsui.makoto@shizuoka.ac.jp

Abstract. Xenon has been typically used as a propellant for Hall thrusters. Its characteristics are its high relative density and low ionization cost. On the other hand, there are disadvantages such as increased tank weight owing to high storage pressure and high costs. Therefore, perspective of ISRU (In-Situ Resource Utilization) during Mars exploration, JAXA and Nagoya University have proposed CO_2 as an alternative propellant. However, its plasma characteristics have not been fully understood, and it is necessary to understand by numerical calculation. Since CO_2 is a molecule, various ions are produced when it is used as a propellant. Ionization other than $CO_2{}^+$ is costly and causes energy loss, reducing thrust performance. Thus, it is necessary to perform collision calculations using the electron temperature as a parameter to estimate the region where a large amount of $CO_2{}^+$ is produced. In this paper, we perform a non-dimensional PIC simulation to estimate the electron temperature to maximize the degree of ionization of $CO_2{}^+$. As a result, it was found that the optimum electron temperature for CO_2 operation is about 15 eV.

Keywords: Electric Propulsion · Hall Thruster · Alternative Propellant · PIC Calculation

1 Introduction

Electric propulsion is a technology designed to provide thrust at high exhaust velocities. In particular, Hall thrusters provide higher specific impulse (1,500 s) and propulsive efficiency (50%) than DC arcjet thrusters and higher thrust density (about 1 mN/cm^2) than ion thrusters. For these reasons, Hall thrusters are used for orbit transition of satellites, north-south Maneuver of space structures. Xenon is typically used as the propellant for Hall thrusters. In addition to its high relative density, it has the advantages of being non-toxic and of having low ionization cost. But must be filled into the tank at high

Y. Ono and J. Kondoh (Eds.): Inter-Academia 2023, LNNS 939, pp. 218–223, 2024.
https://doi.org/10.1007/978-3-031-54450-7_24

pressure and increasing the tank weight. Therefore, it is not suitable for small satellites. Comparison between CO_2 and xenon is tabulated in Table 1. In this table, it is found that CO_2 has some advantages compared with xenon. Furthermore, it can also purify CO_2 from the Martian atmosphere during space exploration, which is a significant advantage compared with other propellants [1]. But the thrust by using CO_2 is lower due to small atomic weight. Ionization cross sections of CO_2 are also lower.

The triple point supply system and the use of Aeroflex as insulation have been found to provide stable storage and supply for hundreds of days [1]. Although Hall thrusters using CO_2 have been tested by JAXA, the plasma characteristics have not been fully understood [2]. In addition, Hall thrusters have non-neutral regions (sheaths) and the region of low electron density. Then, numerical calculation is necessary to understand plasma characteristics, design and optimize operating conditions for CO_2, and these are objectives of this study. To achieve this, it is necessary to analyze the plasma characteristics at the channel, and model that can resolve the sheath is required. Therefore, the Full-PIC (Full Particle-In-Cell) model, which does not require the assumption of electric charge quasi-neutrality, is considered be suitable for this study.

Table 1. Comparison between dry ice and xenon [2].

Propellant	Cost (Normalized) /kg	Tank storage density g/cm^3	Ionization cost eV
Dry ice	0.003	1.56 (at triple point)	13.8 (CO_2^+)
Xenon	1	4.8	12.1
Krypton	0.3	3.7	14.0

When using CO_2, various types of ions are generated in the channel. Because ionization other than CO_2^+ causes energy loss, it is necessary to understand the conditions where CO_2^+ is efficiently produced using the electron temperature as a parameter. In this paper, as a preliminary phase of two-dimensional Full-PIC, we model the collision cross sections of elastic scattering, ionization, and excitation of CO_2 and estimate the optimum electron temperature using non-dimensional PIC simulations.

2 Non-dimensional Discharge Simulation

The simulation in this study is a simple model developed as a preliminary phase of the two-dimensional Full-PIC model. This is a non-dimensional PIC simulation in which electrons and neutrals collide regardless of their position. Also, it has one dimensional velocity (0D1V). Therefore, the current is not affected by the magnetic field, and the electric field is assumed to be uniform. Furthermore, since it is impossible to calculate electron behavior in the magnetic field in this model, the electron density is always assumed to be equivalent to the ion density. The temperature is also kept at a constant. The flowchart is shown in Fig. 1.

To calculate collisions, it is necessary to have a criterion for determining whether there are collisions. In this study, we used a method to decide these by comparing the generated random numbers with the collision probability expressed by Eq. 1.

$$\mathrm{P} = n_e \langle \sigma v_e \rangle \Delta t \tag{1}$$

The collision calculation is performed only when the random number is less than the collision probability. This method of probabilistic processing for collision is called DSMC (Direct Simulation Monte Carlo method). However, since there are various collision types, a process to decide their type is required. In this study, since the value of the collision cross section is proportional to the probability, the ratio of these is used to decide these.

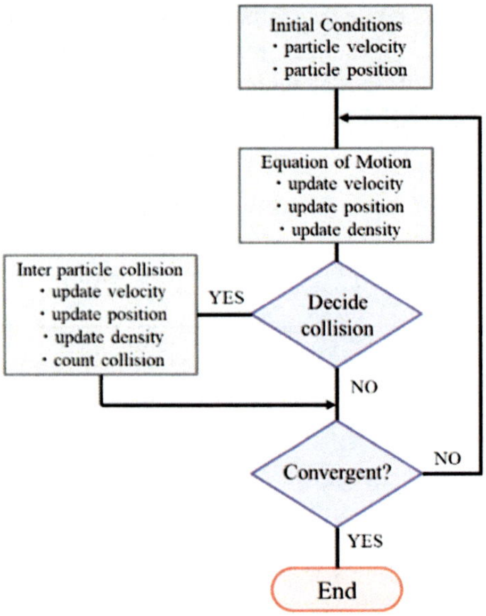

Fig. 1. Flowchart of non-dimensional PIC simulation.

3 Conditions

3.1 Hall Thruster Used in the Simulation

This simulation is performed for a 100 W class magnetic layer Hall thruster. The appearance is in Fig. 2, and the specifications is shown in Table 2.

3.2 Simulation Conditions

We performed simulation under the conditions shown in Table 3. Since this paper focuses only on energy losses, the electron temperature is assumed to be an independent parameter, and the discharge voltage is assumed to remain unchanged with electron temperature.

Fig. 2. 100 W class magnetic layer Hall thruster [3].

Table 2. Specifications of 100 W class magnetic layer Hall thruster [3].

Anode material	Molybdenum
Outer diameter of channel, mm	20
Channel width, mm	6.0
Channel length, mm	8.0
Channel material	Boron nitride

Table 3. Simulation conditions.

Parameters	Values
Time step s	1.0×10^{-10}
CO_2 flow rate mg/s	0.98
Neutral temperature eV	0.10
Discharge voltage V	2.0×10^2

The collision cross sections are shown in Fig. 3. Also, based on the mean free path analysis, the collision types to be included are shown in Table 4.

4 Result and Discussion

Discharge characteristics in the channel are affected by various parameters. In particular, the electron temperature is an important plasma parameter that depends strongly on the discharge voltage that accelerates the particles. In this section, certain number of neutrals are emitted into the channel while varying the electron temperature, and we analyze the ratio of the energy loss. From this, we can estimate the optimum electron temperature in the channel during CO_2 operation. The results are shown in Fig. 4.

From this, the energy consumed for CO_2^+ ionization increases with increasing electron temperature for 15 eV and below. On the other hand, above 15 eV, the energy decreases. Ionization other than CO_2^+ has a higher appearance energy than that of CO_2^+.

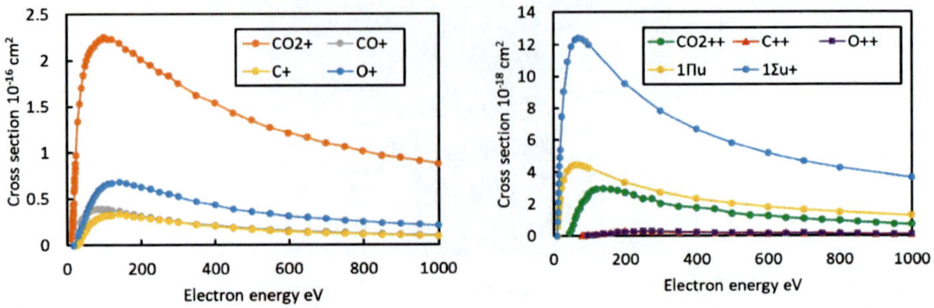

(a) First ionization (b) Double ionization and excitation

Fig. 3. Collision cross sections[4, 5]

Table 4. Collision types in the Hall thruster.

Type	Include
Ionization ($CO_2{}^+$)	○
Ionization (CO^+)	○
Ionization (C+)	○
Ionization (O^+)	○
Ionization (CO^{2+})	×
Ionization (C^{2+})	×
Ionization (O^{2+})	×
Excitation ($^1\Sigma_u{}^+$)	○
Excitation ($^1\Pi_u$)	○
Elastic scattering	○

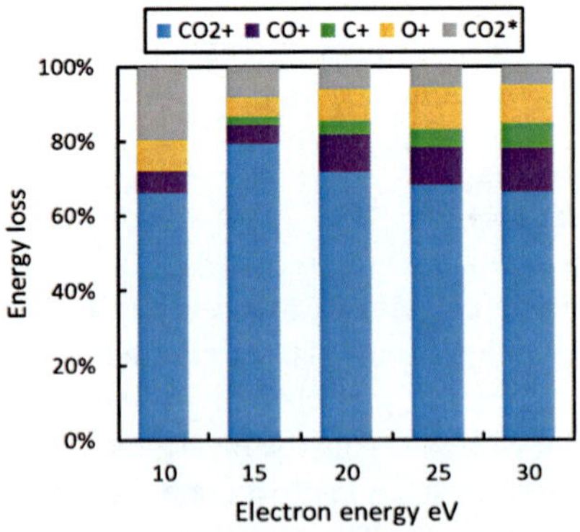

Fig. 4. Energy loss vs electron energy

Therefore, it causes a loss in the operation of the Hall thruster. Thus, a value around 15 eV is desirable, where the energy consumed for CO_2^+ ionization is the highest.

5 Conclusion

A numerical calculation of ionization efficiency in a Hall thruster using CO2 as propellant has been performed based on a non-dimensional PIC simulation that used a selection of collision types. Based on the simulation results, the optimum electron temperature for a 100 W class Hall thruster in CO_2 operation was estimated to be around 15 eV.

Acknowledgement. This work was supported by JSPS KAKENHI Grant Number JP21K04492.

References

1. Maki, T., Kinefuchi, K., Cho, S., Watanabe, H.: electric propulsion system with dry ice propellant. In: Proceedings of Space Transportation Symposium FY2020, STEP-2020-024, Japan (2020)
2. Cho, S., Watanabe, H., Kinefuchi, K.: Discharge characteristics of a small hall thruster using CO_2 propellant. In: 66th Proceedings of the Space Sciences and Technology Conference, 2K08, Kumamoto, Japan (2022)
3. Watanabe, H., Cho, S., Kubota, K.: Performance evaluation of a 100-W class hall thruster. In: 36th Electric Rocket Propulsion Society, IEPC-2019-447, Vienna, Austria (2019)
4. Itikawa, Y.: Cross sections for electron collisions with carbon dioxide. J. Phys. Chem. **31**(3), 749–767 (2002)
5. Kawahara, H., Kato, H., Hoshino, M., Tanaka, H., Campbell, L., Brunger, M.J.: Integral cross sections for electron impact excitation of the $^1\Sigma^+{}_u$ and $^1\Pi_u$ electronic states in CO_2. J. Phys. B: At. Mol. Opt. Phys. **41**(8), 085203 (2008)

Experimental and Numerical Investigation of Body-Cushion Interface Pressure Distribution

Takashi Mashiko[1](✉), Yasushi Kakimoto[1], and Junichi Takano[2]

[1] Shizuoka University, Hamamatsu 432-8561, Shizuoka, Japan
mashiko.takashi@shizuoka.ac.jp
[2] Takano Co. Ltd., Hamamatsu 433-8121, Shizuoka, Japan

Abstract. To design high-functional cushions, it is necessary to elucidate how the body-cushion interface pressure distribution is affected by the specifications of the body as well as that of the cushion. We have experimentally and numerically investigated the pressure distributions for various body-cushion pairs. The experimental results, obtained for relatively simple cushions and bodies, show that the space mean pressure is proportional to the body mass and inversely proportional to the body-cushion contact area. Also, the maximum pressure observed at the body prominence proved to increase with the body surface curvature. Furthermore, it is confirmed that the use of an air-cell cushion can flatten the pressure profile, if the internal air pressure is adjusted appropriately, while the pressure profile has sharp peaks at the body prominences when a polyurethane cushion is used, especially when the body curvature is large. The numerical simulations, in which basic specifications of body and cushion were taken into account, produced pressure distributions that qualitatively agreed with the measurements. The results will contribute to establish the design of optimal cushions which realize desired pressure distributions with consideration of the body specifications.

Keywords: Cushion · Body Pressure · Measurement · Simulation

1 Introduction

The development of high-functional cushions is of increasing importance in recent years. For example, the prevention of pressure ulcer is one of the most desired functions of a cushion in aged societies, where elderly people and bedridden patients are increasing, while caretakers are decreasing with their burden increasing. As the main cause of pressure ulcer is long-time application of high pressure at bony prominences, regular body change is required to prevent its episode, which leads to heavy burden of caretakers. If the body pressures at the prominences are dispersed by the use of a suitable cushion, pressure ulcer can be prevented without heavy burden of caretakers.

Most of the cushion functions, including the prevention of pressure ulcer, could be realized by controlling the distribution of the body-cushion interface pressure or the body pressure. Pain alleviation, motion support, postural maintenance, massage, customization of mattresses, seats, wheelchairs, and so on can be given as examples of cushion functions enabled by the control of the body pressure distribution.

Y. Ono and J. Kondoh (Eds.): Inter-Academia 2023, LNNS 939, pp. 224–231, 2024.
https://doi.org/10.1007/978-3-031-54450-7_25

The air cell, an air-filled cell made of plastic films, is one of the promising cushion materials. The air-cell cushion, composing arrayed air cells, has two main advantages over traditional cushion materials such as flexible polyurethane foam; the pressure dispersibility and the pressure controllability, in addition to many other merits such as the lightness, compactness, low riskiness (against corrosion, leakage, freezing). The pressure dispersibility is that the interface pressure is expected to be the same everywhere in the body-cushion contact area, irrespective of the body shape, due to Pascal's law because air is a fluid. The pressure controllability is that the local body pressure can be changed by adjusting the internal air pressure of the corresponding cell.

Therefore, the effective usage of air-cell cushions has been sought, in particular for the purpose of pressure ulcer prevention [1]. In addition to the measurement of the body-cushion interface pressure [2], physiological measurements of transcutaneous gas pressures [3, 4] or skin perfusion [3, 5, 6] have been performed, which reported that the periodic change of internal air pressure increases the skin perfusion, for example [5, 6]. Also, a finite element study reported the lower human body stresses with the use of an air-cell cushion in comparison with the case of a foam cushion [7].

However, the air-cell cushion still has much to be investigated. Most of the previous studies have focused on "how to use air-cell cushions", using commercial air-cell cushions, rather than "how to design air-cell cushions", which must be more important to achieve desired functions. Also, the importance of consideration of individual anatomy (curvature of bony prominences, mass, composition and mechanics properties of soft tissues) in assessing the risk of pressure ulcer has been pointed out [1]. We believe that the consideration of body properties is important in the design of cushions also for purposes other than pressure ulcer prevention.

Therefore, we aim to clarify the effect of basic specifications of the cushion and of the body on the body pressure distribution, and then to design ideal cushions that realizes the desired pressure distributions, taking the body specifications into account. In the present work, we not only experimentally measure but also numerically compute pressure distributions for simple body-cushion pairs, where we try to grasp the dependence of the pressure distribution on the body and the cushion specifications.

2 Methods

2.1 Measurement

The interface pressure distribution $P(x, y)$ is measured with the use of a sheet-type pressure mapping system (FSA/BodiTrak, Vista Medical, Canada), which is laid on the cushion and below the body, and can measure simultaneously the local pressures at $32 \times 32 = 1024$ points with 14.4 mm pitch and 20 mmHg (2.7 kPa) accuracy (Fig. 1). The cushion is of air-cell or polyurethane. The body has two cylindrical prominences whose curvature radius is R, and its mass M is controlled by the weight put on it. The measurements are performed for different conditions, where the cushion material, internal air pressure p_{air} of air cell, M, and R are varied. The results shown below are data averaged at least over 10 s (at 1 Hz sampling) and over 3 samples for each condition, which is enough for the present measurements in static conditions for non-living bodies.

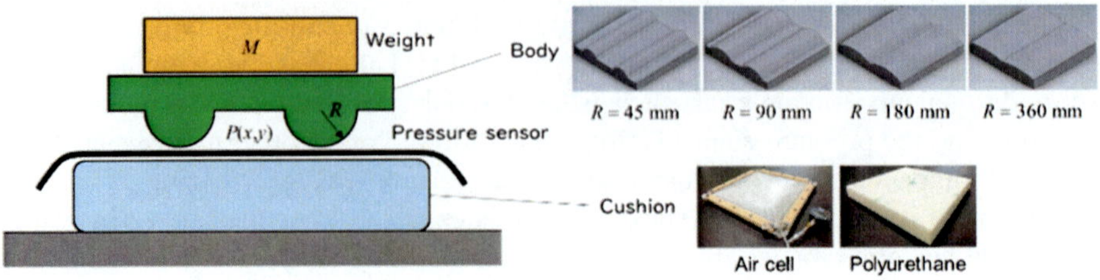

Fig. 1. Experimental setup for the measurement of body pressure distribution $P(x, y)$.

2.2 Simulation

The interface pressure distributions are also obtained in numerical simulations, whose results are compared with the measurements, through which the numerical code will be sophisticated so that it can predict pressure distributions in various situations. The essence of the computation is described below, where the body surface shape is given as the profile of height $b(x, y)$ above some reference plane, and the response properties of cushions and body characteristics are considered (see Fig. 2).

For the air-cell cushion, the unknown quantities are the body pressure $P(x, y)$, height $h(x, y)$, curvature radius $r(x, y)$ of cushion surface (film), internal air pressure p_{air}, film tension σ, and the height z_0 of the body reference plane. On the other hand, we have three local equations for every point (x, y) and three global equations. The former three are the local balance between P, p_{air}, and σ; the geometric relation between r and h; and the film-shape relation between h and b or r. The latter three are Hooke's law for σ; Boyle's law for p_{air}, and the global balance $\iint P(x, y)dxdy = Mg$ (*). Now we have the same number of equations as that of the unknown quantities. For a fixed value of z_0, the body pressure $P(x, y)$ is obtained by solving the simultaneous equations, then the balance (*) is checked. If the left-hand side is smaller than the right-hand side, z_0 is incremented. This procedure is repeated until (*) is satisfied.

For the polyurethane cushion, the unknow quantities are the body pressure $P(x, y)$, pressure of repulsive force from polyurethane $p_{\mathrm{ure}}(x, y)$, strain $\varepsilon(x, y)$ (which is equivalent with $h(x, y)$), and the height z_0 of the body reference plane. We assume that no strain is produced ($\varepsilon = 0$) in non-contact area, which may be a little unnatural. In this case, the computation is easier than the air-cell cushion. As ε is directly obtained by b and z_0, and p_{ure} is related with ε by some stress-strain relation (obtained in a compression test), P is obtained by the local balance $P = p_{\mathrm{ure}}$ immediately. The height z_0 is incremented until the obtained P satisfies (*), as in the case of air-cell cushion.

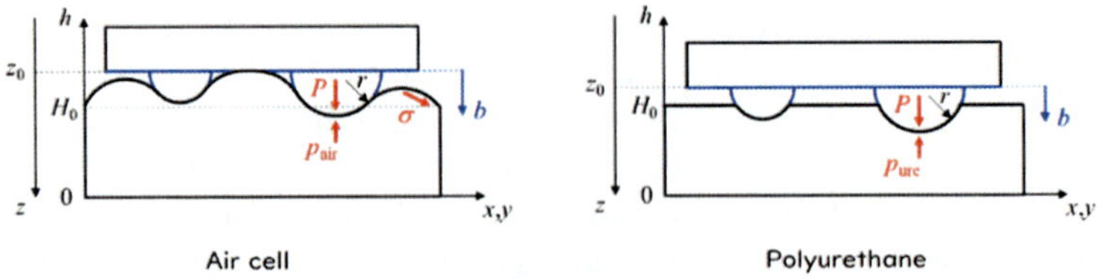

Fig. 2. Basic concept of numerical calculation of body pressure distribution $P(x, y)$.

3 Results and Discussion

Figures 3 and 4 show typical distributions of the body pressure P obtained in measurements and corresponding simulations, respectively, where single air-cell and polyurethane cushions, and 360-mm-square bodies of cylindrical surfaces ($M = 20$ kg, $R = 45, 90, 180, 360$ mm) whose axes are at $x = \pm 90$ mm are used (with the origin at the center of the body). The one-dimensional profiles of $\langle P \rangle_y(x)$ are obtained by averaging the two-dimensional distributions $P(x, y)$ in the central area of $-115 < y < 115$ mm.

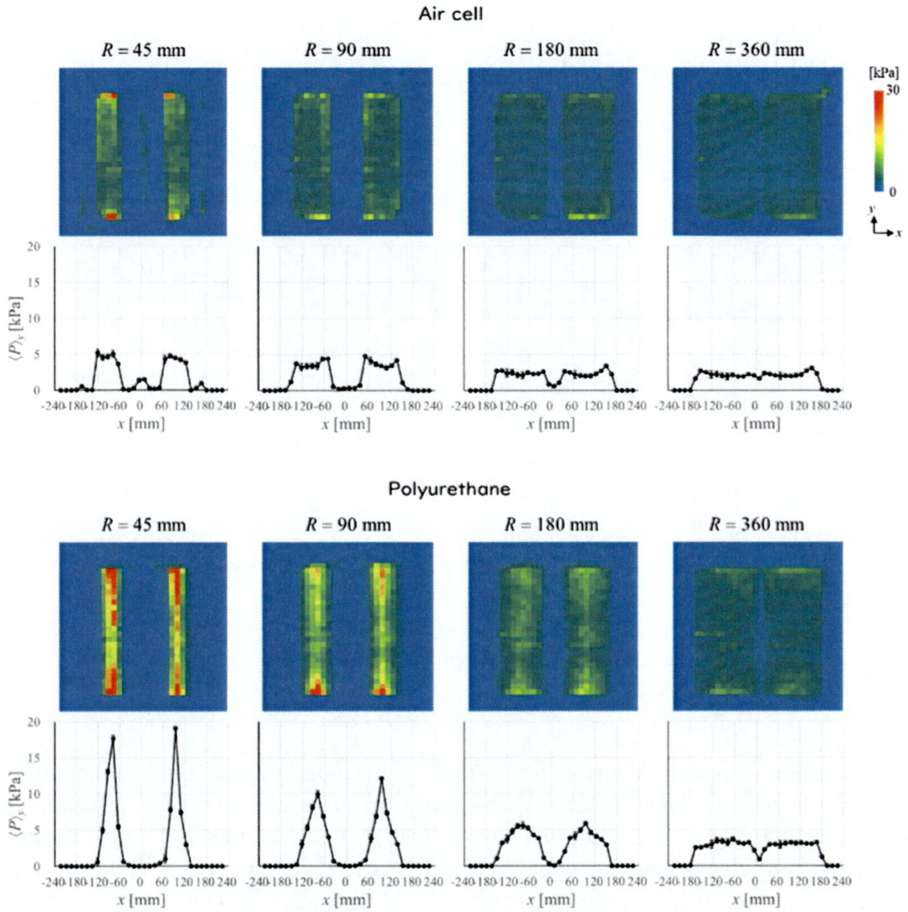

Fig. 3. Typical body pressure distributions of $P(x, y)$ and cross-sectional profile of $\langle P \rangle_y(x)$ obtained in measurements for air-cell and polyurethane cushions, and cylindrical bodies of $M = 20$ kg, $R = 45, 90, 180, 360$ mm.

First, it is clearly seen in Fig. 3 that the air-cell cushion is effective for pressure dispersion. In the case of polyurethane cushion, P is apparently large at prominences ($x = \pm 90$ mm), especially for smaller R, since at these positions the body presses the

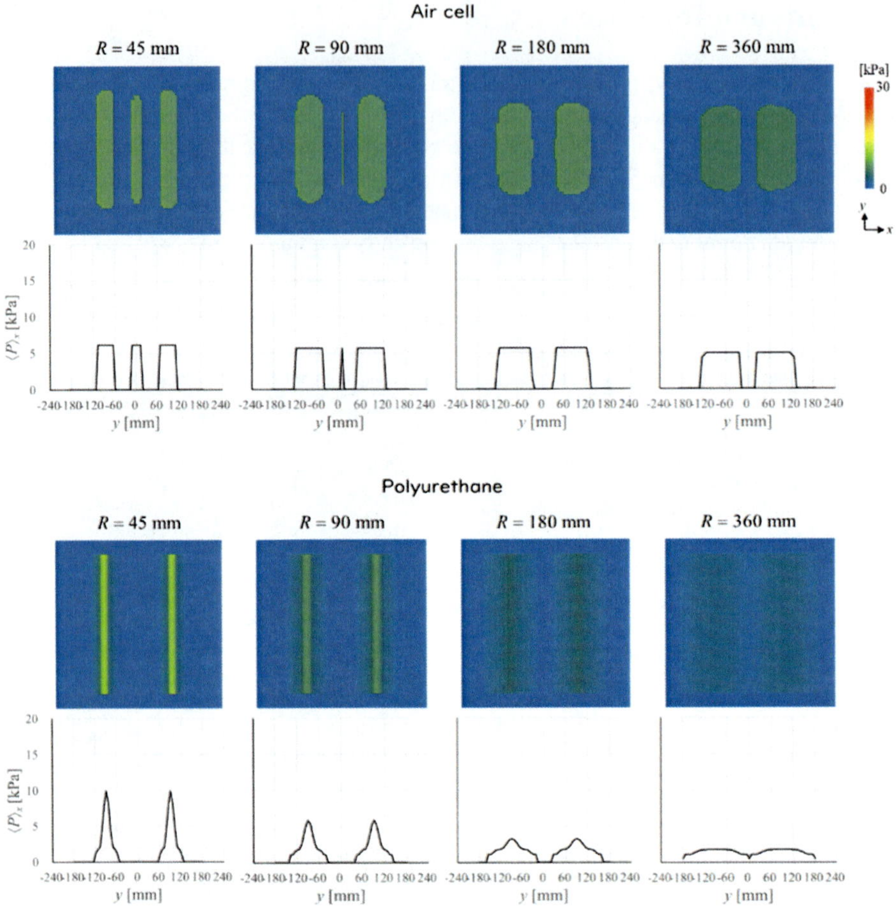

Fig. 4. Typical body pressure distributions of $P(x, y)$ and cross-sectional profile of $\langle P \rangle_y(x)$ obtained in simulations for air-cell and polyurethane cushions, and cylindrical bodies of M = 20 kg, R = 45, 90, 180, 360 mm.

cushion deeply, thus the repulsive force from the cushion becomes large, although the peaks get less outstanding for larger R (*i.e.*, when the prominences are less distinct). On the other hand, in the case of air-cell cushion, the P profile is almost flat over the area where the body and cushion are in contact, which expands in x direction with increasing R. Also, for both cushions, the magnitude of P decreases with increasing R, while it naturally increases with increasing M, as shown later.

The numerical results in Fig. 4 also exhibit the above-mentioned properties. That is, the air-cell cushion shows flat P profiles in the contact area, whereas the polyurethane cushion shows profiles with peaks at the body prominences. Also, as R increases, the P magnitude decreases, while the body-cushion contact area widens. Furthermore, the pressure P proved to increase with increasing the mass M (data not shown). Thus qualitative agreement between the experimental and numerical results was obtained, but the refinement of numerical code is required to achieve quantitative agreement.

We should note that the above-mentioned pressure dispersion effect of the air-cell cushion is an outcome of the adjustment of the internal air pressure p_{air}. Figure 5 shows the P distributions for the air-cell cushion and the body of cylindrical surface of $R =$ 90 mm, $M = 20$ kg, with different values of p_{air}. At $p_{air} = 0.0$ kPa, the profile has sharp peaks, reflecting the bottoming out of the body, *i.e.*, the body is directly supported by the solid support surface below the cushion, since the air is completely deflated. As p_{air} increases, the internal air begins to support the body, and the body-cushion contact area increases. At $p_{air} = 2.0$ kPa, the P profile is almost flat in the contact area (*i.e.*, the pressure is dispersed). With further increase in p_{air}, the contact area decreases, while the body pressure increases in the contact area, as the inflated cushion becomes like a solid. The adjustment of the internal air pressure plays an important role in controlling the body pressure of the air-cell cushion.

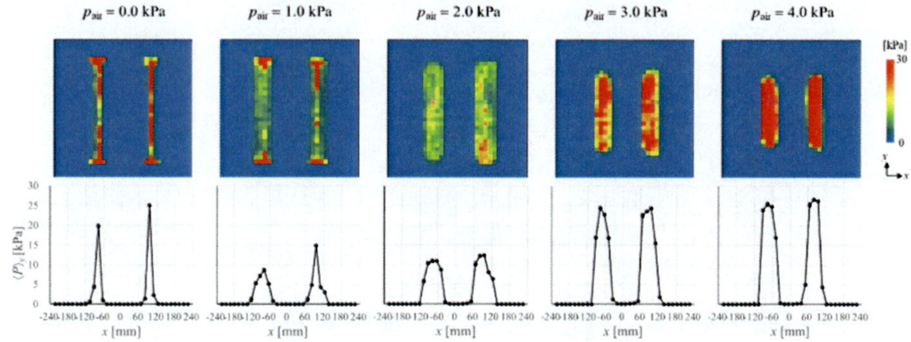

Fig. 5. Body pressure distributions of $P(x, y)$ and cross-sectional profile of $\langle P \rangle_y(x)$ for air-cell cushion with different internal air pressures p_{air} and cylindrical body of $M = 20$ kg, $R = 90$ mm.

We now show in Fig. 6 the space mean value $\langle P \rangle_{xy}$ of P obtained in measurements for the air-cell cushion and the cylindrical body. Here $\langle P \rangle_{xy}$ is plotted as a function of the contact area S, where S is defined by the area for which $P > 0.4$ kPa was recorded. We can see that the data are distributed along (but a little above) the theoretical curves of $\langle P \rangle_{xy} = Mg/S$. Although increasing M leads to increasing S, $\langle P \rangle_{xy}$ is basically proportional to M. We also notice that the increasing R increases S and thus $\langle P \rangle_{xy}$ decreases. Likewise, Fig. 7 shows that the maximum value P_{max} of P increases with increasing M and decreasing R. This implies higher local body pressures for a heavier body and steeper prominences, which seems to agree with real situations.

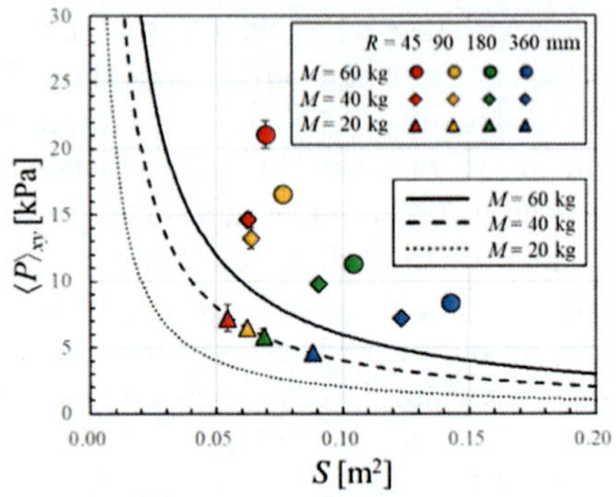

Fig. 6. Mean body pressure $\langle P \rangle_{xy}$ as a function of contact area S for air-cell cushion and cylindrical bodies. The solid, dashed, dotted lines are the curves of $\langle P \rangle_{xy} = Mg/S$ for $M = 20, 40, 60$ kg, respectively.

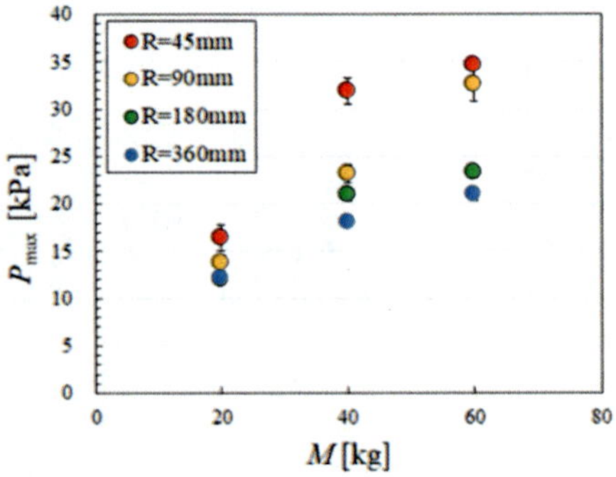

Fig. 7. Maximum body pressure P_{max} as a function of body mass M for air-cell cushion and cylindrical bodies.

4 Conclusion

Body pressure distributions were experimentally measured and numerically simulated for simple pairs of air-cell or urethane cushions and bodies with cylindrical surfaces. Basic effects of the cushion and body specifications, such as the body mass, body surface curvature, internal air pressure of air cell, as well as the pressure dispersion effect of the air-cell cushion, were confirmed in measurements. Also, the numerical results qualitatively agreed with the measurements.

Further measurements and simulations of the pressure distributions are in progress, for the acquisition of more data and the refinement of simulation. Also, more realistic body-cushion pairs, *e.g.*, the air-cell-polyurethane combined cushions and mannequins will be investigated, through which the simulation will be utilized to design ideal cushions that realize desired pressure distributions, *e.g.*, as flat a distribution as possible, whose validity will be verified by measurements.

References

1. EPUAP/NPIAP/PPPIA, Prevention and Treatment of Pressure Ulcers/Injuries: Clinical Practice Guideline (2019)
2. Phillips, L.: Interface pressure measurement: Appropriate interpretation of this simple laboratory technique used in the design and assessment of pressure ulcer management devices. Primary Intention **15**, 106–113 (2007)
3. Rithalia, S.V.S.: Evaluation of alternating pressure air mattress: one laboratory-based strategy. J. Tissue Viability **14**, 51–58 (2004)
4. Chai, C.Y., Bader, D.L.: The physiological response of skin tissues to alternating support pressures in able-bodied subjects. J. Mech. Behav. Biomed. **28**, 427–435 (2013)
5. Jan, Y.-K., Brienza, D.M., Boninger, M.L., Brenes, G.: Comparison of skin perfusion response with alternating and constant pressures in people with spinal cord injury. Spinal Cord **49**, 136–141 (2011)
6. Karg, P., Ranganathan, V.K., Churilla, M., Brienza, D.: Sacral skin blood flow response to alternating pressure operating room overlay. J. Tissue Viabil. **28**, 75–80 (2019)
7. Levy, A., Kopplin, K., Gefen, A.: An air-cell-based cushion for pressure ulcer protection remarkably reduces tissue stresses in the seated buttocks with respect to foams: finite element studies. J. Tissue Viability **23**, 13–23 (2014)

Assuring Reliability of Real-Time Particle Measurement in Flowing Media Using Light Scattering Method

Ryszard Jablonski(✉)

Warsaw University of Technology, Warsaw Sw. A. Boboli Str. 8, 02-525 Warsaw, Poland
ryszard.jablonski@pw.edu.pl

Abstract. The paper presents the optimization of configuration and calibration of special system for measuring the size and quantity of particles in flowing medium. The system is designed for real-time measurement of particles in the range up to 10 μm, moving with velocity up to 15 m/s and under pressure up to 20 bar and while stationary methods of particle size measurement are quite well mastered, their measurement in real time, at higher flow rates and higher pressures, is still a challenge for metrology.

The applied method is based on measuring light scattered by single particle using photomultipliers (single photon counters). We focused on studying the output from photomultiplier and finding relationships between pulse shape and particle size and also the time intervals between pulses.

Keywords: Particle Size · Light Scattering · Particle Seeding

1 Introduction

The measurement of particle size and quantity is routinely carried out across a wide range of industries and is often a critical parameter in the manufacture of many products. Most conventional methods, like microscopy or sieving techniques, are very accurate and reliable, but not suitable for real-time measurements [1].

Measurement of the size and number of particles in situ, at high flow rates with high accuracy and high pressure - forces a very complex construction of the measurement system and calibration of each of its elements. In nanometrology, when the limit of accuracy is reached, the reported results often contain no uncertainty, but a result without measurement uncertainty may raise doubts as to its reliability [2].

The chosen method is based on measuring light scattered by single particle and special emphasis is placed on the calibration of essential elements of the measurement system. The system is designed for real-time measurement of particles in the range up to 10 μm, moving with velocity 15 m/s and under pressure 20 bar.

Y. Ono and J. Kondoh (Eds.): Inter-Academia 2023, LNNS 939, pp. 232–238, 2024.
https://doi.org/10.1007/978-3-031-54450-7_26

2 Measurement System

The methods used to determine particle size can be divided into 5 groups: methods based on the interaction of light and matter, the use of the electrical properties of particles, sorting and classification, microscope methods and sedimentation methods.

Among them, definitely the most suitable for on-line measurement with high flow rate are those based on interaction of light and matter. To fulfill these conditions, we decided to realize special stand based on Multi Angle Light Scattering combined with Time of Flight and with the use of photomultipliers working in Single Photon Mode [3]. The measurement system, based on the above methods and supplemented with a block of standard particles seeding, is shown in Fig. 1. The term "measuring head" is limited to the units enclosed within the "clean air" area containing: lasers, optics, detecting system, nozzle (for particles supply), mechanical construction and clean air construction. The air, after passing though the fine filter, is pumped into the closed compartment containing the measuring head units. The generated small overpressure prevents dust from entering space containing optical systems.

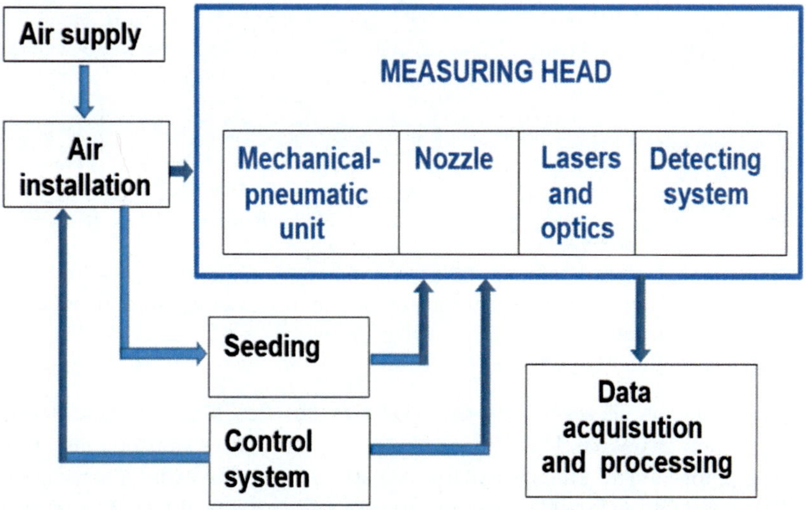

Fig. 1. Block diagram of a system

It was decided that the realized measuring head will contain two lasers of different wavelengths (each with special beam forming optics) and three detectors for collecting back scattered light, forward scattered light and side scattered light. We assume that in the future, measurements may concern particles made of different materials and their interaction with light of a different wavelength may have a different nature - which will help in their identification.

Two separate laser beams form a kind of parallel light curtains located very near each other (the widths and distance between them is 0.3 mm). Figure 2. Shows the schematic of the measuring head. Data from all transducers were connected in the HART

Communication Protocol and software applications for data acquisition and data analysis have been written in LabView [4].

Detectors measure scattered light in 3 directions. Our intention was to add all detector signals in order to increase signal level. Theoretically, the detector measuring forward scattered light receives the strongest signal, but it is also the direction of main laser beams. A special light extinction arrangement was designed. Finally, the measurement signal could be obtained from chosen detector or as the sum of two or three detectors.

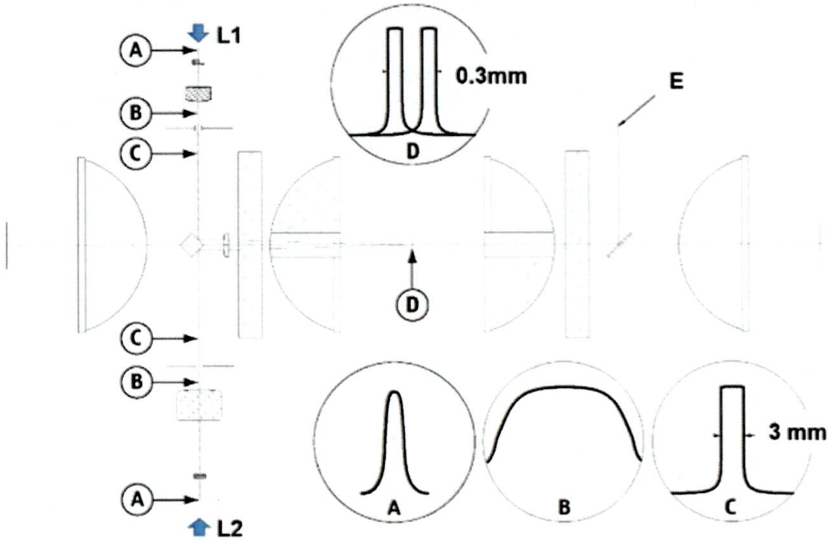

Fig. 2. Measuring head, beam forming: **L1** and **L2** - lasers, **A** – laser beam, **B** – expanded laser beam, **C** – obscured laser beam on 3 mm square aperture, **D** – laser beams in the measurement area, **E** - beam extinguisher.

The first series of measurements, using 0.8 μm particles, had a measurement uncertainty of as much as 0.9 μm [5]. The next stage was the analysis of the most important influencing factors and modification of the measurement system in order to reduce partial errors. The modification included the optimization of the shape of laser beam (Sect. 2.1) and the flow patterns (Sect. 2.2) in the measuring area and also system calibration by the simulation of particles flow (Sect. 2.3). The obtained results are presented in Sect. 3.

2.1 The Shape of Laser Beam

The shape of laser beam was measured directly in the waist. The system was operated automatically measuring the total cross-section of laser beam. The obtained results showed that the shape was influenced by diffraction effects on the edges of diaphragms. The "flat top" of laser beam was achieved by expanding the beam to 3mm, as it is shown in Fig. 3 and focusing it by cylindrical lens to 20 μm (Fig. 4).

Particles passing through (intersecting) two laser beams generate a pairs of measuring signals of time duration and separation depending on the particle velocity, laser beam waist size and displacement between them (Fig. 2D). In order to avoid random (noise) signals data processing algorithm accepts only signals appearing in pairs, besides the velocity of particles in the real time is measured. Different power and wavelength of laser beams was helpful to optimize laser beam parameters. Sampling time can be adjusted; the shortest is 1 μs.

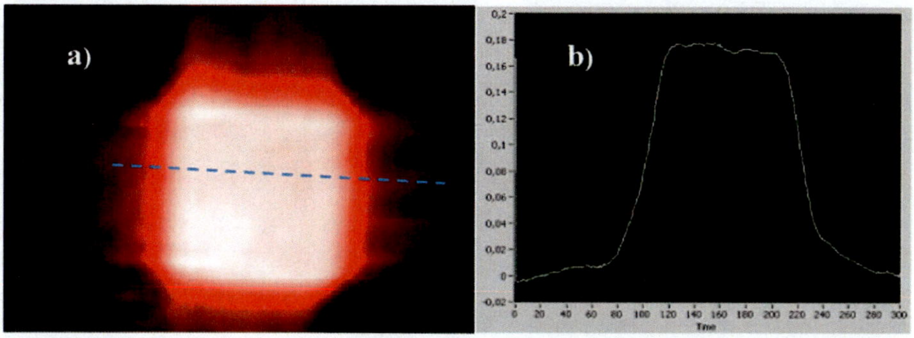

Fig. 3. The 3 × 3 mm shape of laser beam: a) view, b) intensity distribution in cross-section

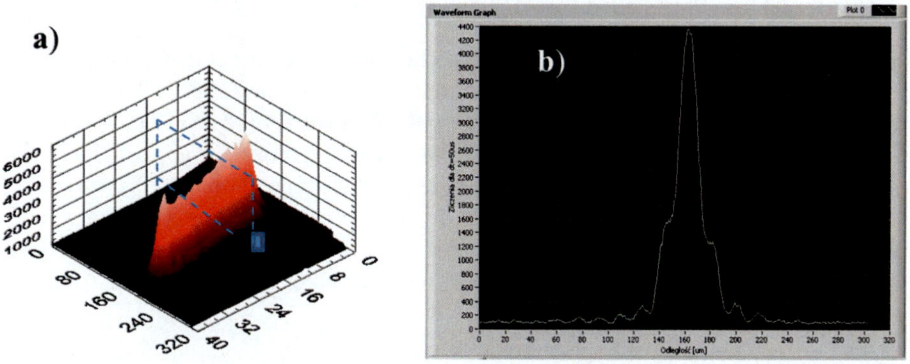

Fig. 4. Shape of a laser beam after focusing: a) view, b) intensity distribution in cross-section

2.2 The Simulation of Particles Flow

Obviously, the best proof of proper photodetectors-software performance is the calibration with particles of known size and known moving speed. This can be done by arranging single particle scanning periodically across both laser beams at the certain velocity. Figure 5 shows the principle of operation. The solid particle is placed on the flat, polished glass disc at the distance R from the center of rotation. The disc is fixed to the shaft of miniature DC motor with the control of its rotational velocity.

During rotation the particle is crossing both laser beams at defined distance R from the laser beam axes (also the nozzle axes) and the real distance between laser beams and the size (cross-section) of each laser beam is defined (calibrated). The experiments were carried with commercially available 2 μm dry glass microspheres, calibrated and certified with NIST traceable mean diameters of size distribution s = 0.6 μm (particle size distribution, PSD, for the most accurate particles reaches the level of 0.2 μm).

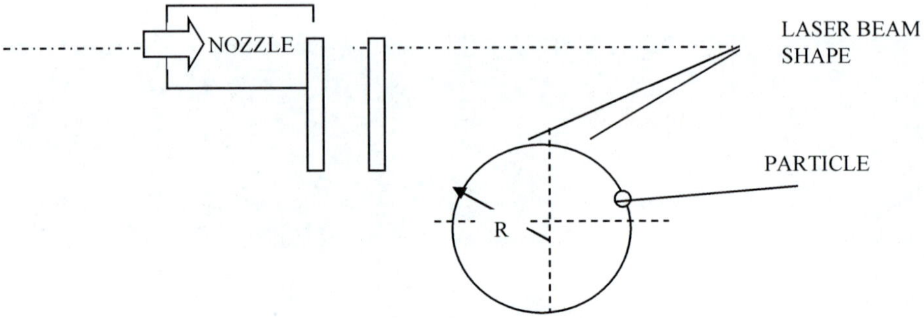

Fig. 5. The simulation of particles flow

2.3 Desired Flow Patterns in the Measuring Area

The velocity of the stream should be equal to the velocity of the surrounding clean air to minimize the turbulent mixing of both central and surrounding stream. Moreover, the velocity profile of the central stream, seeded with particles, should be as flat as possible. In order to obtain such flow characteristics, the converging nozzle is applied. Its profile is close to the profile of the Witoszyński nozzle [6]. This type of nozzles is used widely in flow metrology to obtain flat velocity profile in the primary standards. The flow paths downstream the nozzle exit (Fig. 6) show, that at a distance of 10 mm the mixing with surrounding clean air is still not intensive. The calculated velocity fields is shown in Fig. 7.

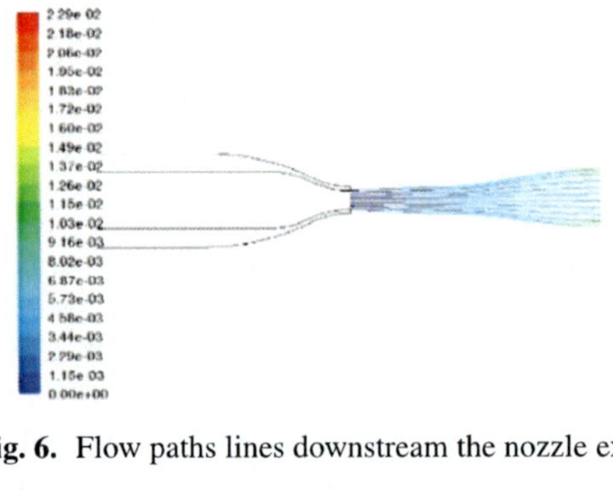

Fig. 6. Flow paths lines downstream the nozzle exit

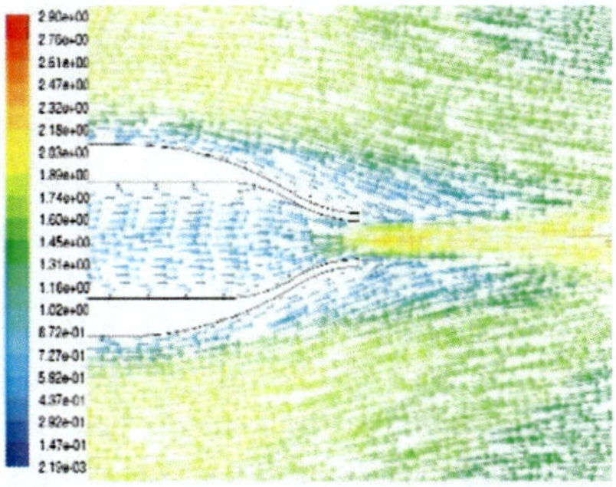

Fig. 7. Velocity field in the nozzle region

3 Results

The project required solving a number of independent threads. Each can be treated as an independent research problem of an experimental or theoretical nature. The theoretical part concerns in fact metrological analysis of the results. The modifications (Sects. 2.1–2.3) made it possible to reduce significantly all partial errors, represented in Table 1 by standard uncertainty of each input component x_j. PSD and less significant components depend on the devices/assemblies/elements used and may change depending on the needs and possibilities. The reliability of the measuring system, expressed in term of relative standard uncertainty is 0.165 (for 2 μm particles).

Table 1. Partial errors

Source of error (input estimate x_j)	Uncertainty $u(x_j)$ [μm]
PSD	0.2
Different particle velocity	0.15
Non-perpendicularity of beam and flow direction	0.15
Crossing the beam by more than one particle at the same time	0.1
Scattering from not filtered out impurities and particles	0.08
Dark current	0.05
Sampling resolution	0.05
Crossing beam at different heights	0.05
Fluctuation of laser direction	0.0409
Instability of laser power	0.01
Software approximation	0.01
Photomultiplier response time	0.0018
Relative combined standard uncertainty $u_r(y)$	**0.165**

4 Conclusion

The uncertainty value $\mathbf{u_r(y)}$ given in Table 1 is obtained for the measurement system calibrated using particles of size 2 μm. These are the smallest, available, calibrated solid particles (smaller particles are amorphous SiO_2 particles suspended in pure water, which is extremely difficult to evaporate). The partial errors included in Table 1 show that the PSD inaccuracy has a dominant share in the total measurement uncertainty. Modification of the measurement system by applying better components or improving its metrological parameters, could reduce the total uncertainty by only about 30%. We therefore assume that the proposed method based on measuring light scattered by single particle using photomultipliers operating in the single photon counting mode is correct and the obtained result can be considered satisfactory.

References

1. Evaluation of measurement data — Guide to the expression of uncertainty in measurement. CGM 100:2008 GUM 1995 with minor corrections (2008)
2. Henk G.M.: Particle Size Analysis. Springer, Dordrecht (2009). https://doi.org/10.1007/978-94-011-1238-3
3. Photomultiplier Tubes. Basics and Application. Fourth Edition. Hamamatsu Photonics K.K. (2017)
4. Woodford, C., Phillips C.: Numerical Methods with Worked Examples: Matlab Edition. Springer, Dordrecht (2011). https://doi.org/10.1007/978-94-007-1366-6. Accessed 16 Sept 2020
5. Jablonski R., Nowak R.: A new system for real time measurement of nano- and micro particles in flowing gas. Design and calibration. Adv. Mater. Res. Trans. Tech. Publ. **222**, 74–77 (2011)
6. Witoszyński Cz.: Selected works. PWN, Warszawa (1957)

Analytical Study of the Distribution of Grip Forces Dependence on the Coefficient of Friction for a Hand Prosthesis

Ewelina Drelich[1] and Jan Tracz[2](✉)

[1] Institute of Micromechanics and Photonics, Warsaw, Warsaw University of Technology, Św. Andrzeja Boboli 8, 02-525 Warsaw, Poland
[2] Institute of Aeronautics and Applied Mechanics, Warsaw University of Technology, Nowowiejska 24, 00-665 Warsaw, Poland
jan.tracz.dokt@pw.edu.pl

Abstract. Data on the distribution of forces acting on the phalanges in individual hand grips are sparse and are usually based on expensive human experiments or time-consuming computer calculations. However, they are essential for the proper design of prosthetic hands. Our goal was to create a simplified model to determine the dependence of these forces on the static coefficient of friction between the cylindrical cylinder held by the hand model, replicating the edgewise objects of everyday use, and the inner surface of the prosthetic phalanges. We created a CAD model of human hand kinematics based on literature data. The cylindrical grip was tested for a 70 mm diameter cylinder mimicking a 1 L bottle. The results showed that the relationship between the tested static friction coefficient and finger-cylinder forces is nonlinear. In the next stages, we plan to validate the obtained force results and use the data in the mechanical design of the prosthesis.

Keywords: Hand prosthesis · Grip force · Friction coefficient

1 Introduction

The human hand is considered one of the most complex mechanisms due to its ability to grasp, manipulate objects and make gestures. Therefore, a lack or a loss of it is a life changing dilemma. There are 541,000 amputees in the United States alone, even though amputation of the upper limb is approx. 4–5 times rarer than that of the lower [1]. Approximately 90% of upper limb amputations are caused by injury to the limb [2]. Severe infections or trauma also led to amputations.

There is a need for anthropomorphic hand prostheses to enable return to normal functioning in society. The process of selecting a prosthesis should begin as soon as possible after amputation to facilitate adaptation to the new situation [3]. Different types of prostheses are currently available, which vary in functionality and price. The group most at risk of amputation are people of low social status who cannot afford expensive prostheses [4, 5]. Recently, costs have come down thanks to 3D-printed prostheses such as the Hero Arm and True Limb [6, 7] and have contributed to a drop in prototyping costs to as low as $500 [8].

Y. Ono and J. Kondoh (Eds.): Inter-Academia 2023, LNNS 939, pp. 239–246, 2024.
https://doi.org/10.1007/978-3-031-54450-7_27

In order to make a prosthesis device, designers have to consider many aspects, such as the design of the prosthesis sockets, the number of actuated fingers, the capacity of the wrist or manufacturing technologies [9, 10]. Patients' needs are important, including the number of grips depending on the degrees of freedom, which ranges from four to even eight [11–14]. The use of new materials and methods, such as soft robotics, allows prostheses to be improved [15–17].

In order to design a prosthesis performing different grasps, it is necessary to determine the forces needed experimentally or analytically. Documentation gives a prosthesis force range of 70–140 N [11–14], but this refers to the whole device. There are a few experimental studies [18–20] performed with patients, however, the real hands are not prosthetic ones, and they differ in both geometry and prehension. There are engineering tools such as Finite Element Methods, which can determine forces, but usually expensive software is required to perform calculations. We decided to calculate the forces analytically as the first step in the design process.

2 Model and Methods

The model (Fig. 1) consists of a metacarpal part, four fingers made up of three phalanges and an inversion/adduction part, and a thumb made up of two phalanges and a metacarpal part, as well as a thumb counter element. The metacarpal part was modelled as a rigid prism on which the axes of the finger joints were fixed at fixed points. The thumb attachment axis was placed on the metacarpal tendon. The fingers were modelled with uniform lengths. The thumb was deflected 120° from the vertical for functionality, omitting full anthropomorphism.

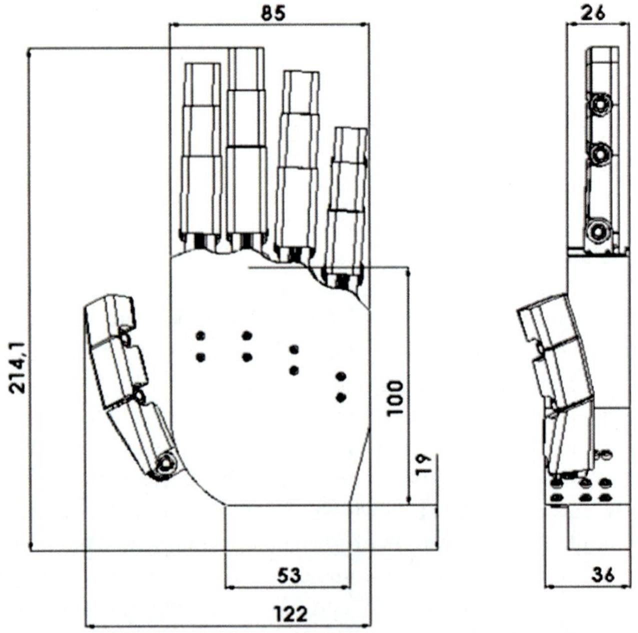

Fig. 1. Computer Aided Design model of the right-hand prosthesis

The interphalangeal joints were given the possibility of uniaxial rotation, while the metacarpophalangeal joints of fingers II–V, i.e. (index, middle ring and little), were given two degrees of freedom through the use of an adduction part. The modelling of the metacarpal bone of the thumb and an additional opposing element made it possible to reproduce the mobility of the thumb. To increase the anthropomorphic nature of its kinematics, an MCP metacarpophalangeal joint rotated 30° inwards was used. The model consists of open kinematic chains for each finger/thumb attached to the metacarpal housing. Flexion and inversion angles were defined for the fingers; additional angles defined the thumb opposition. The ranges of movement correspond to physiological values [21].

A cylindrical grip including the thumb, index finger and middle finger was modelled based on preliminary calculations. The phalanges were positioned so that the lateral side of the phalanges was parallel to the base of the cylinder, resulting in phalangeal thrust forces on the cylinder acting in horizontal parallel planes. In the kinematic model, the tangent bonds between the phalanges' surfaces and the cylinder were given and then the coordinates of the contact points were read out. There were 7 contact points in total, and they were numbered from 1 to 7. The 1st and 2nd corresponded to the thumb phalanges (distal and proximal), the 3rd and 4th are assigned to index phalanges (distal and middle) whereas 5th and 6th are for the middle finger phalanges (distal and middle). The metacarpal is assigned to the 7th point and it is in the middle of hands width. The 2D view is visible in Fig. 2.

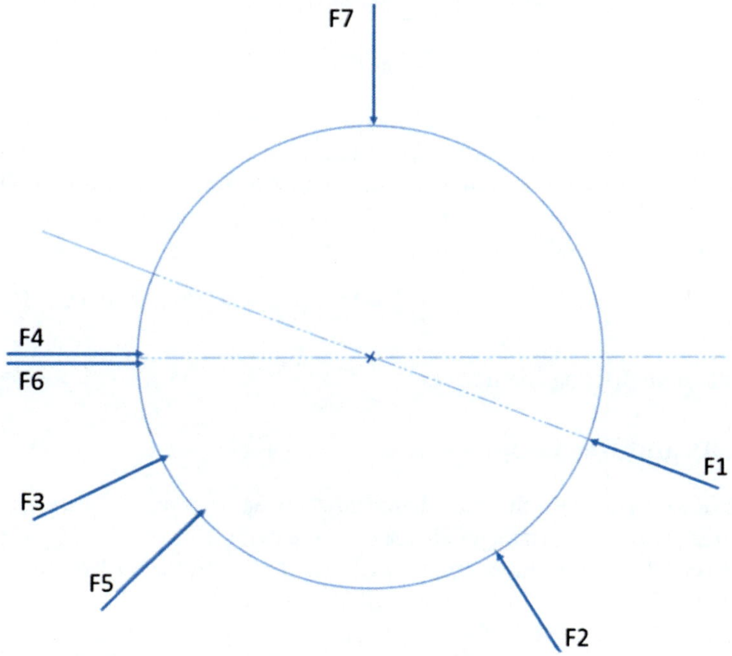

Fig. 2. Free body diagram of forces acting on the system

Each of the fingers included in the grip modelling had 2 points of contact with the cylinder. The contact of the metacarpal part was modelled as a line of contact with the side surface of the cylinder and for the calculations the force was concentrated in the centre of this line.

On the basis of the data read out, the assumed body weight and the parameterized friction coefficient, the equilibrium equations of force in the direction parallel to the cylinder axis were established (Eqs. 1–3).

$$F_{1X} + F_{2X} + F_{3X} + F_{4X} + F_{5X} + F_{6X} + F_{7X} = 0 \tag{1}$$

$$F_{1Y} + F_{2Y} + F_{3Y} + F_{4Y} + F_{5Y} + F_{6Y} + F_{7Y} = 0 \tag{2}$$

$$F_{1Z} + F_{2Z} + F_{3Z} + F_{4Z} + F_{5Z} + F_{6Z} + F_{7Z} = Q \tag{3}$$

where:

Q – is load equal to mass times the gravitational acceleration (m*g).

F – Force in the given direction (x,y,z).

Moments in x,y,z (Eqs. 4–6) were also considered around the axis of the coordinate system placed in the base of the cylinder, the z-axis of which coincided with the cylinder axis.

$$\sum Mx = 0 \tag{4}$$

$$\sum My = 0 \tag{5}$$

$$\sum Mz = 0 \tag{6}$$

To reduce the number of unknowns, the force moduli in the phalanges of one finger were assumed to be equal. The equilibrium equations were written for a cylinder with 4 unknown modulus forces (thumb, index finger, middle finger, metacarpal). The forces acting vertically are friction forces (Eq. 7). This reduced the number of unknowns to 4. The equations were solved numerically.

$$2\mu Fx_{THUMB} + 2\mu \mathrm{Fx}_{\mathrm{INDEX}} + 2\mu \mathrm{Fx}_{\mathrm{MIDDLE}} + \mu Fx_{METACARPAL} = Q \tag{7}$$

where:

-μ is the static friction coefficient.

3 Results and Discussion

The results (see Fig. 3) show that the relationship between the coefficient of static friction and the force acting between the phalanges and the cylinder is not linear. It can be seen that increasing the coefficient of friction (μ) by 0.3 (from 0.3 to 0.6) causes a 50% decrease in the force needed to hold the cylinder, but if friction is increased by another 0.3, the force will decrease by only 17%. To put it in broader perspective, the 0.9 friction coefficient corresponds to the rubber tires to concrete contact and the human hand has about 0.5 static friction coefficient. To achieve a high static friction coefficient, one would require rubber glove to be used over the device.

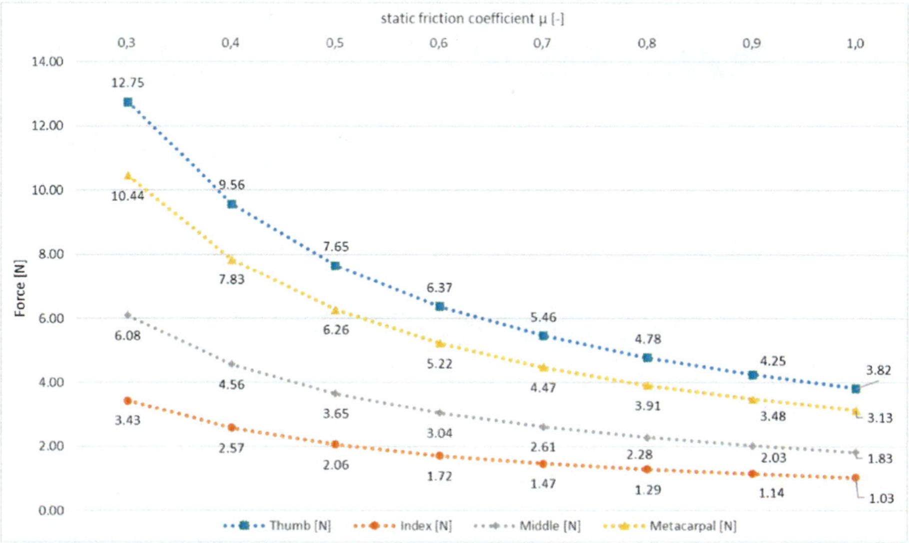

Fig. 3. Force acting on the phalanges and metacarpal as a function of friction coefficient.

The study [22] showing that it is friction not texture, which dictates grip force, has reported that a low friction coefficient increases the force required to lift an object nonlinearly. In the study [23], where dry and soapy objects were grabbed, the maximum grip force did not linearly follow the friction coefficient, although the relation was not inverse like in our case, but logarithmic. It might be possible that the extremities of the model (i.e. the 0.3 coefficient and below) are not correct. Nevertheless, we conclude that the method is physically correct due to the fact that for $\mu = 1$ the sum of the forces is equal to the weight (9.81 N).

However, that does not mean that the presented model is accurate. A comparison with literature data has been performed. The reports contained experiments performed on participants grabbing objects with diameters of 50, 65 and 88 mm, having mass of 600, 1000 and 1620 g respectively. Since setups were different and the friction coefficients were not reported, the percentage distribution of forces can be useful to compare the results. They have been calculated by dividing the force reported for a given finger by the total algebraic sum of forces for the grip. In the analytical model, the ratio stays constant with respect to static friction coefficient. The comparison can be seen in Fig. 4

We can observe that the relative values reported in the literature are different from the ones obtained in our method. The difference may come from different setups of the grip. There are also similarities - the force is always the highest on the opposing digit as it is balancing the forces from the rest of the fingers. In our case, it is the thumb. Although the results are different, we can draw the conclusion that in order to perform a grip the forces must be distributed over the fingers such that for a given condition forces stay in equilibrium. This might be obvious since it was assumed that the forces are in equilibrium, and the cylinder stays at rest. However, in experiments subjects usually have sensory feedback which enables them to judge how much force they are exerting, which isn't true for prosthesis driven by engines.

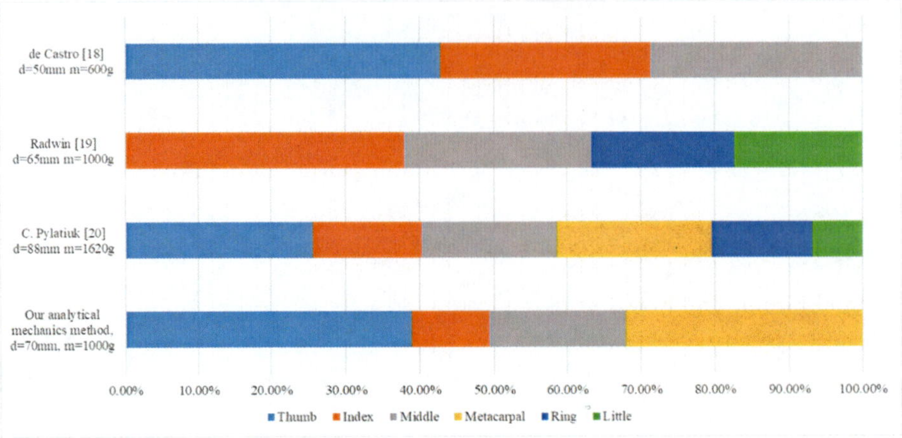

Fig. 4. Comparison of percentage force distributions on fingers in cylindrical grips

4 Conclusion

The study shows that the thumb contributes most of the grip force and therefore it's important for the device to have a fully functional mechanism that will replicate its function. Moreover, contact patch/area will also be crucial when performing the grip. Usually, softer materials provide better friction and are deformable, thus adapting to a given shape. This comes at the cost of mechanical properties and rigidity.

The values calculated are different from the ones reported by other studies, but they show that the cylindrical grip can be performed with various percentages of involvement of the individual fingers. However, the grip must be uniform so that the forces acting on the object stay in equilibrium. The final value of forces depends on the geometry and setting of fingers on the object. We think that the model is physically correct, but it is possible that for lower friction coefficients it loses accuracy, due to the simplifying assumptions that had to be used to solve it. Our conclusion is that future tests are warranted, but for now, further design can be based on the model with some safety margin.

Based on the results obtained, it was assumed that during the prototyping phase, a polymer resin will be chosen to create the model, which will be printed using SLA 3D printing technology giving a coefficient of friction of about 0.6. The resin has average mechanical properties, but has good printability and the 3D prints will have better tolerances than in FDM/FFF. The prototype will be able to be covered with an additional surface providing increased friction. Future work will include validating the forces and using the resulting values for mechanical design of the prosthesis.

References

1. Ziegler-Graham, K., MacKenzie, E.J., Ephraim, P.L., Travison, T.G., Brookmeyer, R.: Estimating the prevalence of limb loss in the United States: 2005 to 2050. Arch. Phys. Med. Rehabil. **89**(3), 422–429 (2008). https://doi.org/10.1016/j.apmr.2007.11.005. PMID: 18295618
2. Pezzin, L.E., Dillingham, T.R., MacKenzie, E.J.: Rehabilitation and the long-term outcomes of persons with trauma-related amputations. Arch. Phys. Med. Rehabil. **81**(3), 292–300 (2000). https://doi.org/10.1016/s0003-9993(00)90074-1. PMID: 10724073
3. Hubbard, S., Bush, G., Kurtz, I., Naumann, S.: Myoelectric prostheses for the limb-deficient child. Phys. Med. Rehabil. Clin. N. Am. **2**(4), 847–866 (1991)
4. Resnik, L., Meucci, M.R., Lieberman-Klinger, S., et al.: Advanced upper limb prosthetic devices: implications for upper limb prosthetic rehabilitation. Arch. Phys. Med. Rehabil. **93**, 710–717 (2012)
5. van der Riet, D., Stopforth, R., Bright, G., Diegel, O.: An overview and comparison of upper limb prosthetics. In: AFRICON, 2013. IEEE, New York (2013)
6. Open bionic prosthetics Hero Arm. https://www.youtube.com/watch?v=JuLRBcW8DsY
7. True limb available online. https://www.youtube.com/watch?v=N01QN-3OB3s
8. ten Kate, J., Smit, G., Breedveld, P.: 3D-printed upper limb prostheses: a review, disability and rehabilitation. Assist. Technol. **12**(3), 300–314 (2017). https://doi.org/10.1080/17483107.2016.1253117
9. Setty, K., van Niekerk, T., Stopforth, R.: Design considerations of the Touch Hand 4. Procedia CIRP. **91**, 494–502 (2020). https://doi.org/10.1016/j.procir.2020.02.205
10. Demofonti, A., Carpino, G., Tagliamonte, N.L., Baldini, G., Bramato, L., Zollo, L.: Design of a modular and compliant wrist module for upper limb prosthetics. The Anatomical Record. **306**, 764–776 (2022). https://doi.org/10.1002/ar.24911
11. Otto bock Healthcare GmbH: "Michelangelo Hand" (2017). https://www.ottobock.com/en-us/product/8E500
12. Touch Bionics: I-Limb Ultra Revolution Brochure. No. 694 5462, pp. 1–2 (2013)
13. Vincent Systems, VINCENTevolution4. https://www.vincentsystems.de/vincent-evolution4?lang=en
14. Taska Prosthetics: Taska Hand (2018). http://www.taskaprosthetics.com
15. Cuellar Lopez, J.S., Plettenburg, D., Zadpoor, A.A., Breedveld, P., Smit, G.: Design of a 3D printed hand prosthesis featuring articulated bio-inspired fingers. Proc. Inst. Mech. Eng. Part H: J. Eng. Med. **235**(3), 336–345 (2021). https://doi.org/10.1177/0954411920980889
16. Dunai, L., Novak, M., García, E.C.: Human hand anatomy-based prosthetic hand. Sensors. **21**(1), 137 (2020). https://doi.org/10.3390/s21010137.PMID:33379252;PMCID:PMC7795667
17. Capsi-Morales, P., Piazza, C., Catalano, M.G., et al.: Comparison between rigid and soft poly-articulated prosthetic hands in non-expert myo-electric users shows advantages of soft robotics. Sci. Rep. **11**, 23952 (2021). https://doi.org/10.1038/s41598-021-02562-y
18. de Castro, M.C.F., Cliquet, A.: An artificial grasping evaluation system for the paralysed hand. Med. Biol. Eng. Comput. **38**, 275–280 (2000). https://doi.org/10.1007/BF02347047
19. Radwin, R.G., Oh, S., Jensen, T.R., Webster, J.G.: External finger forces in submaximal five-finger static pinch prehension. Ergonomics **35**(3), 275–288 (1992). https://doi.org/10.1080/00140139208967813. PMID: 1572337
20. Pylatiuk, C., Kargov, A., Schulz, S., Döderlein, L.: Distribution of grip force in three different functional prehension patterns. J. Med. Eng. Technol. **30**, 176–182 (2006). https://doi.org/10.1080/03091900600565217

21. Netter, F.H.: Atlas of Human Anatomy: Latin Terminology Elsevier. 7th edn. Elsevier, Amsterdam (2019)
22. Cadoret, G., Smith, A.M.: Friction, not texture, dictates grip forces used during object manipulation. J. Neurophysiol. **75**(5), 1963–1969 (1996). https://doi.org/10.1152/jn.1996.75.5.1963
23. O'Meara, D.M., Smith, R.M.: Functional handgrip test to determine the coefficient of static friction at the hand/handle interface. Ergonomics **45**(10), 717–731 (2002). https://doi.org/10.1080/00140130210159247

Photonics, Optoelectronics and High-Energy Physics

Study of Femtosecond Laser Pulse-Induced Carbonization of Polyimide

Masaki Uematsu[1], Saulius Juodkazis[2], Junko Morikawa[3], and Vygantas Mizeikis[1,4(✉)]

[1] Graduate School of Integrated Science and Technology, Shizuoka University, Shizuoka, Japan
mizeikis.vygantas@shizuoka.ac.jp
[2] Centre for Optical Sciences, Swinburne University of Technology, Hawthorn, Australia
[3] School of Materials and Chemical Technology, Tokyo Institute of Technology, Tokyo, Japan
[4] Research Institute of Electronics, Shizuoka University, Shizuoka, Japan

Abstract. Spatially localized carbonization of cheap and widely available insulating polymeric films, such as polyimide (PI), into electrically and thermally conductive graphitic carbon is important for the development of photonic, electronic and thermal devices. We use tightly focused femtosecond laser pulses to locally convert PI into carbon at the laser beam focus, and by translating the focus draw extended 2D and 3D structures in PI. By embedding micro-regions of electrically and thermally conductive carbon into PI using laser irradiation, we create a composite material having modified physical properties. In the future such modification can be used to create new materials and devices for control over optical, electrical, and thermal microdevices phenomena.

Keywords: Laser Material Processing · Femtosecond Pulses · Laser-Induced Carbonization · Polyimide · Kapton · Carbon

1 Introduction

Polyimide (PI) was developed by DuPont in 1960 and is also known by its trademark Kapton. This material can be mass-produced cheaply and is attractive for applications in various fields of science and technology. Among its notable physical properties are high heat resistance retained within the temperature range of −269 °C to 400 °C, excellent electrical and thermal insulation capability, and high mechanical, chemical, and ionizing radiation robustness. This commercially affordable and cheap polymer is widely used in microelectronics, electrical and electronic equipment, automobiles, aircraft and is typically prepared in the form of thin or thick film or molded parts. At the same time, the possibility to embed PI with other materials having higher electrical and thermal conductivity, may further widen its applicability in technology and engineering. It was found that subjecting PI to temperatures higher than the threshold value of 850 °C converts it fully or partially to graphitic carbon [1]. Irradiating PI by a focused laser beam can be used to achieve a highly spatially localized photothermal effect with temperature at the focus exceeding the carbonization threshold, and spatial scan of the focus on the surface

Y. Ono and J. Kondoh (Eds.): Inter-Academia 2023, LNNS 939, pp. 249–255, 2024.
https://doi.org/10.1007/978-3-031-54450-7_28

or in the bulk of PI can be used to create complex 2D or 3D carbonization patterns. The resulting composite material may have electrical and thermal properties somewhat different from those of pristine PI, controllable via parameters of the laser beam and the spatial structure of the recorded pattern. It was found in 1991 that ultraviolet (UV) KrF laser irradiation of PI surface increased its electrical conductivity by 15 orders of magnitude [2]. Subsequently it was clarified using Raman scattering spectroscopy that laser-irradiated contain high density of C-C bonds [3]. J. Lin et al. [4] found that carbonization of PI surface can also be induced using irradiation by an infrared (IR) CO_2 laser. IR radiation, especially that belonging to the near-infrared (NIR) spectral region has one important advantage, namely its relative optical transparency, which allows it to propagate into the bulk of PI without absorption. Focused NIR laser beams, especially those delivering ultrashort picosecond (ps) and femtosecond (fs) pulses can still induce two-photon and multi-photon absorption (TPA and MPA) selectively at the focus, leading to highly localized carbonization of PI, and allowing fabrication of complex 3D carbonization patterns embedded in the bulk of PI [5]. In this study we have used fs laser processing of PI to fabricate simple graphitic carbon structures, and report on their physical properties. We were able to confirm fabrication of carbon micro-patterns in the bulk and on the surface of PI, and provide evidence of modified electrical and thermal properties in these regions. Similar approach may be used for fast, non-contact prototyping of carbon-based microstructures in PI, and possibly also other optically transparent organic materials.

2 Experimental Details

In this section we describe opto-mechanical setup used to fabricate the samples and the initial materials used in the experiments. Figure 1 shows schematic picture of the laser processing setup used in the experiments. The light source was an Yb:KGW femtosecond laser (Carbide, Light Conversion) emitting laser pulses having a central wavelength of 1030 nm, a pulse duration of 400 fs at a repetition rate of 200 kHz. The laser beam was expanded by a 5× telescope to a diameter of 6.5 mm which was sufficient to fill the entrance pupil of the microscope objective lens which was used for focusing the laser beam into the sample. In our experiments we used two lenses having magnification and numerical aperture (NA) of (20×, NA = 0.46) and (50×, NA = 0.80). Transverse diameter of the focal spot d obtained using these lenses were calculated using the well-known relationship $d = 1.22\lambda/\mathrm{NA}$, and are $d = 2.7$ μm and $d = 1.57$ μm, respectively. The laser beam was polarized linearly along the x-axis direction.

During the fabrication the samples were translated using a high-precision nanopositioning system (Aerotech ANT95) controlled by a specialized software (Poli3D, Femtika). The fabrication process was observed in situ using an LED illuminator and a video camera. During the observation the laser-damaged areas appear as dark (low-transparency) points or lines surrounded by bright (high-transparency) regions of non-irradiated pristine PI. The samples used were sheets of commercially available PI and had a thickness of 20, 50, 100, and 200 μm.

The laser-fabricated structures were subsequently characterized using various techniques. Optical microscopy was used for visual evaluation of the laser-processed regions,

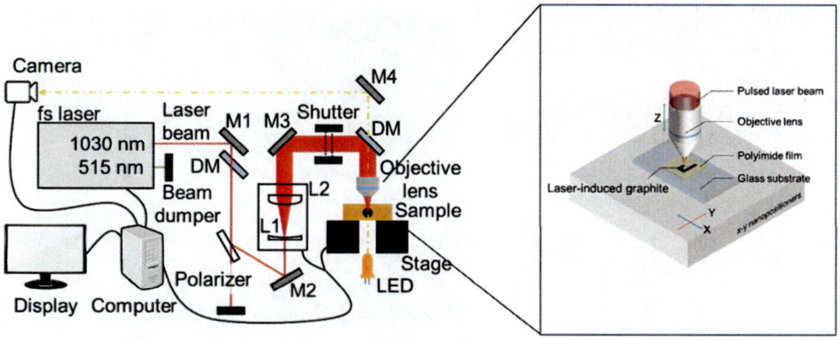

Fig. 1 Opto-mechanical setup used for laser-induced carbonization of PI. Components denoted by abbreviations are M1…M4 – mirrors, L1, L2-lenses, DM-dichroic mirrors, LED – light emitting diode illuminating the sample

while Scanning Electron Microscopy (SEM) was used to observe fine features in these regions. In order to detect the laser-induced generation of carbon, Energy-dispersive X-ray spectroscopy (EDS) and micro-Raman spectroscopies were employed. Modification of electrical properties was observed using two-terminal electrical resistance measurements, while modification of thermal conductivity was visualized using thermal imaging [6].

3 Results and Discussion

3.1 Bulk Modification of PI

Modification of PI by tightly focused fs laser pulses can be visualized using optical microscopy. Typically, the photomodified regions are seen in transmission images of the sample as dark spots in a bright background consisting of non-irradiated PI. The dark appearance does not prove that generation of carbon by laser irradiation has occurred; the darkening may be caused by strong refractive index modulation, and presence of carbon was confirmed using other methods as described in Sect. 3.2. Figure 2(a,b) shows optical top-view (a) and side-view (b) images of an area containing periodic array of photomodified spots. Each spot was irradiated using average laser power of 10 mW for 200 ms, resulting in the total number of 40,000 pulses.

The laser beam was focused to the depth of about 100 μm under the surface in order to generate carbonized regions in the bulk of PI, since bulk photomodification would allow formation of 3D structures fully embedded in the bulk of PI, which is more interesting for possible applications. The bulk nature of fabrication is illustrated by the side-view image in Fig. 2(b). In order to aid the side-view observation the sample was cut as close as possible to the fabricated area, and the edge was mechanically polished. Detailed images of the laser-induced photomodification were obtained using SEM observation. Figure 2(c) shows a modified surface region which consist of a laser-induced damage pit with fine pattern of periodic ripples, or laser-induced periodic surface structures (LIPSS) having the period of 230 nm. The ripples are perpendicular to the polarization orientation E of the incident laser beam, emphasized by arow in the figure. Surface nanogratings

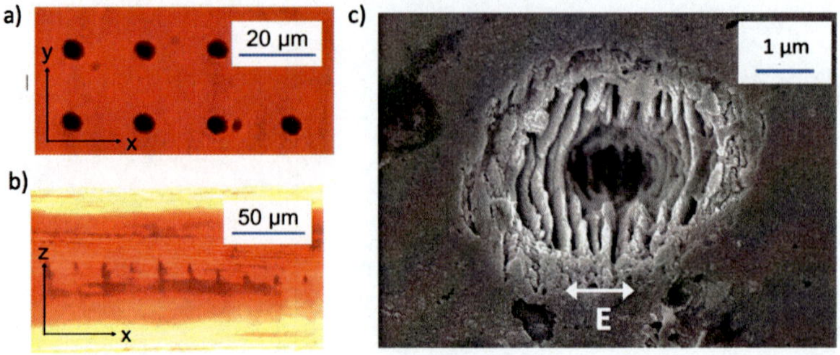

Fig. 2 Optical top and side-view images of laser-induced modifications in the bulk of PI film (a,b), detailed SEM image of surface damage featuring periodic LIPSS structure (ripples) oriented perpendicular to the polarization of the laser beam. The double arrow indicates orientation of the electric field vector *E* of the incident laser beam (c)

due to LIPSS are of interest to a wide range of application areas [7]. Their formation on the surface of PI and possible applications will be addressed in our future publications.

3.2 Evidence for Carbon Generation

To confirm the laser-induced carbonization of PI, surface structures were investigated using EDS technique, which enables one to analyze the elemental composition of the selected area. The results are shown in Fig. 3(a). The EDS data is presented as shows the dependence of spectral emission intensity (in counts per second per eV) on the energy (in keV). Comparison between the pristine (left) and laser-irradiated (right) spectra clearly reveal increased carbon concentration due to irradiation. For this, the ratios between magnitudes of C peak to N and O peaks were estimated and are shown in the figure. As can be seen, the estimates show relative increase of C response by 88% (with respect to N) and 20% (with respect to O). Although at present it is difficult to determine which element (N or O) is a better reference, these results qualitatively confirm the carbonization on the surface of PI. It can be expected that similar conditions are also achieved during bulk irradiation (the local temperature at the focus might be even higher, since the irradiated region is surrounded by highly thermally-resistant PI).

Further evidence for carbonization was obtained from Raman data. The measurements were performed using a micro-Raman spectrometer (NRS-7100, Jasco) at the excitation wavelength of 532 nm. The results are shown in Fig. 3(b). Pristine PI regions not irradiated by the laser exhibit a broad Raman spectrum without significant spectral features. On the other hand, laser-irradiated regions clearly exhibit the characteristic graphite D and G peaks at 1300 and 1600 cm^{-1}, respectively. The peak assignment can be confirmed using the reference spectrum of graphite taken from a core of 10B pencil, which is also shown in the figure.

Hence, the data shown so far suggests the following qualitative picture of C generation on PI. Laser-induced photomodification at the wavelength of 1030 nm, where PI is optically transparent, starts with TPA or MPA process, instantaneously producing a limited amount of initial free electrons, which become excited by the same laser pulse,

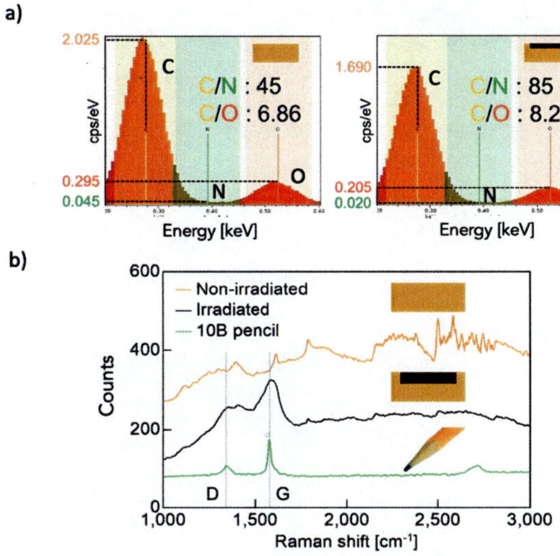

Fig. 3 EDS spectra taken from pristine and laser-irradiated areas of PI(a), Raman spectra taken from pristine and laser irradiated areas of PI together with the spectrum taken from 10B pencil(b). Optical top and side-view images of laser-induced modifications in the bulk of PI film (a,b)

leading to photoionization of PI constituent atoms, and local increase of the temperature above the threshold value of 850 °C required for carbonization. The carbonization then occurs by breaking of C–O and C–N bonds in the PI, and forming new C–C bonds as indicated by the data in Fig. 3(a). As noted above, one can expect that this process can take place both on the surface and in the bulk of PI.

3.3 Modification of Electrical and Thermal Properties

Electrical and thermal properties of PI and carbon are vastly different, with carbon being much more conductive than PI. Using laser-induced carbonization in the bulk of PI one can create conductive microstructures embedded in a dielectric. In order to verify this possibility, linear channels having a length of 3 mm were fabricated below the PI surface. At the same time both ends of the channel were pointed vertically towards the surface. At these points silver paste was applied, and electrical contacts were formed permitting resistivity measurements. The measured channel resistivity for several laser scanning (writing) velocities are shown in Fig. 4(a). The inset shows schematic configuration of the experiment. As can be seen, the resisitivity decreases with laser power. Higher laser power leads to a higher local temperature, which affects physical properties of the laser-generated carbon. It was suggested previously that laser-irradiated regions contain a nanocrystalline nanostructure consisting of polycrystalline and amorphous carbons [8]. At higher temperature a more connected percolation network for the flow of electric current may result, thus leading to decreased resistivity. However, using excessively high laser power would obliterate the carbonized region, and more studies are needed in order to determine the optimum exposure conditions for fabrication of electrically conductive structures.

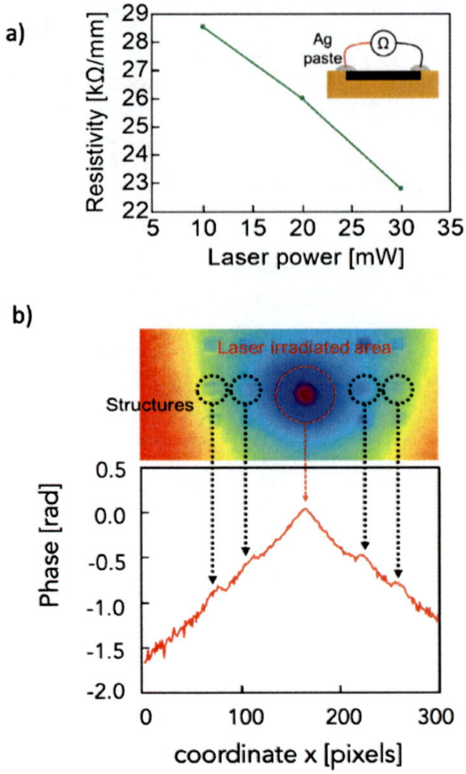

Fig. 4 Measured dependence of the electrical resistivity of laser-generated carbon channels buried in the bulk of PI film versus the average laser power (a), thermal imaging of a heat wave in a 200-μm-thick PI film containing laser-processed areas. The top image shows 2D phase profile of the heat wave, the bottom plot shows the profile along a horizontal straight line traversing the fabricated areas (b). Vertical arrows emphasize the laser processed regions and their thermal properties notably different from those in pristine PI

Modification of thermal properties due to fs laser irradiation was qualitatively investigated using thermal imaging technique. The experimental setup and measurement conditions used for investigations were similar to those used in Ref. [6]. Temperature localization and a heat flow in the samples locally heated by a focused beam of a laser diode were in situ monitored using an infrared camera. The results are shown in Fig. 4(b). As can be seen, laser-structured regions can create strong thermal phase gradients of a few degrees over the tens of micrometers distance, distinctly different from that in the pristine PI film. Detailed interpretation of the data will be reported elsewhere in the future. Nevertheless, these qualitative results clearly reveal the consequences of laser-induced carbonization, and indicate the possibility create spatially anisotropic heat transport using anisotropic carbon-based structures in PI.

4 Conclusions

The results reported here illustrate versatility of fs laser processing technique for the realization of PI-carbon composite material using commercialy available PI films. We were able to realize simple carbon-based microstructures embedded in the bulk of PI film, to confirm the increased carbon density in the laser-processed regions, and to obtain qualitative evidence of modified electrical and thermal properties in these regions. These findings promise fast laser prototyping of electrical and photonic devices using cheap and widely available materials.

Acknowledgements. This work was supported by the JST CREST Grant Number JPMJCR19I3, Japan.

References

1. Küper, S., Brannon, J., Brannon, K.: Threshold behavior in polyimide photoablation: single-shot rate measurements and surface-temperature modeling. Appl. Phys. A **56**, 43–50 (1993)
2. Schumann, M., Sauerbrey, R., Smayling, M.C.: Permanent increase of the electrical conductivity of polymers induced by ultraviolet laser radiation. Appl. Phys. Lett. **58**, 428 (1991)
3. Gu, X.J.: Raman spectroscopy and the effects of ultraviolet irradiation on polyimide film. Appl. Phys. Lett. **62**, 1568–1570 (1993)
4. Lin, J., Peng, Z., Liu, Y., Ruiz-Zepeda, F., Ye, R., Samuel, E.L.G., Yacaman, M.J., Yakobson, B.I., Tour, J.M.: Laser-induced porous graphene films from commercial polymers. Nat. Commun. **5**, 5714 (2014)
5. Gattass, R.R., Mazur, E.: Femtosecond laser micromachining in transparent materials. Nat. Commun **2**, 219–225 (2008)
6. Morikawa, J., Hayakawa, E., Hashimoto, T., Buividas, R., Juodkazis, S.: Thermal imaging of a heat transport in regions structured by femtosecond laser. Opt. Expr. **19**, 21 (2011)
7. Bonse, J.: Quo Vadis LIPSS?—recent and future trends on laser-induced periodic surface structures. Nanomaterials **10**, 1950 (2020)
8. Dorin, B., Parkinson, P., Scully, P.: Direct laser write process for 3D conductive carbon circuits in polyimide. J. Mater. Chem C **5**, 4923–4930 (2017)

Reflectance of Black Silicon with Sol-Gel ZnO Passivation Films: Optical Simulation and Experimental Correlation

Alina Semchenko[1(✉)], Dmitry Kovalenko[1], Gagik Ayvazyan[2], and Levon Hakhoyan[2]

[1] Skorina Gomel State University, Sovetskaya 104, 246019 Gomel, Belarus
alina@gsu.by
[2] National Polytechnic University of Armenia, 105, Teryan, 0009 Yerevan, Armenia

Abstract. The results of a study of the optical properties of a black silicon (BS) layer passivated with ZnO films obtained by the sol-gel technique are presented. The BS layer was formed by the reactive ion etching process using an SF_6/O_2 plasma. The reflectance spectra of the BS/ZnO structures were simulated using the finite difference time domain method. In this simulation, the structure is considered a thick silicon substrate with regular near-surface straight circular nanocones uniformly coated with a thin ZnO film. The reflectance of the experimental samples was measured using a spectrometer with an integrating sphere. The ZnO sol-gel film, in addition to surface passivation of the BS layer, noticeably improves the reflectance in the entire spectral range necessary for the operation of a solar cell. BS/ZnO structures exhibit a low reflectivity in the wavelength range of 400–800 nm and at radiation incidence angles up to 50°. Thicker films are most preferred. Studies have shown that the simulation results are in good agreement with experimental data in the case of thin films. For structures with thick ZnO films, it is necessary to take into account the differences in the morphology of the model and real surfaces.

Keywords: Black Silicon · Sol-Gel Film · ZnO · Reflectance · Morphology · Solar Cells

1 Introduction

Nanostructured silicon-based surfaces feature unique electronic and optical properties, high surface-to-volume ratios, and facile surface modification [1–4]. Such subwavelength surfaces are a promising approach to reducing broadband light reflection. In recent years, black silicon (BS) has been of the greatest interest as an antireflection nanostructured surface for single junction [5–7] and tandem [8] silicon solar cells. This near-surface layer of crystalline silicon consists of an array of randomly distributed and densely packed nanoneedles. It is important to note that the BS layer formation process is self-organized without additional external techniques.

Y. Ono and J. Kondoh (Eds.): Inter-Academia 2023, LNNS 939, pp. 256–263, 2024.
https://doi.org/10.1007/978-3-031-54450-7_29

The large surface area of the BS layer leads to a high surface recombination rate, and therefore, efficient surface passivation is of utmost importance in employing BS layers in solar cells. Additionally, the BS layer tends to be gradually oxidized and contaminated by impurities in contact with air, which negatively affects the performance of solar cells. An effective way to overcome these problems is to deposit thin passivating films on the surface of BS [9–11]. Besides providing good surface passivation, applying a film onto a BS surface can further increase the antireflection ability.

Previously, we investigated the passivation properties (comfortability of the coating and charge carrier lifetime) and photoelectric characteristics (current-voltage and capacitance-voltage) of thin sol-gel films on the BS surface [12]. It was shown that sol-gel ZnO films provide effective passivation and stability of the structural properties of BS layers over time. In our other work, we studied the optical properties (transmission and absorption spectra) of sol-gel ZnO films on glass and silicon planar (non-textured) substrates. It was shown that they are almost transparent in the spectral range of 400–1000 nm [13].

This work is aimed at experimental investigation and simulation of the optical properties of BS layers passivated with sol-gel ZnO films to determine the possibility of their combined use in solar cells. The reflectance of BS/ZnO structures was estimated as a function of the incidence angle of solar radiation.

2 Simulation Results

The optical properties of BS/film structures depend on the geometrical parameters (diameter, period, height, or depth) of the nanoneedle arrays [5, 6], as well as on the refractive film index and thickness [9, 10]. Optical simulation of these structures can effectively optimize their parameters at the initial stage of solar cell design. Such an analysis eliminates the need for numerous experimental and technological reworks later.

Various rigorous methods, such as the finite difference time domain method (FDTD), the effective index technique, the rigorous coupled-wave analysis, the transfer matrix and finite element methods, allow optical simulation of sub-wavelength surfaces [14–17]. The FDTD method is known for its accuracy and simplicity in modeling the antireflective properties of nanostructured silicon-based surfaces.

For simulation, we used the Rough Surface module of the commercial software package FDTD Solutions by Lumerical Co. This software package allows us to determine the optical properties of periodic sub-wavelength surfaces depending on their shape and geometric dimensions, as well as on the wavelength (λ) and incidence angle (θ) of solar radiation. The numerical calculation is based on the parameters of stochastic nanostructured surfaces, namely, the values of the root mean square (RMS), correlation length (LC) and spatial resolution (δ).

Based on the results of studying the morphology of the BS/ZnO structures [12], their optical models can be represented as a thick silicon substrate with regular near-surface straight circular nanocones uniformly coated with a thin ZnO film (Fig. 1). Such periodic structures can be characterized by the following basic parameters: in-plane period (p), out-of-plane height (h) and diameter (d) of nanoneedles, and thickness of substrate (H) and film (t). Based on these geometric parameters, the initial parameters RMS, δ, and LC needed for simulation are uniquely determined.

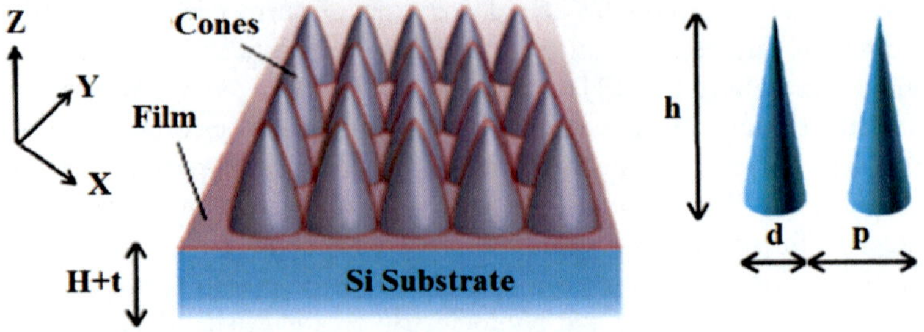

Fig. 1. Schematic diagram of the simulated BS/ZnO structure.

The simulated reflectance spectra of BS/ZnO structures with different film thicknesses at $\theta = 0°$ are shown in Fig. 2. For comparison, the reflectance spectrum of the initial BS layer (without the film) is also shown. Note also that the average reflectance of a planar (non-textured) silicon wafer is ~30% at $\lambda = 300–800$ nm.

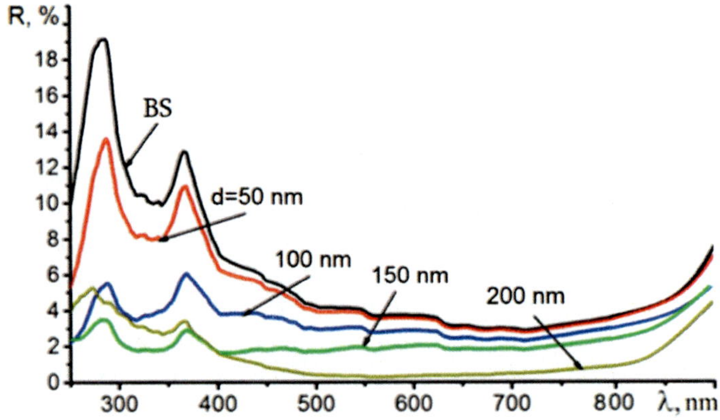

Fig. 2. Simulated reflectance spectra of the BS/ZnO structures.

Regardless of the film thickness, the BS/ZnO structures demonstrate a decrease in the reflectance compared to the initial BS layer. This means that the ZnO film contributes to the improvement of the antireflection ability of the BS layer. With an increase in the film thickness, the simulated reflectance decreases. For ZnO films with a small thickness (t = 50, 100 nm), the reflectance is relatively high in the most important spectral range $\lambda = 400–700$ nm, corresponding to the maximum solar radiation. At film thicknesses t = 150 and 200 nm, the reflectance in the considered spectral region is relatively small and weakly depends on the wavelength.

The 3D dependence of the simulated reflectance of BS with a 100 nm thick ZnO film on the incident angle and wavelength is presented in Fig. 3. As shown, the BS/ZnO structure exhibits a low reflectance (~3%) in the visible range for incidence angles up to 50°. However, for higher angles, the reflectance increases up to 5% at 60° and is even higher than 10% at 70°. The omnidirectional light trapping capability of the simulated structures is very important for photovoltaic stations to generate electricity in the morning and evening [18]. Note that this does not apply to stations with a sun-tracking system using light-sensitive sensors.

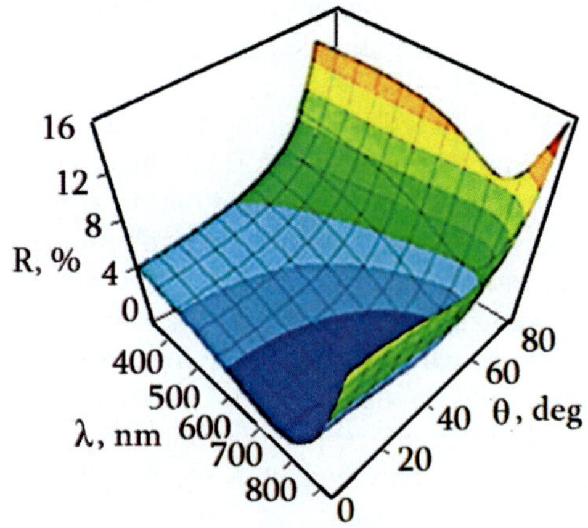

Fig. 3. 3D dependence of the simulated reflectance of the BS/ZnO structures.

3 Experimental Results

The p-type 400 μm-thick Si wafers with a resistivity of 3.0 Ω cm were used as the initial experimental samples. The BS layers were produced by the reactive ion etching (RIE) process using an SF_6/O_2 plasma. In this study, those values of the RIE process parameters that were previously chosen as the best for single-junction c-Si solar cells were used [19]. The process pressure was 55 mTorr, and the gas flow rates were 75 cm^3/min and 40 cm^3/min for SF_6 and O_2, respectively. The etching time was kept constant at 10 min. After plasma processing, the samples were washed with distilled water and dried in air. The average period, height and diameter of the BS needles were 150, 640 and 100 nm, respectively.

Homogeneous ZnO films were obtained by the sol-gel technique using the spin-coating method [12]. The needed amount of zinc acetate [$Zn(CH_3COO)_2{\cdot}2H_2O$] was poured with absolute isopropyl alcohol, dimethylformamide, and 2-methoxyethanol and mixed. The resulting solutions were then stirred for 30 min in an ultrasonic bath. To mature the solution, it was kept at ambient temperature for two days. The sol solution was spin-coated onto the BS layer, rotating at 2000 rpm for 20 s. The thickness of the sol seed layer was approximately 50 nm. The thin films were pre-heated at 350 °C for 10 min to evaporate the solvent and remove the organic residuals. The spin-coating and pre-heating process was repeated until the desired thickness was obtained. In the last stage, the samples were placed in an oven and heated in steps at an interval of 20 to 550 °C. The film thicknesses were 48, 146 and 207 nm.

A JEOL JSM-6700F Scanning Electron Microscope (SEM) and Solver Pro Atomic Force Microscopy (AFM) were used to characterize the morphology of the obtained samples. The analysis of the obtained data was carried out using the Gwyddion software package, which makes it possible to calculate the arithmetic mean deviation of all points of the roughness profile from the mean line along the evaluation length, as well as the number and size of grains. The adhesion of ZnO thin films was investigated by the ramp loading scratch test. The omnidirectional optical reflectance was measured using a UV-3101 PC spectrometer with an integrating sphere. The sample was rotated through an incidence angle in the range of 8°–70°.

Figure 4 shows typical cross-sectional and top view SEM images of BS layers with and without sol-gel ZnO films. The highly conformal film can be distinguished from the cross-sectional image as a lighter, narrow coating on top of the BS needles. It can be seen that the BS needles have a distinct conical shape with smooth surfaces. Sol-gel ZnO films are quite conformal and accurately repeat the BS morphology, which is important for effective surface passivation. The synthesized films had high adhesion to the surface of the BS layers. The average particle size on the film surface was about 19 nm, the number of grains was 280, and the roughness was 4.3 nm. It should be noted that during the post-annealing of the films in an oven, the morphology and structural properties of the initial BS layers remain almost unchanged due to the relatively low temperature of such treatment [19].

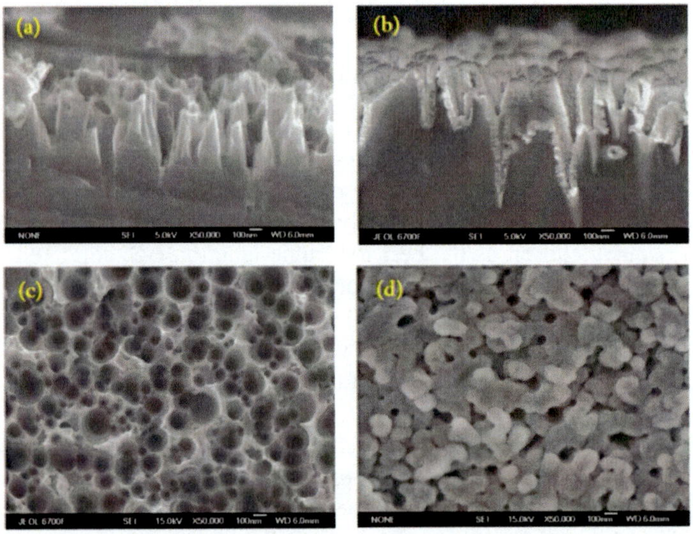

Fig. 4. Typical cross-sectional (a, b) and top view (c, d) SEM images of BS layers with (b, d) and without (a, s) sol-gel ZnO films.

The experimental reflectance spectra of BS/ZnO structures with different film thicknesses at $\theta = 0°$ are shown in Fig. 5. Figure 6 shows the reflectance of the BS layers covered conformally with 146 and 207 nm thick ZnO films as a function of the incident angle. Omnidirectional reflectance was measured at $\lambda = 600$ nm.

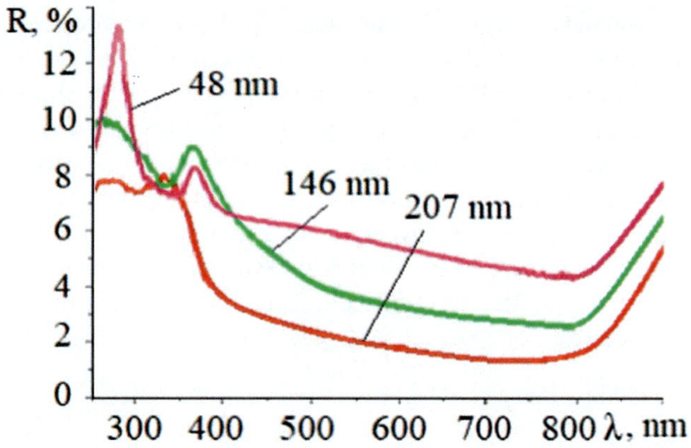

Fig. 5. Experimental reflectance spectra of the BS/ZnO structures.

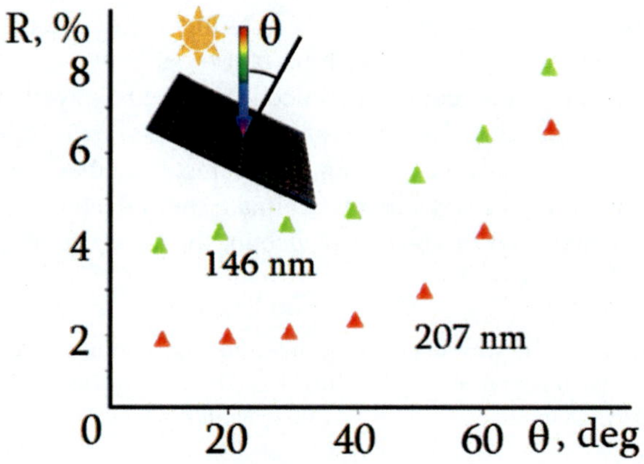

Fig. 6. Experimental reflectance of the BS/ZnO structures as a function of the incident angle.

From the presented results, it can be seen that the fabricated BS/ZnO structures demonstrate low reflectance in the range of $\lambda = 400$–800 nm. In addition, the structures have a wide-angle antireflective characteristic at incidence angles less than 50°, indicating their omnidirectional light-trapping capability. Moreover, the reflectance of structures is noticeably lower than the reflectance of a BS layer without a sol-gel ZnO passivation film (not shown in the figures). Thicker films are most preferred. In particular, at $t = 200$ nm, a reflectance of $R \sim 1$–3% is practically achievable in the entire visible range. In the range of ultraviolet radiation ($\lambda = 250$–400 nm), the reflectance increases but remains less than 8%. Thus, the sol-gel ZnO film, in addition to surface passivation of the BS layer, at least, does not deteriorate the reflectance in the entire spectral range necessary for solar cell operation. Similar results were previously obtained for BS layers coated with thin films of metal oxides by atomic layer deposition [10].

A comparison of the simulated and experimental reflectance spectra demonstrates good qualitative agreement in terms of the dependence of the reflectance on the film thickness and the incidence angle of solar radiation. However, noticeable quantitative differences exist, particularly for thick ZnO films. This is most likely due to the difference in the morphology of the model BS/ZnO structure from the structure formed in the ALD film real deposition process. When modeling reflectance spectra using the Lumerical FDTD Solutions program, the film is introduced into the model by vertically displacing the surface of the BS layer by the thickness of the film, and the relief of the film completely repeats the relief of the BS layer. Conversely, the real film does not follow the surface relief of the BS layer, but its surface is a shell. The layers grow toward each other, causing a "collapse" effect and smoothing out the relief. As a result, real films are thicker than model BS/ZnO structures.

4 Conclusion

An experimental study and simulation of the optical properties of black silicon (BS) layers passivated with sol-gel ZnO films was carried out to determine the possibility of their joint use in solar cells. It is shown that ZnO films are quite conformal and accurately repeat the BS morphology, which is important for effective surface passivation. In addition, these films noticeably reduce the reflection of the BS layers, which is retained at incidence angles less than 50°. The effect of film thickness on the reflectance spectrum of the BS/ZnO structure is analyzed. The simulation results and experimental data show the same trend. However, to quantitatively estimate the reflectance of structures with thick ZnO films, it is necessary to take into account the differences in the morphology of the model and real surfaces.

Acknowledgments. This work was supported by the Higher Education and Science Committee of the Republic of Armenia under project 21AG-2B011 and by the Belarusian Republic Foundation for Fundamental Research by grants T21ARMG-004, T22UZB-074, and T22MLDG-003.

References

1. Saive, S.: Light trapping in thin silicon solar cells: a review on fundamentals and technologies. Prog. Photovolt. **29**(10), 1125–1137 (2021)
2. Ayvazyan, G., Hakhoyan, L., Ayvazyan, K., Aghabekyan, A.: External gettering of metallic impurities by black silicon layer. Phys. Status Solidi A **220**(5), 2200793 (2023)
3. Dong, L., et al.: Fabrication of hierarchical moth-eye structures with durable superhydrophobic property for ultra-broadband visual and mid-infrared applications. Appl. Opt. **58**(24), 6706–6715 (2019)
4. Maniya, N.H.: Recent advances in porous silicon based optical biosensors. Rev. Adv. Mater. Sci. **53**, 49–73 (2018)
5. Liu, X., Coxon, P., Peters, M., Hoex, B., Cole, J., Frayc, D.: Black silicon: fabrication methods, properties and solar energy applications. Energy Env. Sci. **7**(10), 3223–3263 (2014)
6. Jian, L., Zhang, T., Zhang, P., Li, S.: Review application of nanostructured black silicon. Nanoscale Res. Let. **15**, 1–10 (2018)
7. Otto, M., et al.: Black silicon photovoltaics. Adv. Opt. Mater. **3**, 147–164 (2015)

8. Ayvazyan, G.Y., Kovalenko, D.L., Lebedev, M.S., Matevosyan, L.A., Semchenko, A.V.: Investigation of the structural and optical properties of silicon-perovskite structures with a black silicon layer. J. Contemp. Phys. (Armenian Acad. Sci.) **57**(3), 274–279 (2022)
9. Repo, P., Savin, H.: Effect of different ALD Al_2O_3 oxidants on the surface passivation of black silicon. Energy Procedia **69**, 381–385 (2016)
10. Ayvazyan, G.Y., et al.: Anti-reflection properties of black silicon coated with thin films of metal oxides by atomic layer deposition. J. Contemp. Phys. (Armenian Acad. Sci.) **56**(3), 240–246 (2021)
11. Kabalan, A.A.: Comparative study on the effects of passivation methods on the carrier lifetime of RIE and MACE silicon micropillars. Appl. Sci. **9**(9), 1804 (2019)
12. Semchenko, A.V., et al.: Characteristics of nanocomposite sol-gel films on black silicon surface. In.: Khakhomov, S., Semchenko, I., Demidenko, O., Kovalenko, D. (eds.) INTER-ACADEMIA 2021. LNNS, vol. 422, pp. 213–220. Springer, Singapore (2022). https://doi.org/10.1007/978-981-19-0379-3_21
13. Rogachev, A., Semchenko, A., Sidsky, V., Gaishun, V., Kovalenko, D.: Structural, optical and electrical properties of ZnO: Al thin films synthesized by sol-gel method. IFMBE Proc. **55**, 111–114 (2016)
14. Rahman, T., Boden, S.A.: Optical modeling of black silicon for solar cells using effective index techniques. IEEE J. Photovolt. **7**, 1556–1562 (2017)
15. Ma, S., et al.: A theoretical study on the optical properties of black silicon. AIP Adv. **8**, 035010 (2018)
16. Han, K., Chang, C.-H.: Numerical modeling of sub-wavelength anti-reflective structures for solar module applications. Nanomaterials **4**, 87–128 (2014). https://doi.org/10.3390/nano4010087
17. Ayvazyan, G., Gasparyan, F., Gasparian, V.: Optical simulation and experimental investigation of the crystalline silicon/black silicon/perovskite tandem structures. Opt. Mat. **140**, 113879 (2023)
18. Davidsen, R.S., et al.: Angle-resolved characterization of nanostructured and conventionally textured silicon solar cells. Sol. Energy Mat. Sol. Cells **140**, 134–140 (2015)
19. Ayvazyan, G., Vaseashta, A., Gasparyan, F., Khudaverdyan, S.: Effect of thermal annealing on the structural and optical properties of black silicon. J. Mater. Sci. Mater. Electron. **33**, 17001–17010 (2022)

Photosensitivity of ZnO_x:MgO_x Nanocomposite Sol-Gel Films on the Surface of Silicon Depending on Temperature

Alina Semchenko[1] (✉), Viktoriya Malyutina-Bronskaya[2], Anna Kuzmitskaya[2], Anton Nestsiaronak[2], Konstantin Danilchenko[1], Vladimir Pilipenko[3], Dmitrii Zhigulin[3], Alexander Pyatlitski[3], and Tatiana Piatlitskaya[3]

[1] Skorina Gomel State University, Sovetskaya 104, 246019 Gomel, Belarus
alina@gsu.by
[2] State Scientific and Production Association of Optics, Optoelectronics and Laser Technology, Nezavisimost ave 68-1, 220092 Minsk, Belarus
[3] JSC "INTEGRAL", Kazintsa Str. 121A, 20108 Minsk, Belarus

Abstract. The photoelectric characteristics of sol-gel ZnO_X:MgO_X layers on an n-Si substrate, measured in the temperature range from 5 to 50 °C, has been analyzed. For photodetector applications, the current must flow across the ZnO_X:MgO_X/n-Si/InGa structure and it should operate in pre-avalanche or avalanche mode (U of the order of −20 V). A strong effect of temperature on the photoelectric characteristics has been established, associated with the inclusion in the current transfer of various local energy levels due to electrically active impurities and defects at the interface of the structure.

Keywords: sol-gel · photodetectors · nanocomposite films · photosensitivity

1 Introduction

From the point of view of practical applications of ZnO films, the influence of structural features on the photoelectric properties of diode structures based on the ZnO/Si heterojunction is of interest. In such designs, the transparent-conducting ZnO layer can provide the transmission of radiation to the Si substrate and high sensitivity in a wide spectral region both due to the generation of minority charge carriers in the wide-gap ZnO layer and the potentially low recombination rate in the near-surface region. It is possible to use doped ZnO films as a photosensitive material, which at the same time can act as a transparently conducting electrode. There is a number of works devoted to the use of the n-ZnO/p-Si heterojunction to create UV and visible photodetectors [1, 2]. Therefore, it is of interest to create a photodiode operating in a wide spectral range and having a sensitivity not lower than that for a silicon photodiode. Recently, there have been lots of reports on good-quality ZnO films fabricated by the well-known techniques such as metal-organic chemical vapor deposition (MOCVD), molecular beam epitaxy (MBE), radio frequency magnetron sputtering, pulsed laser deposition (PLD), etc. [3]. So far, there are few messages on UV detectors using chemically synthesized ZnO composite film [4, 5].

Y. Ono and J. Kondoh (Eds.): Inter-Academia 2023, LNNS 939, pp. 264–272, 2024.
https://doi.org/10.1007/978-3-031-54450-7_30

An expansion of the practical application of composites is limited by the lack of detailed information on the process of electrotransfer in the structure at various intensities and frequencies of the applied electric field, and the temperature dependence of the electrical-physical characteristics. Despite intensive studies of the properties of composite multilayer structures in recent years, the mechanism of charge carrier transfer in them is still insufficiently investigated. In particular, the mechanism of carrier capture, retention, and emission from localized states, which determines the parameters of some effects in composite films, has not been elucidated. Also, the role of the intermediate layer, i.e., the interface between phases and the processes which occur at the interface, is not fully understood due to the specifics of the change in the electronic states on the surface of the composite. The electronic structure is noticeably affected by the appearance of space charges at localized surface levels, the formation of an electrostatic barrier, a shift in the levels of filled and unfilled electronic states in the composite, etc.

Currently, research is being actively conducted to find new materials for sensors with sensitivity to ultraviolet (UV) radiation. A good candidate for such sensors is zinc oxide (ZnO). It is characterized by high electron mobility, good thermal conductivity, and a wide and straight band gap (Eg ~ 3.37 eV) [6]. Undoped ZnO thin films exhibit photosensitivity in the visible region of the spectrum due to the presence of intrinsic defects. In order to vary the optical and electrical properties, ZnO is doped with Group III metals of the periodic table [7]. To increase the band gap, such an alloying element is chosen, so that, as a result, a compound with small distortions of the crystal lattice is formed. The radius of the Mg^{2+} ion (0.57 Å) is comparable to the radius of the Zn^{2+} ion (0.60 Å), making Mg suitable for substitution in the Zn lattice and facilitating the increase in the band gap [8]. In their previous work, the authors of the paper showed that the introduction of Mg makes it possible to obtain a composite based on zinc oxide with an optical band gap of about 5 eV, and these structures exhibit selective photosensitivity to UV radiation when a voltage of more than 10 V is applied [9].

The main purpose of this work is to evaluate the effect of temperature on the photoelectric characteristics of ZnO_x:MgO_x/Si with an optical band gap of about 5 eV ZnO_x:MgO_x.

2 Experimental Details

Layers based on ZnO_X:MgO_X films were obtained by the sol-gel technique using separate hydrolysis. The spin-coating technique has been used as the deposition method. The starting materials used were zinc acetate dihydrate (ZnAc) [Zn $(CH_3COO)_2 \times 2H_2O$]; magnesium acetate (Mg $(CH_3COO)_2$); isopropyl alcohol; distilled water [H_2O]; diethanolamine (DEA) [$HOCH_2CH_2NHCH_2CH_2OH$]. The film-forming solution has been prepared as follows: zinc acetate and magnesium acetate were separately dissolved in isopropyl alcohol and stirred at 60 °C for 10 min. When the solution was transformed into an emulsion, diethanolamine has been added to it at a DEA/ZnAc molar ratio of 1:1 and H_2O/ZnAc 2:1, respectively. A sol based on magnesium acetate was prepared in a similar way. Thereafter, the sols were mixed in component ratios 1:5, hydrochloric acid has been additionally added to sol. The deposition of the film was carried out by the spin-coating technique at a rotation speed of 2000 rpm for 40 s, followed by drying each layer at 60 °C for 10 min (2 layers). The final heat treatment has been realized at temperatures of 250 °C and 450 °C for 30 min. The sol was applied by the spin coating (five layers), which corresponds to a film thickness of 0.4 μm. Silicon wafers KEF-20 (n-type) have been used as substrates. After each layer application, the films were pre-dried in an oven (SNOL) for 5 min at a temperature of 250 °C. After applying the fifth layer, annealing has been carried out at a temperature of 550 °C for an hour. The temperature conditions for processing sol-gel layers were determined experimentally. The process of application and drying has been repeated until a film of the required thickness with the required structural and optical characteristics was obtained. The temperature of 250 °C is used at intermediate stages of film formation to remove sol organic residues, and the temperature of 550 °C is used to form the crystalline structure of the film.

The structural properties of the ZnO_x:MgO_x structures were studied using the Hitachi S-4800 Scanning Electron Microscope (SEM) (Hitachi, Tokyo, Japan). The top view and cross-section of the samples have been characterized.

Measurement of photoelectric characteristics (current-voltage characteristics) was carried out on the automated basic laser test complex, according to the schemes presened in Fig. 1. The measurement of photoelectric characteristics has been carried out according to two schemes: longitudinal, when the probes were installed on top of the ZnO_x:MgO_x film (scheme 1); and transverse, when one probe was mounted on top of the ZnO_x:MgO_x film, and another one was mounted from below to the Si substrate, through the InGa contact (scheme 2).

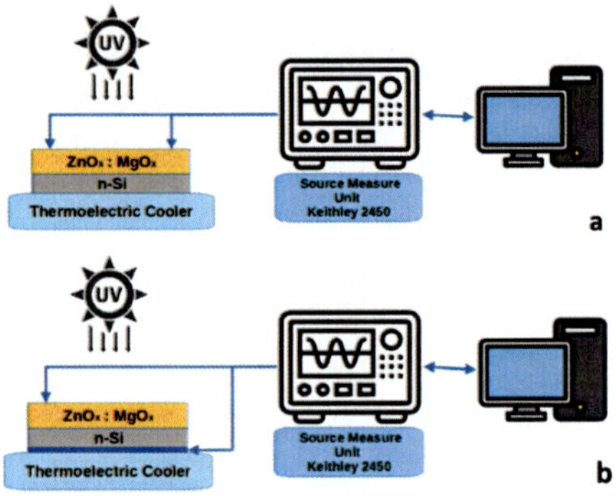

Fig. 1. Block diagram of measurements of photoelectric characteristics (longitudinal (a) and transverse (b) ones).

Measurements of photoelectric characteristics was carried out at temperatures of 5, 25 and 50 °C without and under the influence of radiation. On this concern the multispectral laser radiation source has been used, which is a set of 9 laser diodes with wavelengths of 405, 450, 520, 660, 780, 808, 905, 980 and 1064 nm with a common fiber output and control boards, as well as with a calibrated radiation power of about 2 mW. The multispectral source uses 9 lasers of different wavelengths in order to produce a spectral range from ultraviolet to near-infrared. At these wavelengths, various types of photodiode structures often have the greatest spectral sensitivity. As a source of

UV (278 nm), a 300 μW TO-3535BC-UVC265-30-6V-E LED located in a specialized tooling was used.

3 Results and Discussion

Figure 2 shows the typical top and cross-sectional SEM images of the obtained samples. On the SEM images, one can easily distinguish between the silicon layer and the ZnO_X:MgO_X film. The thickness of the ZnO_X:MgO_X layer is 185 nm. The layer contains close-packed structures without porosity, with an almost smooth surface, and dense packing of crystallites, which is suitable for the fabrication of high-efficiency UV-sensors.

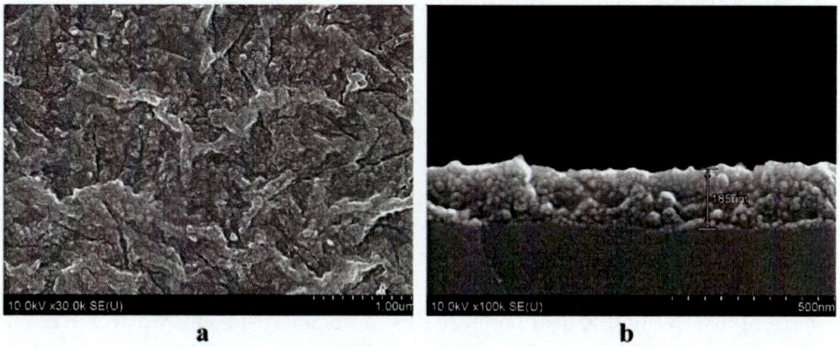

Fig. 2. Typical SEM images of the ZnO_X:MgO_X/Si structures: (a) top view; (b) cross-sectional view.

Sol-gel ZnO_X:MgO_X layers, annealed at 500 °C for 60 min, are characterized by light attenuation in the visible range of (5–10) %. In the absorption spectrum of some films two maxima have been observed, which corresponded to an optical band gap of the order of 3.21 and 4.68 eV. An Eg value in tile range of 3.21–3.4 eV indicates the presence of zinc oxide in the thin-layer system. This value coincides or it is slightly below the tabular value typical for ZnO (3.37 eV). For sols with a high magnesium content, the band gap value of ZnO_X:MgO_X is in the range from 4.68 to 5.47 eV [8]. The investigation of the photoelectric characteristics was carried out on the ZnO_X:MgO_X/n-Si/InGa structures.

The dark current-voltage (I—V) characteristics measured according to the scheme 1 and to the scheme 2 (Fig. 1) at different temperatures are presented on Fig. 3. As can be seen from Fig. 3, the current value when measured in a transverse circuit is an order of magnitude higher than that in a longitudinal circuit. In semiconductor photodetectors the work is based on the internal photoelectric effect, in which the photon energy must be large enough for the photoelectron to overcome the band gap and pass into the conduction band. Photons are absorbed inside the bulk of the material, creating electron-hole pairs and the corresponding conductivity. The current-voltage characteristics measured according to scheme 1 (Fig. 3a) show that with increasing temperature, the current at positive voltages is decreasing, and at negative voltages it first decreases and then increases. The current-voltage characteristics measured according to scheme 2 (Fig. 3b) show that with increasing temperature the current at negative voltages decreases, and at positive voltages it first increases and then falls.

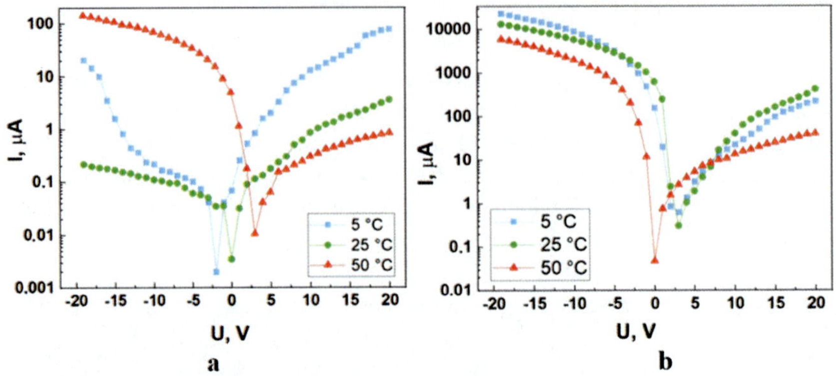

Fig. 3. Current-voltage characteristics of the ZnOx:MgOx/Si structure measured by the longitudinal scheme (a) and the transverse scheme (b) at different temperatures.

Typically, the internal resistance of photoresistors increases with decreasing temperature, which is not observed in our work. This behavior of the conduction processes may be due to the percolation effect. The ZnO_x:MgO_x film has a microcrystalline structure and the current flow along the grain boundaries of microcrystals may differ in each measurement. Additionally, in such systems, a large number of surface states are observed, which strongly affects the processes of current flow as well [9].

Analyzing the current-voltage characteristic with a positive voltage applied to the ZnO_x:MgO_x film, measured according to scheme 2, one can observe three sections with currents obeying the dependence of the I ~ U^n type. This dependence is typical for space-charge-limited current (SCLC). For a measurement temperature of 5 °C: n = 2.6 from 2 V to 12 V, n = 4.2 from 12 to 16 V and n = 2.8 from 16 to 20 V; for 25 °C: n = 3.0 from 3 V to 5 V, n = 4.0 from 6 to 12 V and n = 2.7 from 13 to 20 V; for 50 °C: n = 1.2 from 0 V to 6 V, n = 1.0 from 6 to 8 V and n = 1.5 from 10 to 20 V. In the region of negative voltages, a power law has been observed as well, however, at 5 °C n = 1.5 from 0 V to −19 V, at 25 °C there are two sections with n = 1.0 from 0 V to −8 V and n = 1.3 from −8 V to −20 V; at 50 °C n = 1.7. An increase in the current, at n > 2, indicates the filling of a separate group of monoenergetic traps characteristic for injection currents in the ZnO_x:MgO_x film and at the ZnOx:MgOx/Si interface. An exponent n close to 2 indicates the presence of SCLC in the case of injection of charge carriers into the dielectric in the absence of a compensating charge. An exponent n close to 1 indicates the presence of an Ohmic conduction mechanism. Analysis of the current-voltage characteristics showed that the main mechanism of conductivity in the ZnO_x:MgO_x/Si structure under study is the SCLC in the presence of trap levels [9].

The spectral characteristics of photodiode structures with deep energy levels, caused by the surface states, are changing with temperature depending on the operating mode.

As shown by the current-voltage characteristics measured in the photovoltaic mode (U = 0 V) the field in the space charge region is so small that the time of flight of carriers there can be commensurate and it is sometimes even longer than the lifetime of minority carriers. As a result, most of the carriers created by light recombine with each other without affecting the photocurrent. The spectral sensitivity in this case is negligible. In photodiode structures an external electric field is applied to overcome the potential barrier, as a result of which the field in the space-charge region increases strongly. The carriers created by light move under the action of the field without having time to recombine, and the sensitivity increases. Therefore, when calculating the sensitivity, voltages of −5, −10, −20, +5, +10, +20 V were chosen, when an increase in photosensitivity was observed in certain regions of the spectrum. The maximum change in photosensitivity, especially in the UV part of the spectrum, was observed in both measurement schemes at a bias voltage of −20 V.

The photosensitivity of the ZnO_X:MgO_X/Si structure obtained from the current-voltage characteristics measured (at U = −20 V) according to the scheme 1 (Fig. 1a) and the scheme 2 (Fig. 1b) at different temperatures is presented in Fig. 4. The value of photosensitivity, in the case of measurements in the longitudinal scheme, is an order of magnitude less than in the case of measurements in the scheme 2 (Fig. 1,b).

As can be seen from Fig. 4, a (measurement along the scheme 1) at a temperature of 5 °C, the photosensitivity maximum is observed at a wavelength of 980 nm. As the temperature rises to 25 °C, it drops to almost zero, and as the temperature rises to 50 °C, negative photosensitivity appears in the UV region (from 278 nm to 450 nm) and it is insignificant in the IR region (at about 1000 nm).

When measured in the scheme 2 an increase in photosensitivity is observed in the entire spectral range. At a temperature of 5 °C, two maxima are observed with high photosensitivity at wavelengths of 278 nm and 780 nm.

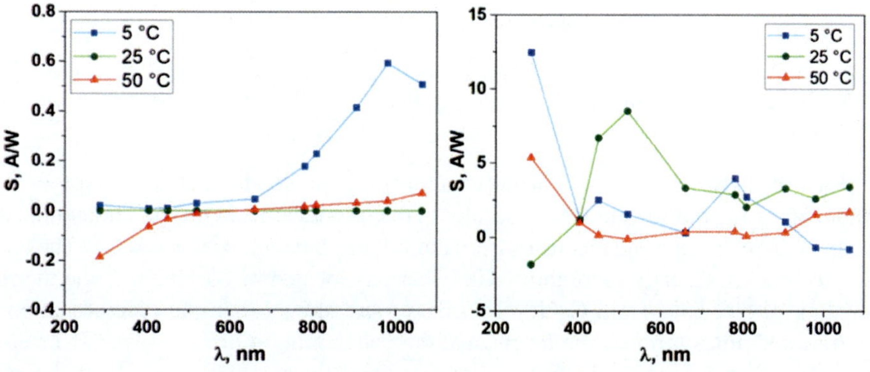

Fig. 4. The sensitivity of the ZnO_X:MgO_X/Si structure obtained from the current - voltage characteristics measured (at a voltage of −20 V) in the longitudinal scheme (a) and the transverse scheme (b) at different temperatures.

With an increase in temperature to 25 °C, the spectral sensitivity maxima are shifting to the long-wavelength part of the spectrum (450 nm and 880 nm). At a temperature of

50 °C, the spectral sensitivity maximum is observed at a wavelength of 278 nm and 905 nm, and the sensitivity has become insignificant in the visible wavelength range.

The obtained high values of the spectral photosensitivity are possible if the structure under investigation operates in the pre-avalanche or avalanche mode, which is observed at −20 V when measured according to the scheme 2. This operation mode contributes to a more complete collection of nonequilibrium charge carriers and accelerated movement of minority charge carriers to the transition.

The observed temperature dependence of photosensitivity, as well as negative photosensitivity, are caused by electrically active impurities and structural defects, in the presence of which shallow and deep local energy levels are formed in the band gap of the semiconductor, which play an important role in current transfer processes, including those under the influence of optical radiation [9].

Qualitatively, the phenomenon of negative photoconductivity can be explained by the presence of two groups of local energy trap levels located near the bottom of the conduction band (G1) and the top of the valence band (G2), as shown in Fig. 5. The distribution of states at the interface is not very important, since any dense set of states at the interface near or below the mid-bandgap of Si creates a double depletion layer near the junction, resulting in rectification or saturation in both directions.

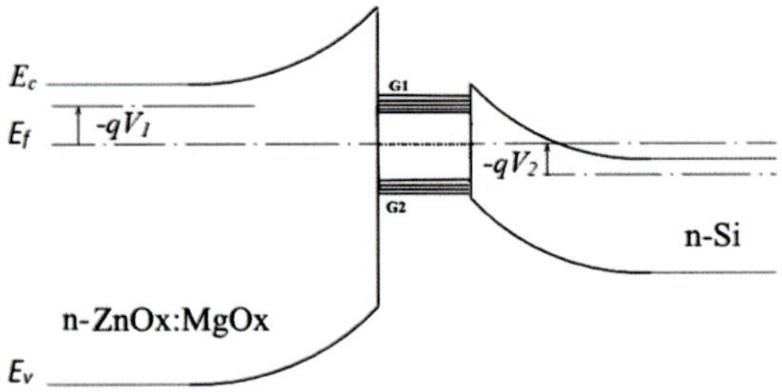

Fig. 5. Energy band diagram in an n-n heterojunction, taking into account states at the interface.

However, in such a junction, current can flow from one semiconductor to another even without being in equilibrium with the states at the interface. In an n-type semiconductor, light quanta $h\nu$ are exciting electrons from the valence band to local levels G1. The holes G2 levels cannot directly recombine with electrons located in G1 levels. If the thermal excitation of electrons from G1 levels occurs more slowly than the recombination of electrons and holes through the levels, and the concentration of centers of G1 levels is not too low, then the recombination of electrons from the conduction band with holes of G2 levels leads to a decrease in the dark concentration of intrinsic charge carriers. The concentration of free holes in this case increases insignificantly, less than the decrease in the concentration of free electrons. Therefore, with such light irradiation, the conductivity of the semiconductor decreases.

4 Conclusions

The photoelectric characteristics of ZnO_X:MgO_X/n-Si with an optical band gap of ZnO_X:MgO_X of the order of 5 eV were measured at temperatures from 5 to 50 °C. Analysis of the I-V characteristics of the ZnOx:MgOx/Si structure, measured at different temperatures, showed that the current, at both voltage polarities, obeys the $I \sim U^n$ type dependence. Thus, the main mechanism of conduction is the space-charge-limited current. In some areas of the I-V dependence the exponent n is greater than 2.0, which indicates the filling of a separate group of monoenergetic traps in ZnOx:MgOx and at the ZnOx:MgOx/Si interface. It has been found that the temperature dependence of the photoelectric characteristics, as well as the observed negative photosensitivity, are due to electrically active impurities and structural defects. If the current in the structure has been flowing along the ZnO_X:MgO_X film, the photosensitivity (at $U = -20$ V) was insignificant in a larger wavelength range, only at a temperature of 5 °C a sensitivity values of 0.6 A/W at 980 nm and −0.2 A/W at 278 nm have been observed. For photodetector applications, the current should flow across the ZnO_X:MgO_X/n-Si/InGa structure and it must operate in pre-avalanche or avalanche mode (at U of the order of −20 V). In this mode, at a temperature of 5 °C, the maximum spectral sensitivity at a wavelength of 278 nm is 12 A/W. At the same time, there is a slight photosensitivity in the visible and IR regions of the spectrum. At 25 °C in the UV range, the photosensitivity value becomes negative and amounts to −2.5 A/W. In the visible and IR regions of the spectrum it decreases. With an increase in temperature to 50 °C, the UV sensitivity increases to a positive value of 5 A/W. In the IR region it is ~2 A/W, and in the visible region it becomes close to 0.

Acknowledgments. This work was supported by Belarusian Republic Foundation for Fundamental Research by grants No. T22UZB-074 and T22MLDG-003.

References

1. Liu, K., Sakurai, M., Aono, M.: ZnO-based ultraviolet photodetectors. Sensors **10**(9), 8604–8634 (2010)
2. Jose, J., Ravindran, A., Nair, K.: A review on ZnO heterojunction photodetector for UV application. Ictact J. Microelectron. **02**(04), 305–310 (2017)
3. Jagadish, C., Pearton, S.: Zinc Oxide Bulk, Thin Films and Nanostructures. Elsevier Science (2006)
4. Basak, D., Amin, G., Mallik, B., Paul, G.K., Sen, S.K.: Photoconductive UV detectors on sol–gel-synthesized ZnO films. J. Cryst. Growth **256**(1–2), 73–77 (2003)
5. Tsay, C.-Y., Yu, S.-H.: Optoelectronic characteristics of UV photodetectors based on sol–gel synthesized GZO semiconductor thin films. J. Alloy. Compd. **596**, 145–150 (2014)
6. Sabayev, V., Aronov, D., Oster, L., Rosenman, G.: Induced surface reactivity modification in Zinc oxide-based thin films. Appl. Phys. Lett. **93**, 144104 (2008)
7. Priscilla, S.J., Daniel, R., Dhakshayani, Y., Caroline, S.C., Sivaji, K.: Effect of magnesium dopant on the structural, morphological and electrical properties of ZnO nanoparticles by sol–gel method. Mater. Today: Proc. **36**, 793–796 (2021)

8. Sidsky, V.V., Malyutina-Bronskaya, V.V., Soroka, S.A., Danilchenko, K.D., Semchenko, A.V., Pilipenko, V.A.: Photoactive and structure properties of ZnO:XMgO. In: Khakhomov, S., Semchenko, I., Demidenko, O., Kovalenko, D. (eds.) INTER-ACADEMIA 2021. LNNS, vol. 422, pp. 227–235. Springer, Singapore (2022). https://doi.org/10.1007/978-981-19-0379-3_23
9. Malyutina-Bronskaya, V., Zalesski, V., Zhyhulin, D., Mudryi, A.: Structural, optical and photoelectric properties of Tb doped ZnO thin films for device applications. Opt. Mater. **127**(60), 112305 (2022)

ITO-Based Sol-Gel Layers Containing Rare Earth Complexes for Solar Cells and Optoelectronics

Sergei Khakhomov[1](✉), Alina Semchenko[1], Vladimir Gaishun[1], Dmitry Kovalenko[1], A. A. Maevsky[1], K. D. Danilchenko[1], Viktoriya Malyutina-Bronskaya[2], Shavkat U. Yuldashev[3], and Victor Verlan[4]

[1] Francisk Skorina Gomel State University, Sovetskaya 104, 246028 Gomel, Belarus
{akhakh,semchenko}@gsu.by

[2] SSPA "Optics, Optoelectronics and Laser Technology", Nezavisimost ave 68-1, 220092 Minsk, Belarus

[3] Center of Nanotechnologies Development, National University of Uzbekistan, University str. 4, 100174 Tashkent, Uzbekistan

[4] Institute of Applied Physics, State University of Moldova, Str. Academiei, 5, 2028 Chisinau, Republic of Moldova

Abstract. This paper presents the development of a technique for synthesizing ITO sol-gel films with embedded YAGG:Cr^{3+}, Yb^{3+} nanocrystals that exhibit a long-lasting afterglow on the surface of a photoelectric solar cell. The properties of ITO layers for further application in optoelectronics and solar energy have been studied. The Pechini method provides the luminescence intensity of ITO films with YAGG:Cr^{3+}, Yb^{3+} nanocrystals 4 times higher than the chemical deposition method. Solar cells with an additionally deposited ITO coating with YAGG:Cr^{3+}, Yb^{3+} luminescent nanocrystals maintained a residual voltage of up to 0.51 V for 30–60 min.

Keywords: sol-gel · luminescent nanocrystals · nanocomposite films

1 Introduction

Transparent conducting films (TCF) have been used for decades in optoelectronic devices as a transparent conductor due to their high optical transmission and electrical conductivity [1–3]. Each TCF material has different optical and electrical properties that define the characteristics of the device: high transparency combined with low resistivity, such as indium. The properties of indium tin oxide (ITO) have attracted tremendous interest in this material. ITO thin films are used in optoelectronics and microelectronics as a transparent conductive contact for a range of optoelectronic products, including solar cells, LEDs, laser diodes, and flat panel displays. In_2O_3: SnO_2 (ITO) is a semiconductor with a band gap of 3.7–4.3 eV, which provides high transmittance in the visible region of light. Oxygen intrinsic vacancies and doping impurities of Sn^{-4} ions arising during ITO

Y. Ono and J. Kondoh (Eds.): Inter-Academia 2023, LNNS 939, pp. 273–279, 2024.
https://doi.org/10.1007/978-3-031-54450-7_31

formation explain the high electrical conductivity of the material. The properties of ITO films in optoelectronic devices strongly depend on the film deposition technique. One of the successful technologies is the sol-gel method, which makes it possible to obtain various functional materials [3–5]. This method is chosen by the authors of this work.

Persistent luminescent materials can be used to increase the efficiency of solar cells. This type of luminescence can last up to several hours, usually at room temperature, after the excitation sources have ceased. The most acceptable mechanism can be qualitatively explained by an electron capture-release process that is excited by ultraviolet or visible light.

A technique was developed for obtaining film-forming compositions with the optimal concentration of doping additives leading to the maximum quantum yield of luminescence in thin films of transparent conducting oxides. Nanocrystals with a long-lasting afterglow of various compositions were chosen as declared additives.

The introduction of YAGG:Cr^{3+}, Yb^{3+} nanocrystals with stable luminescent properties into the sol at low temperatures of heat treatment allows for the preparation of coatings showing persistent luminescence on the surface of a solar cell, and the coatings have sufficient mechanical resistance to be used in solar cells [6]. To obtain such coatings, nanocrystals YAGG:Cr^{3+}, Yb^{3+} were introduced into the indium tin oxide films.

2 Experimental

Chemically pure divalent tin anhydride chloride ($SnCl_2$), ammonium fluoride (NH_4F), hydrochloric acid (HCl), and propanol were used to prepare indium tin oxide (ITO) films. In the first step, $SnCl_2$ (0.07 M) and NH_4F (0.03 M) were dissolved in 7.5 mL of propanol and 2.5 mL of deionized water to form a mixed solution, and 3 drops of hydrochloric acid (HCl) were added to obtain a clear solution. For doping, carbon (0.5 wt.%) was added to the above-prepared solution and stirred. For the homogeneous precursor solution, the solution was kept in an ultrasonic bath until complete dissolution. After that, the solution was kept at room temperature (about 20 °C) for several days until complete hydrolysis of the components. YAGG:Cr^{3+}, Yb^{3+} nanocrystals were prepared by the following method [7, 8]: gallium oxides (Ga_2O_3) and yttrium oxides (Yb_2O_3) were dissolved in nitric acid (HNO_3) and then boiled until evaporated, after which the salt residue was diluted in distilled water. This solution was evaporated and dried at 110 °C to form a xerogel. Xerogel was calcined in a muffle furnace for five hours at a temperature of 800 °C. Further amorphous powder was annealed in a furnace at 1000 °C for ten hours. The resultant powder of nano-crystals YAGG:Cr^{3+}, Yb^{3+} was cooled and processed in a mill to a homogeneous state. Then the powder of nanocrystals was incorporated into fluorine- and carbon-doped tin oxide. The procedure for obtaining nanocrystals is illustrated in Fig. 1.

YAGG:Cr^{3+}, Yb^{3+} nanocrystals were introduced into the sol at a concentration of 5 wt% after hydrolysis of the main components. Subsequently, the resulting composition was applied to the surface of the solar cell using centrifugation at a substrate rotation speed of 6000 rpm. The deposited coating underwent annealing in a muffle furnace at a temperature of 250–300 °C for 60 min. YAGG:Cr^{3+}, Yb^{3+} nanocrystals were introduced into the sol at a concentration of 5 wt % after hydrolysis of the main components.

Subsequently, the resulting composition was applied to the surface of the solar cell using centrifugation at a substrate rotation speed of 6000 rpm.

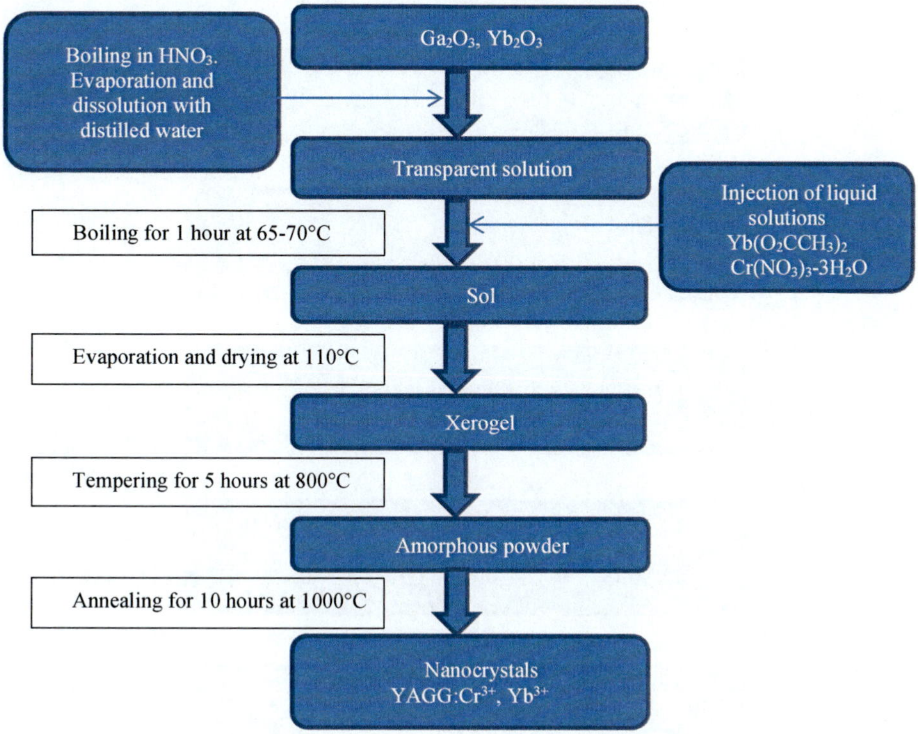

Fig. 1. Schematic of nanocrystal production YAGG:Cr^{3+}, Yb^{3+}

The deposited coating underwent annealing in a muffle furnace at a temperature of 250–300 °C for 60 min. The annealing of the interlayer was conducted at a temperature of 250 °C. The composition was applied in the amount of 5 layers. Figure 2 demonstrates the procedure that was employed in the formation of sol-gel coatings with nanocrystals YAGG:Cr^{3+}, Yb^{3+}.

The transmission spectra were recorded on a Cary-500 spectrophotometer λ. The steady-state luminescence spectra and luminescence excitation spectra were measured on an SFL-1211A spectrofluorimeter and were then corrected for the spectral sensitivity of the recording system and spectral density distribution of exciting radiation, respectively. The microstructure of the films was evaluated by method of high-resolution atomic force microscopy (AFM) SOLVER RRO 47 («NT-MDT»).

3 Results and Discussion

Figure 3 shows the optical transmission spectra (T) of ITO films deposited on a glass substrate containing different numbers of layers annealed at different temperatures.

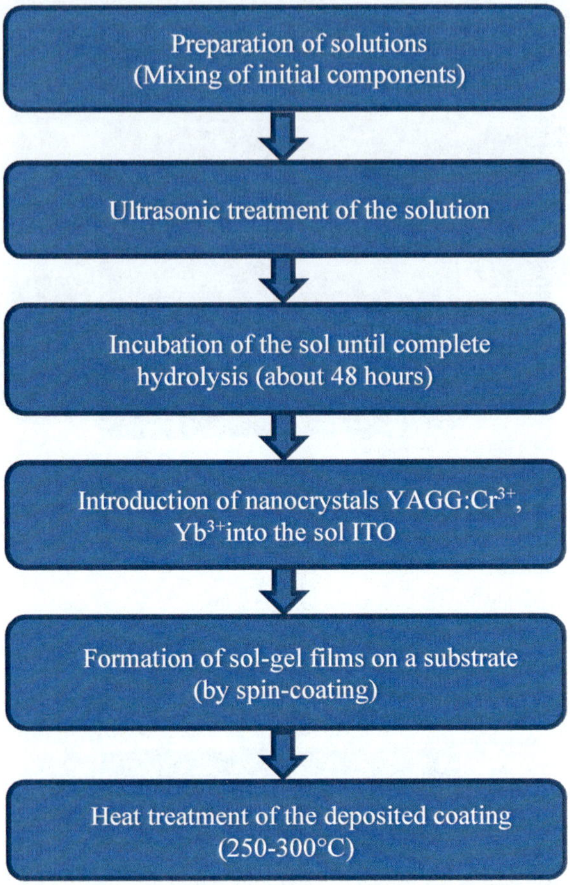

Fig. 2. Technique for creating sol-gel coatings through the introduction of nanocrystals

ITO films deposited on glass have good transmittance (75–85%) in the visible and near-IR ranges. Longer annealing at 550 °C leads to a slight shift of the edge of the fundamental absorption band to the short-wavelength side due to the densification of the film and the improvement of its crystal structure. Also, the transmission spectra have an oscillating character, which is characteristic of the emerging interference effects in multilayer thin-film structures. The optical properties of ITO films do not change after additional vacuum annealing (Fig. 4).

Figure 3 shows typical AFM images of the surface of an ITO sample (topography, phase contrast, grain marking), the scanning area was 4×4 μm.

The ITO layer is formed by evenly distributed spherical grains with clear boundaries between them. Before additional vacuum annealing, a dense layer with a characteristic grain size of 120 nm and a subroughness of 5.5 nm is formed on the surface of the multilayer ITO film. For comparison, we note that the subroughness of the ITO sample layer (6 layers) on the polished silicon surface does not exceed 3 nm. After additional vacuum annealing, the structure becomes denser, the grain size decreases to 70 nm, and the

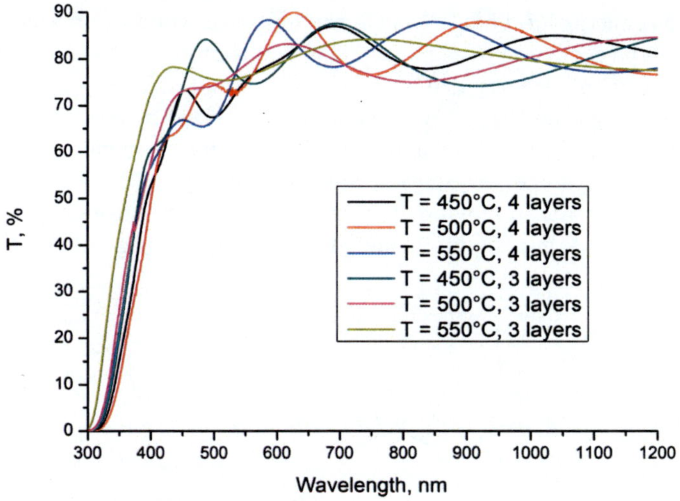

Fig. 3. Transmission spectra of ITO layers

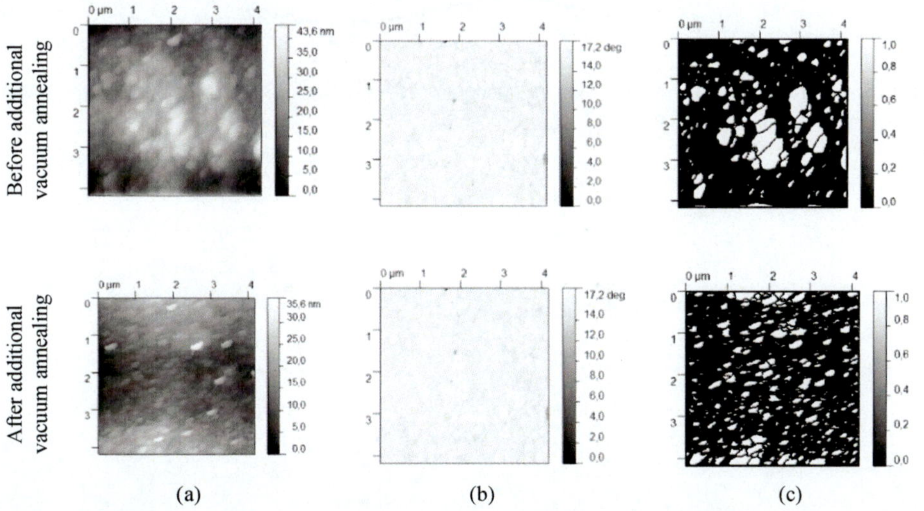

Fig. 4. AFM images of the ITO sample surface: a - topography, b - phase contrast, c - grain marking

subroughness to 4.2 nm, which is in good agreement with the resistivity measurements. The synthesized layers had high adhesion to the surface of the substrates.

Persistent ceramic phosphors with composition YAGG doped with Ce^{3+}, Cr^{3+} and Yb^{3+} ions were prepared, then they were into ITO matrix embedded and their spectroscopic properties of composite materials were measured. Bright green emissions are produced as a result of the Ce^{3+}:$5d_1 \rightarrow 4f$ transition that continues for several hours after ceasing the blue light excitation. In this material, the threshold of photoionization between the 5d$_1$ energy level of $Ce^{3+}$and the bottom of the conduction band with photostimulation can be decreased by Ga^{3+}. At room temperature, Cr^{3+} ions can function as electron traps with an ideal trap depth for persistent luminescence.

Powders made by the Pechini method are characterized by more intense luminescence in the ranges of 510–580 nm and 700–750 nm, which is apparently explained by the stronger coupling of the activator ion with the matrix.

It was found that after removing the external light source, solar cells with an additional applied luminescent coating maintained a residual voltage of 0.5 V for 60–90 min.

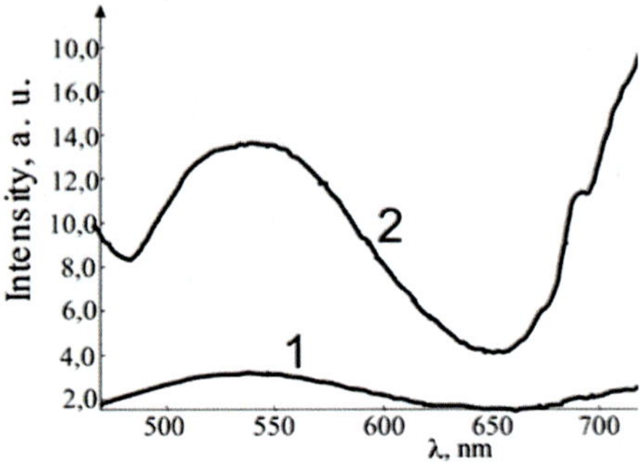

Fig. 5. Luminescence spectra of ITO composites with YAGG:Cr^{3+}, Yb^{3+} nanocrystals (chemical precipitation): λ_{exc} = 325 nm (1); λ_{exc} = 435 nm (2)

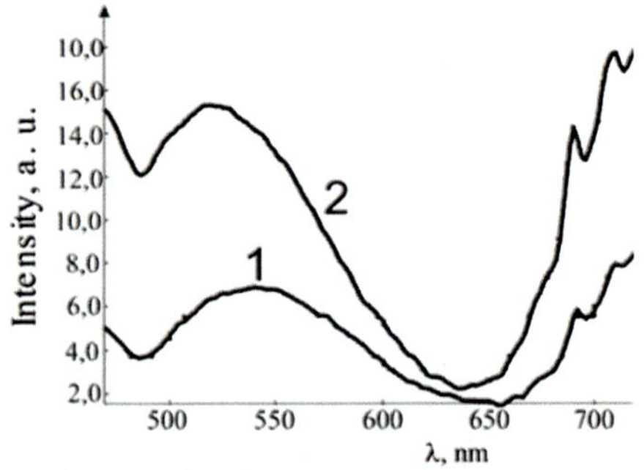

Fig. 6. Luminescence spectra of ITO composites with YAGG:Cr^{3+}, Yb^{3+} nanocrystals (λ_{exc} = 435 nm): chemical precipitation (1); Pechini (2)

The powders produced by the Pechini method are characterized by more intense luminescence in the ranges of 510–580 nm and 700–750 nm. This phenomenon can be presumably explained by stronger binding of the activator ion to the matrix. Figures 5 and 6 show the luminescence spectra of composites with YAGG:Cr^{3+}, Yb^{3+} nanocrystals. Increasing the concentration of doping nanocrystals above 0.5 wt% reduced the transmittance of the sol-gel film to less than 80%, which is unacceptable for a transparent conducting electrode. When the external light source was removed, solar cells with an additionally deposited ITO coating with YAGG:Cr^{3+}, Yb^{3+} luminescent nanocrystals maintained a residual voltage of up to 0.51 V for 30–60 min.

4 Conclusions

The technique for synthesizing ITO sol-gel films with embedded YAGG:Cr^{3+}, Yb^{3+} nanocrystals that exhibit a long-lasting afterglow on the surface of a photoelectric solar cell has been developed. Increasing the concentration of doping nanocrystals above 0.5 wt% reduced the transmittance of the sol-gel film to less than 80%, which is unacceptable for a transparent conducting electrode. The Pechini method provides the luminescence intensity of ITO films with YAGG:Cr^{3+}, Yb^{3+} nanocrystals 4 times higher than the chemical deposition method. Solar cells with an additionally deposited ITO coating with YAGG:Cr^{3+}, Yb^{3+} luminescent nanocrystals maintained a residual voltage of up to 0.51 V for 30–60 min.

Acknowledgments. This work was supported by Belarusian Republic Foundation for Fundamental Research by grants T22UZB-074, T22MLDG-003 and by ANCD of R. Moldova through projects 20.80009.5007.14, 22 80013.5007.6BI.

References

1. Ayvazyan, G.Y., Kovalenko, D.L., Lebedev, M.S., Matevosyand, L.A., Semchenko, A.V.: Investigation of the structural and optical properties of silicon-perovskite structures with a black silicon layer. J. Contemp. Phys. (Armenian Acad. Sci.) **57**(3), 274–279 (2022)
2. Ayvazyan, G.Y., Hakhoyan, L.A., Kovalenko, D.L., Semchenko, A.V.: Nanostructured thin films and surfaces: formation, research and application: scientific monograph. In: Ayvazyan, G.Y., Hakhoyan, L.A., Kovalenko, D.L., Semchenko, A.V. (eds.) NPUA, 240 p. Chartaraget, Yerevan (2023)
3. Semchenko, A.V., et al.: Photoactive properties of transport sol-gel layers based on strontium titanate for perovskite solar cells. Photonics **10**, 845 (2023). https://doi.org/10.3390/photonics10070845
4. Semchenko, A.V., et al.: Nanoscale piezoelectric properties and phase separation in pure and la-doped $BiFeO_3$ films prepared by sol–gel method. Materials **14**, 1694 (2021). https://doi.org/10.3390/ma14071694
5. Sidsky, V.V., Malyutina-Bronskaya, V.V., Soroka, S.A., Danilchenko, K.D., Semchenko, A.V., Pilipenko, V.A.: Photoactive and structure properties of ZnO:XMgO. Nanocomposite sol–gel films on the surface of silicon. In: Khakhomov, S., Semchenko, I., Demidenko, O., Kovalenko, D. (eds.) INTER-ACADEMIA 2021. LNNS, vol. 422, pp. 227–235. Springer, Singapore (2021). https://doi.org/10.1007/978-981-19-0379-3_23
6. Dai, Z., et al.: Optical studies of $Y_3(Al,Ga)_5O_{12}$:Ce^{3+},Cr^{3+},Nd^{3+} nano-phosphors obtained by the Pechini method. J. Rare Earths **37**(11), 1132–1136 (2019). https://doi.org/10.1016/j.jre.2019.04.006
7. Saladino, M.L., Caponetti, E., Martino, D.C., Enzo, S., Ibba, G.: Effect of the dopant selection (Er, Eu, Nd or Ce) and its quantity on the formation of yttrium aluminum garnet nanopowders. Opt. Mater. (Amst) **31**(2), 261–267 (2008)
8. Chen, D., Wang, Y., Hong, M.: Lanthanide nanomaterials with photon management characteristics for photovoltaic application. Nano Energy **1**, 73–90 (2012). https://doi.org/10.1016/j.nanoen.2011.10.004

Computational Science

Check for updates

Deep Learning for Relevant Findings in Colonoscopy

Mihaela Luca[1(✉)], Adrian Ciobanu[1], Radu-Alexandru Vulpoi[2], and Vasile Liviu Drug[2,3]

[1] Institute of Computer Science, Romanian Academy - Iasi Branch, Iasi, Romania
{mihaela.luca,adrian.ciobanu}@iit.academiaromana-is.ro
[2] Institute of Gastroenterology and Hepatology, Iasi, Romania
[3] "Grigore T. Popa" University of Medicine and Pharmacy, Iasi, Romania

Abstract. Colorectal cancer is a leading cause of death in the world, thus, early prevention and treatment became a major priority. Improved detection, on-site evaluation and the accuracy of the medical exploration are essential for this attempt. Automatic detection of polyps, diverticula and membrane bleedings or lesions, might enhance the objectivity, precision and efficiency of this examination, aiding the endoscopy expert. Video colonoscopy files can be split in frames and these can be classified into informative or non-informative images. Applying the technology advancements on the chosen frames, the system can learn to detect and point on the relevant findings. It can assist on-line object detection and off-line classification and segmentation of images. Using diverse structures of deep learning, it is able to help the physicians or to assess the quality of the video colonoscopy procedures. We will further describe some affordable, efficient, solutions that might be used to tackle these problems worldwide.

Keywords: video colonoscopy · deep learning · detection · segmentation · classification

1 Artificial Intelligence in Endoscopy Exams

1.1 Deep Learning in Image Processing

The accelerated development of artificial intelligence, the diversification of the deep learning (DL) architectures and the new hardware dedicated to parallel computing, determined a qualitative improvement in image processing. With the appropriate training on well-structured annotated images databases, they are successfully replacing the classic computing approaches [1–3].

Due to affordable parallel computing devices, DL might now be used for object detection, image classification and semantic segmentation.

Endoscopy, and particularly colonoscopy, is a long, tedious procedure, needing focused attention and skilled training in observing small polyps, therefore automatic image processing for experts assistance is useful and necessary [4].

Y. Ono and J. Kondoh (Eds.): Inter-Academia 2023, LNNS 939, pp. 283–293, 2024.
https://doi.org/10.1007/978-3-031-54450-7_32

Video colonoscopies might contain both clear, informative images (with valuable information) and non-informative images which are blurred, full of reflexes, bubbles or residues on the colon's membrane. Labeling these images and marking the regions of interest is a compulsory, yet laborious step, and using an important amount of annotated images is essential for obtaining good performances with deep learning [5].

1.2 Video-Colonoscopy Image Database

We used our own video-colonoscopy image database which we have progressively augmented from 18 videos at the beginning of our testing to 28 video colonoscopies. This database of video colonoscopies is selected within an internal agreement between the Institute of Gastroenterology and Hepatology, "Grigore T. Popa" University of Medicine and Pharmacy, Iasi, Romania, and the Institute of Computer Science, Romanian Academy - Iasi Branch, with the ethical commission approval and the patients' written consent.

Taking into account almost 50000 frames for a colonoscopy of about 15–17 min (more or less upon the particularities and abnormalities discovered) we worked on hundreds of thousands of images. We selected about 300000 informative frames and narrowed the choice step by step towards the identification of certain particularities, in order to be able to label the most important types of "objects".

2 Object Detection Using MobileNet

The onset of colorectal cancer initiates with the formation of benign polyps on the mucosal surface of the colon or rectum. Over a span of 5 to 15 years, these polyps may undergo transformation into cancer, leading to the invasion of the colonic wall [6].

In video colonoscopies, physicians are trying to detect the abnormal structures that might evolve into cancer, especially the large polyps [6–10]. There are polyps of different sizes (diminutive – less than 5 mm, small - between 6 and 9 mm, or large - more than 10 mm) [8], and structures (pedunculated, serrated, adenomatous, dysplasia, adenocarcinoma), or other structures (bleeding areas, diverticula, lipoma) [7]. Clinical studies [8] found that colon polyps are mainly less than 10 mm in size (85%–90%). The risk of cancer in diminutive lesions is negligible, while the risk in small lesions is less than 1% [7]. Literature, guides and medical blogs are developing a lot these aspects [9, 10].

Artificial intelligence is useful here. Well annotated, in hundreds and thousands of frames, these chosen areas are used to train DL networks on parallel hardware, appropriated for operating millions of convolutions in a record time [5, 11].

We have focused on polyps, because this is a dilemma in colonoscopy: finding and removing them, or letting them on their place, when they are too small? Polyps, pedunculated or sessile, diverticula, bleedings are issues to be found. The other categories we have chosen to detect were in order to better differentiate from the important ones, during the learning process. We observed that this is the better way to avoid confusion, to "teach" more different objects to the deep learning network, not only the main entities.

An automatic colonoscopy analysis system should "know" all the elements that can appear in colonoscopy videos. Therefore we have trained the networks on hundreds

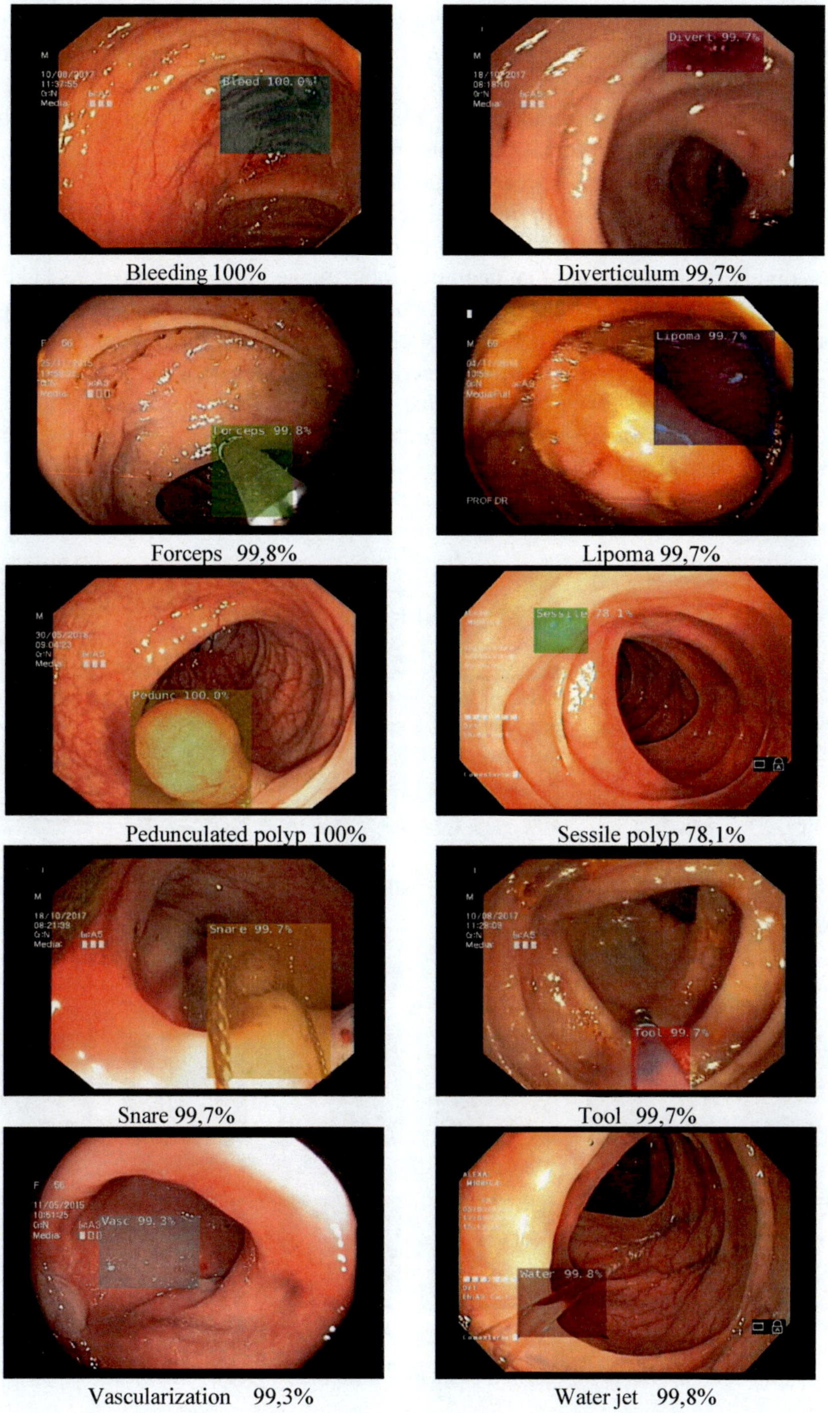

Fig. 1. Ten classes of objects selected to train the DL network for detection

of well-chosen images for each kind of "*object*" to be detected. In fact we denote by "*object*" classes like: *bleeding area, diverticulum, forceps, lipoma, pedunculated polyps, sessile polyps, snare, tool-head, vascularization* and *water jet*, presented in Fig. 1.

MobileNet architecture reduces the number of parameters and computations required, compared to traditional convolutional neural networks (CNNs) while maintaining a good performance. It is less computationally complex than other deep learning architectures and thus interesting for mobile vision applications, as the seminal paper of Andrew Howard's and colleagues [11], is proving. We have obtained excellent on-line results, highly overpassing our classical, previous, image processing attempts [11].

We used a Jetson Xavier NX, NVIDIA computing device [11, 12] and we retrained a MobileNet DL structure [5, 11, 13]. MobileNet is a deep learning model designed for efficient image classification and object detection tasks on mobile and embedded devices, particularly well-suited for applications that require real-time processing. MobileNet architecture is built using depth-wise separable convolutions, which are composed of two main components: depth-wise convolution and point-wise convolution (Figs. 2 and 3).

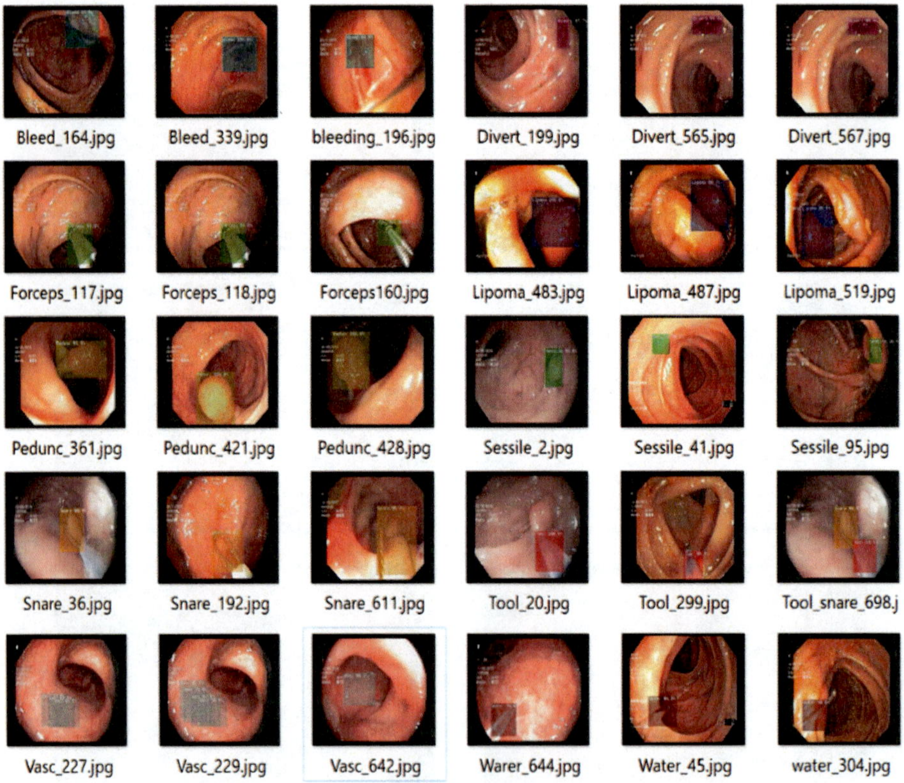

Fig. 2. Random selection of images from the 10 classes of objects.

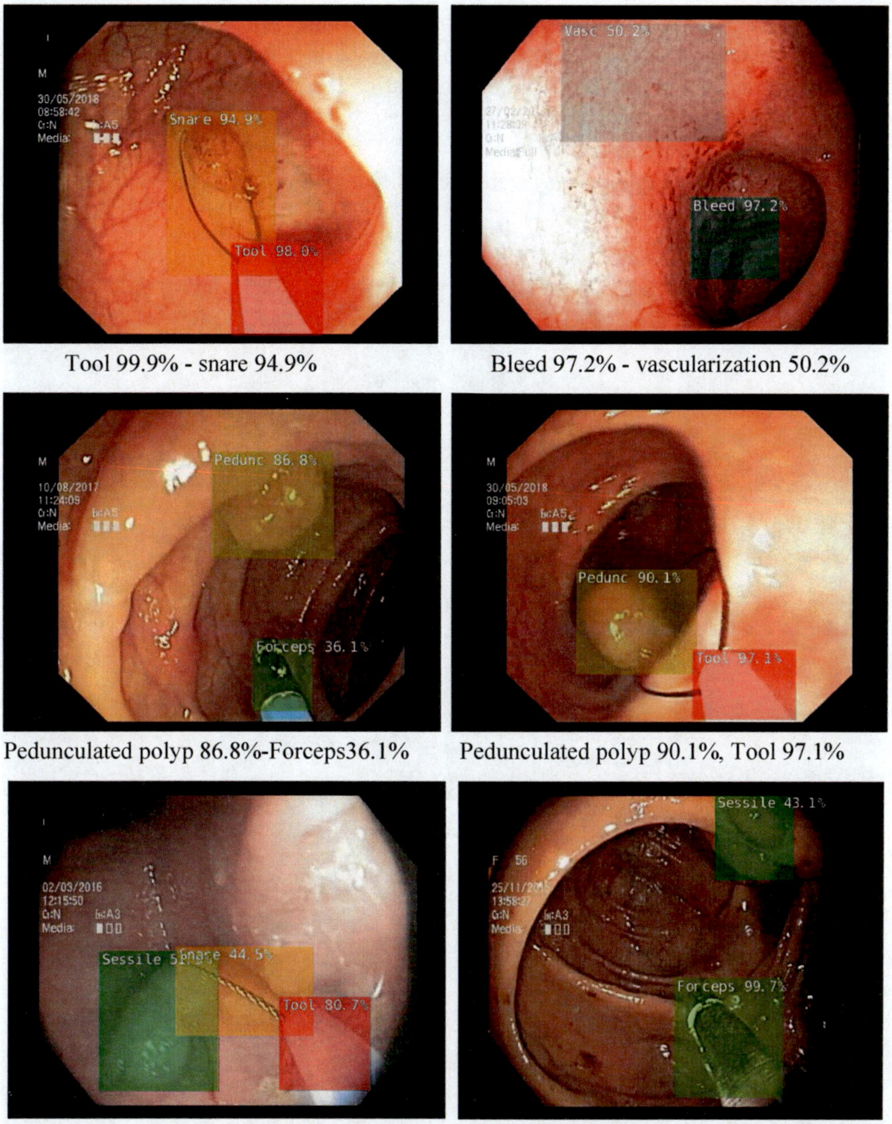

Tool 99.9% - snare 94.9% | Bleed 97.2% - vascularization 50.2%

Pedunculated polyp 86.8%-Forceps36.1% | Pedunculated polyp 90.1%, Tool 97.1%

Tool 80.7%, polyp sessile 51.9%, snare 44.5% | Forceps 99,7%, Sessile 43,1%

Fig. 3. Two or more "objects" detected in the same image and the confidence coefficient

The resulted trained neural network was obtained with a limited number of images per class. Our goal was to demonstrate that it is possible to obtain such a well-trained neural network with limited resources. We used the 18 video colonoscopies in.mp4 format, from which we extracted 322,486 frames (.jpg images). We made several selections, first down to 9,087 relevant frames, then down to 3,226 frames.

Finally, for training we used 700 images, 70 images for each of the 10 classes taken into consideration. We applied the trained neural network on all the 18 video colonoscopies and we extracted 322,486 frames after detection. This was an experiment, therefore we are satisfied with the results, as in fact we do not have a project for a finite product, we just verified our method.

We stated that the large polyps were well detected by MobileNet, with 98%–99% confidence coefficient (see Fig. 4).

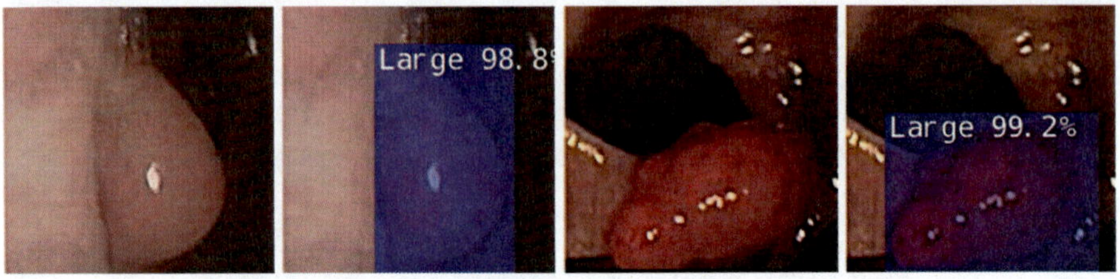

Fig. 4. Large polyps well detected in video colonoscopy using MobileNet

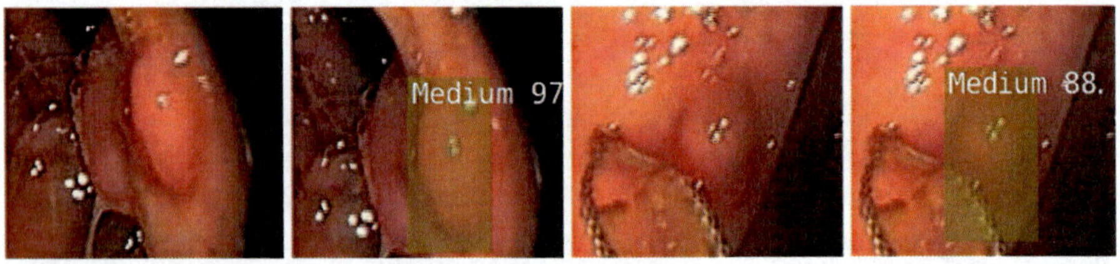

Fig. 5. Medium size polyps detected in video colonoscopy using MobileNet

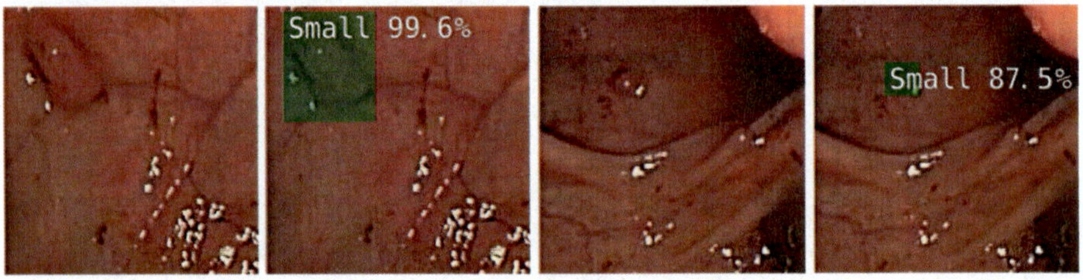

Fig. 6. Small polyps detected in colonoscopy frames with a complex MobileNet structure

The detection confidence for the medium-size polyps was between 88% up to 97% (see Fig. 4).

The diminutive ones were sometimes misinterpreted or detected with a smaller confidence (Fig. 5).

Consequently, we have trained a special DL model (also based on MobileNet) on images with smaller regions of interest, to enhance the accuracy of small polyps detection obtaining good results (see Fig. 6), detailed in a previous paper [13].

3 Image Classification with Deep Learning

To evaluate the quality of the video colonoscopy procedure, as the physician has to reach with the colonoscope till the appendix, the path is passing through the rectum, sigmoid colon, descending colon, splenic flexure, transverse colon, hepatic flexure, ascending colon, cecum, ileocecal valve. Thus, some important loci (or places) have to be visually documented, as it is stated into the colonoscopy guides [14, 15]. We have chosen as landmarks for assessing the correctitude of a colonoscopy four entities: appendix, ileocecal valve, hepatic flexure and splenic flexure. The list of landmarks differs upon the professional society which is agreed in a certain geographical region, usually either European or American guidelines are followed.

These images have their own characteristics that can be taught into the DL system. A special attention has to be focused on the resemblance of these images. It is better to exclude the atypical images from the training database in order not to spoil the results.

Therefore, we had to put the two types of flexure images (hepatic and splenic, which are in fact blueish shadows on the colon membrane) in the same class. Only the temporal order of their appearance distinguishes them, while physicians are obviously knowing that in the normal anatomical succession the ileocecal valve precedes the hepatic flexure that is before the splenic one, which makes them recognize each flexure. The images are very similar. The appendix is just a small aperture towards the small intestine, difficult observables.

We trained on these different three image categories and then compared the results of several DL structures: ResNet50, GoogLeNet, VGG19, and AlexNet [15].

Table 1. Landmarks images classification with four different types of deep learning networks

DL network	Correct classification (over 80%)
ResNet50	90%
GoogLeNet	86,65%
VGG19	98,33%
AlexNet	98,33%

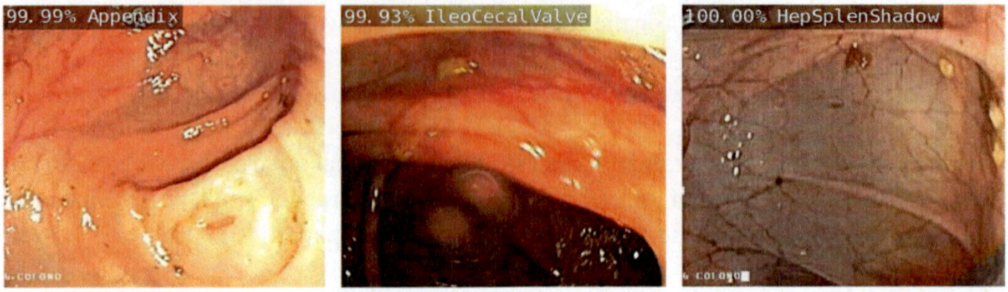

Fig. 7. Appendix (left), ileocecal valve (center), hepatic/splenic flexures (right) image classification using Deep Learning

The best rate of recognition was 98.33% for VGG19 and AlexNet (Table 1, Fig. 7).

4 Semantic Segmentation for the Overall Evaluation

The Boston Bowel Preparation Score (BBPS) is a conventional mark given upon the degree of colon membrane cleansing and visibility. However it is a subjective appreciation depending on the expert's experience and skills. It differs upon individual standards, being relatively difficult to mediate a final score. Two different experts might give two different marks, each of them with its own modality of evaluation; therefore an objective estimation is desirable.

Scarce residues on the tissue surface are time consuming and difficult to annotate in order to feed a DL network for learning, aiming to automatically segment the image.

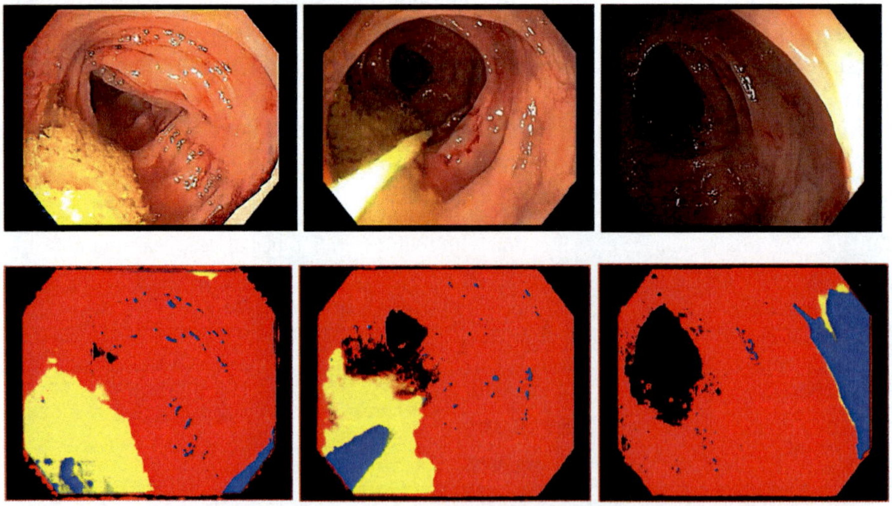

Fig. 8. Membrane surface semantic segmentation for BBPS evaluation (mucosa on red, residues in yellow, lumen in black, artifacts in blue)

We have proposed a CIELAB color-based method of identifying different regions on the video colonoscopy frames and trained a semantic segmentation DL network to recognize them, obtaining valuable results in segmenting mucosa, lumen, residues and artifacts [16]. In a first stage we transformed the RGB images in LAB color space and we extract image fragments characteristic for each of the four regions. Then we created a specific LAB cube for each region (mucosa, lumen, residues and artifacts) and these color features (let's say similar to fingerprints) were applied to the selected colonoscopy frames to obtain the annotated images. We annotated with red the mucosa, with yellow the residues, with blue the artifacts and with black the lumen. We configured in Matlab a semantic segmentation DL network and we trained it with original and annotated colonoscopy frames. Figure 8 presents some results of applying the trained semantic segmentation network to colonoscopy frames, showing that the network was capable to generalize and find related colors, for each region, that were not present in the annotated images.

Automatically identifying the residues covered areas (in yellow) permits counting the total covered surface and deducing a value conventionally comparable with the standard evaluation BBPS.

Having a clean colon is very important, as a well-observable surface, permits reliable diagnosis. Upon the pattern of the polyp surface, recognizing the "texture" on high resolution images, the physician assisted by the artificial intelligence system might decide if the diminutive polyp might be safely left in place, or not, thus diminishing the medical costs [17, 18].

Our methods permit *to detect the important objects* (or regions of interest) that we want to find, *to classify the landmarks images* in order to identify the correctitude of the procedure (automatically parsing the way of the instruments), and use color and DL *to automatically segment images*, generalizing, permitting to objectively compute the covered surface of the colon.

Next, step by step, we will try to learn to use imperfect datasets in image segmentation [19], augmenting the available information in a consistent way with reality, as databases are still scarce, difficult to collect and to correctly annotate.

5 Conclusions

We developed several systems which use deep learning in approaches for assisting the video colonoscopy procedures. We used MobileNet DL model on Jetson Xavier NX for on-line object detection on the video colonoscopies. The system is capable of identifying pedunculated and sessile polyps, as well as diverticula, bleedings etc., in real time, with high degrees of confidence. We mainly focused on polyps' detection. Besides, we propose a few off-line techniques to evaluate the correctness of the colonoscopy procedure. A deep learning classifier for several anatomic landmarks detected during video colonoscopy might verify the compliance with medical guidelines. The methods presented are tests on our database. Larger image databases might allow us to compute statistics, too. The DL semantic segmentation method on the CIELAB color space offers a new method to be applied to the video recordings to estimate the quality of the colon cleansing. These are low cost methods for colonoscopy evaluation that permit interesting applications. For example, in the field of education, a useful application that might be developed in the future is for training students in detecting polyps, in a competing manner with our system. The system might give marks to the students, and evaluate their answers. Another possible and necessary approach for the insurance societies is a system that might evaluate the correctitude of the colonoscopy procedure verifying the documentation of the important landmarks and evaluating the degree of colon cleansing. But, important applications of the methods described in this paper might be developed only in collaboration with the producers of the endoscopic devices, as the images of very high resolution demand only embedded solutions. The colonoscopy images of high resolution are not usually available (we had only Standard Definition videos).

References

1. Taghiakbari, M., Mori, Y., von Renteln, D.: Artificial intelligence-assisted colonoscopy: a review of current state of practice and research. World J. Gastroenterol. **27**(47), 8103–8122 (2021)
2. Vulpoi, R.A., Luca, M., Ciobanu, A., Olteanu, A., Bărboi, O.B., Drug, V.L.: Artificial intelligence in digestive endoscopy-where are we and where are we going? Diagnostics (Basel) **12**(4), 927 (2022). https://doi.org/10.3390/diagnostics12040927. https://pubmed.ncbi.nlm.nih.gov/35453975/. PMCID: PMC9029251
3. Luca, M., Ciobanu, A., Barbu, T., Drug, V.: Artificial intelligence and deep learning, important tools in assisting gastroenterologists. In: Lim, C.-P., Vaidya, A., Jain, K., Mahorkar, V.U., Jain, L.C. (eds.) Handbook of Artificial Intelligence in Healthcare. ISRL, vol. 211, pp. 197–213. Springer, Cham (2022). https://doi.org/10.1007/978-3-030-79161-2_8
4. Luca, M., Barbu, T., Ciobanu, A.: An overview on computer processing for endoscopy and colonoscopy videos. In: Balas, V.E., Jain, L.C., Balas, M.M., Shahbazova, S.N. (eds.) SOFA 2018. AISC, vol. 1222, pp. 3–22. Springer, Cham (2021). https://doi.org/10.1007/978-3-030-52190-5_1
5. Luca, M., Ciobanu, A.: Polyp detection in video colonoscopy using deep learning. J. Intell. Fuzzy Syst. **43**(2), 1751–1759 (2022). https://content.iospress.com/articles/journal-of-intelligent-and-fuzzy-systems/ifs219276
6. Huck, M.B., Bohl, J.L.: Colonic polyps: diagnosis and surveillance. Clin. Colon Rectal Surg. **29**(4), 296–305 (2016). https://doi.org/10.1055/s-0036-1584091. PMID: 31777460; PMCID: PMC6878826
7. World Health Organization. International Agency for Research on Cancer, Colorectal Cancer Screening, IARC Handbooks of Cancer Prevention, vol. 17 (2019). https://publications.iarc.fr/open-access. https://publications.iarc.fr/Book-And-Report-Series/Iarc-Handbooks-Of-Cancer-Prevention/Colorectal-Cancer-Screening-2019. Website address available in November 2023
8. Mahmood, S., Aguila, E.J., ur Rahman, A., Shuja, A., Bollipo, S.: How to approach small polyps in colon: tips and tricks. Tech. Innov. Gastroint. Endosc. **23**(4), 328–335 (2021). https://www.sciencedirect.com/science/article/pii/S2590030721000453
9. Rex, D.K.: Best practices for resection of diminutive and small polyps in the colorectum. Gastrointest. Endosc. Clin. **29**, 603–612 (2019)
10. Medical News Today, Colon polyp size chart: How does size affect cancer risk? https://www.medicalnewstoday.com/articles/colon-polyp-size-chart#size-chart. Accessed 15 Nov 2023
11. Howard, A.G., et al.: MobileNets: efficient convolutional neural networks for mobile vision applications. https://arxiv.org/abs/1704.04861 (2017)

12. https://www.nvidia.com/en-us/autonomous-machines/embedded-systems/jetson-xavier-series/. Accessed 12 Sept 2023
13. Ciobanu, A., Luca, M., Vulpoi, R.A., Barboi, O., Drug, V.L.: Deep learning in colonoscopies: improving small polyps' recognition rate. In: 10th International Conference on E-Health and Bioengineering – EHB 17–19 November 2022 (2022). https://doi.org/10.1109/EHB55594.2022.9991415
14. Marques, S., Bispo, M., Pimentel-Nunes, P., Chagas, C., Dinis-Ribeiro, M.: Image documentation in gastrointestinal endoscopy: review of recommendations. GE Port J. Gastroenterol. **24**(6), 269–274 (2017). https://www.ncbi.nlm.nih.gov/pmc/articles/PMC5731161/pdf/pjg-0024-0269.pdf. Accessed 15 Nov 2023
15. Ciobanu, A., Luca, M., Vulpoi, R.A., Drug, V.L.: Anatomic landmarks detection by deep learning in colonoscopy. In: 11th International Conference on E-Health and Bioengineering - EHB 2023, 9–10 November to appear in the Conference Proceedings, Ed. Springer Nature (2023)
16. Ciobanu, A., Luca, M., Drug, V.L.: Objective method for colon cleansing evaluation using color CIELAB features. In: IEEE International Conference on e-Health and Bioengineering-EHB 2020 (2020). https://doi.org/10.1109/EHB50910.2020.9280110, Corpus ID: 228098537
17. Hassan, C., Balsamo, G., Lorenzetti, R., Zullo, A., Antonelli, G.: Artificial intelligence allows leaving-in-situ colorectal polyps. Clin. Gastroenterol. Hepatol. **20**, 2505–2513 (2022). https://www.cghjournal.org/article/S1542-3565(22)00639-5/pdf. Accessed 15 Nov 2023
18. Hassan, C., et al.: Post-polypectomy colonoscopy surveillance: European Society of Gastrointestinal Endoscopy (ESGE) Guideline – Update 2020 (2020). https://www.esge.com/assets/downloads/pdfs/guidelines/2020_a_1185_3109.pdf
19. Tajbakhsh, N., Jeyaseelan, L., Li, Q., Chiang, J.N., Wu, Z., Ding, X.: Embracing imperfect datasets: a review of deep learning solutions for medical image segmentation. Med. Image Anal. **63**, 101693 (2020). https://doi.org/10.1016/j.media.2020.101693

Fuzzy Inference with Sequential Fuzzy Indexed Search Trees

Balázs Tusor[1,2(✉)], Ondrej Takáč[1], Štefan Gubo[1], and Annamária R. Várkonyi-Kóczy[1,2]

[1] Department of Informatics, J. Selye University, 94501 Komárno, Slovakia
{tusorb,takaco,gubos}@ujs.sk, varkonyi-koczy@uni-obuda.hu
[2] Institute of Software Design and Software Development, Óbuda University, Budapest 1034, Hungary

Abstract. In this paper, a new method is presented for fuzzy inference-based classification. Its base idea lies in the dimensional decomposition of the problem space: the proposed method builds a structure from the training data in which the search among the fuzzy rules is done dimension by dimension, and thus, the number of rules that are needed to be evaluated is gradually restricted. The structure has a layered architecture, where each layer corresponds to a given dimension of the input data and contains a set of fuzzy membership functions, each with a self-balancing binary search tree to quickly identify the relevant fuzzy sets. These are implemented using indexing arrays to enhance the operating speed.

Keywords: Classification · Fuzzy Logic · Fuzzy Inference · Machine Learning

1 Introduction

Fuzzy inference systems [1] are widely used nowadays for classification [2], decision making systems [3], etc. They apply fuzzy logic to map a given input to an output using fuzzy rules that combine predefined fuzzy sets and their fuzzy membership functions. Many implementations have been developed during the past decades, one of which is the Sequential Fuzzy Indexing Tables (SFITs, [4, 5]) classifier that achieves a very fast operation by storing the precalculated *fuzzy membership function* (FMF) values for known parts of the problem space (thus, creating a dimensional decomposition of it), effectively reducing the inference process that would require the evaluation of all fuzzy rules into a series of array accesses. However, a significant disadvantage of SFITs is that they have a non-negligible memory requirement that scales substantially with the size of the problem space.

In this paper, a new fuzzy classifier called *Sequential Fuzzy Indexed Search Trees* (SFISTs) is proposed that is based on the same idea as the SFITs, but instead of storing the precalculated values in large indexing arrays, SFISTs only store the significant parameters of the fuzzy sets (divided into intervals based on the training data) and use self-balancing binary search trees to identify the fuzzy sets that are need to be evaluated for each attribute (narrowing down the corresponding problem space dimension). This

Y. Ono and J. Kondoh (Eds.): Inter-Academia 2023, LNNS 939, pp. 294–309, 2024.
https://doi.org/10.1007/978-3-031-54450-7_33

significantly reduces the size of the structure of the classifier. A secondary goal of the new classifier is to surpass the classification accuracy of the SFIT classifier as well, which is ensured by a more thorough evaluation phase.

The rest of the paper is as follows. In Sect. 2, a brief overview is given of the general architecture and properties of the SFIT classifier. In Sect. 3, the architecture of the proposed SFIST classifier is described, then the training and evaluation algorithms are presented. In Sect. 4, the classification performance of the new classifier is given, alongside a brief analysis considering its computational and spatial complexities. Finally, the Conclusions section summarizes the paper, and presents future work.

2 Sequential Fuzzy Indexing Tables

The SFIT classifier builds and maintains a layered structure (one layer for each attribute of the problem). In each layer, there is a fuzzy value array M (1D in the first layer, and 2D in the rest) that stores the pre-calculated fuzzy membership function values for each element of the array, and an index array Λ of the same size, that determines which row should be regarded in the arrays of the next layer. Each row in a given layer is basically a separate FMF.

Figure 1 presents an illustration for the classifier. To demonstrate its usage, let us consider the input sample $X = (5.0, 3.8)$. First the values are mapped to integers using a simple linear mapping function that converts the input values into a suitable positive integer domain: $\tilde{X} = (5, 4)$. The new values are used as the column coordinate of the arrays. In the first layer, $\Lambda_5 = 2$ index value shows that in the next layer, row 2 should be checked, at column $\tilde{X}_1 = 4$. In the last layer of the structure, class ID numbers are stored in the index array (as positive integer values), of which the output value is selected using the linearly mapped value of the last attribute (as column address) and the index value obtained in the previous layer (as row address). The overall fuzzy membership value belonging to it indicates the confidence of the system in this output, which is calculated from the fuzzy values (gained from M^{L_0} and M^{L_1}, , marked with rounded squares in the figure) along the path the evaluation has taken to reach it. E.g., the overall FMF value $\mu = \text{Min}\left(M_5^{L_0}, M_4^{L_1}\right) = \text{Min}(0.8, 1) = 0.8$ in the example shown in Fig. 1, meaning that the confidence of the system in the output is 80%.

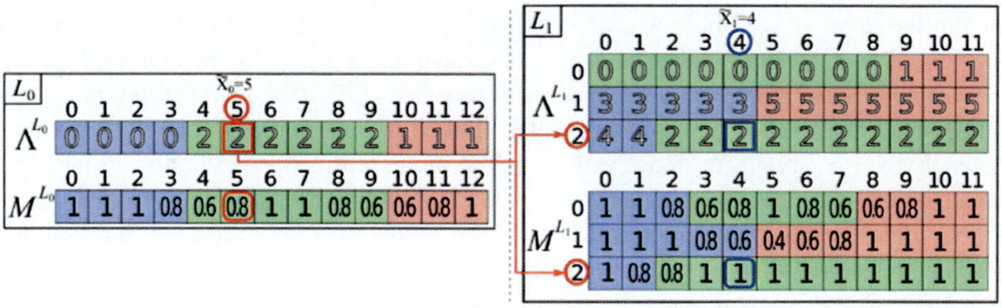

Fig. 1. An illustrative example for the general architecture of the Sequential Fuzzy Indexing Tables classifier.

Remark: The numbering of the sets and intervals is generally given by the order in which they were created from the training data.

Tests shown [5] that although the SFIT classifier provides a slightly lower (by ~1–2%) classification accuracy compared to other contemporary fuzzy classifiers, but it is also much faster considering its evaluation speed. This is because its time complexity is $\sim O(N)$, , meaning that it only depends on the number of attributes (since it simply has to access that many array elements to gain the indices along the path of the evaluation, and the class label in the last layer).

However, its spatial complexity depends on the size of the domains of the training data attributes, aside from the number of attributes. Since its index and class matrices in each layer have as many columns as the scaled domain size (D_i), and as many rows as the number of index values in the previous layer (which in practice can be bounded by the number of training samples, P). Therefore, the spatial complexity of the SFIT is $\sim O\big(\mathrm{N} \cdot \min\big(\overline{D}, P\big)\big)$, where $\overline{D}$ is the average attribute domain size among the layers. Although the size of the attribute domains can be lowered by scaling them accordingly, but that also reduces the classification accuracy of the classifier (as the nuances between data values are getting lost in the scaling process). Furthermore, for larger domains there often are "holes" in the arrays (i.e., valueless elements representing unknown parts of the problem space), so a portion of the used memory space can be wasteful.

Another disadvantage of the SFIT classifier is that it only evaluates one sequence of FMFs: the one that seems to result in the highest fuzzy membership value. This sometimes results in the classifier choosing a class that seems resembles the input data sample in the first few attributes but less so in the rest of the attributes, over one that would fit it more in general.

3 Sequential Fuzzy Indexed Search Trees

3.1 The General Architecture of SFISTs

In order to solve the problem of both the domain-size dependency and the empty regions, a new structure has been designed that uses the same base idea (i.e., creating and maintaining a dimensional decomposition of the problem space), but uses a more compact description of the FMFs in the form of intervals.

Figure 2 shows an example for the structure of the proposed classifier for the same problem as shown previously, roughly representing the same FMFs depicted in Fig. 1. In the first layer ($L_{0,0}$), there is one FMF with 3 *fuzzy sets*, which are represented as 5 *intervals*. An interval can either be a *transition* (an intersection between two neighboring sets) or a *plateau* (where the μ FMF value is constant 1). Each fuzzy set (and thus, each interval that is a part of a given fuzzy set) is linked to the appropriate FMF in the next layer (by a corresponding index value), so the next step in the evaluation in the next layer is determined by which interval the value of the input data (X_0) falls into. E.g., for input data $X = [5.0, 3.8]$, the first attribute value falls into the transition (interval #2) between fuzzy sets 0 and 2 in the FMF of $L_{0,0}$, so in the next layer the FMFs of $L_{1,0}$ and $L_{1,2}$ have to be examined using the second attribute value.

An FMF is divided into multiple intervals by so-called *pins*. Each pin represents a single data value from the data set used to train the classifier. To each pin belongs an

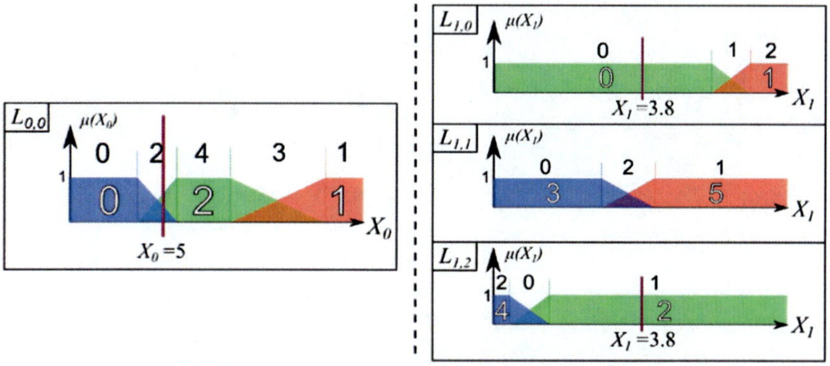

Fig. 2. The general architecture of the SFIST and an example for its usage.

interval that starts at the pin and ends at the next one (with the exception for the rightmost interval, where the upper bound is ∞^+). Furthermore, since the lower boundary value of the leftmost interval is at ∞^-, the leftmost interval also belongs to the leftmost pin. The leftmost and the rightmost intervals in any given FMF are always plateau type intervals.

In the following, let $I_{i,j,k}$ denote the kth interval of the jth FMF in layer i. An interval is defined with the following 5 values:

$$I_{i,j,k}\left(B^L, B^U, \eta^L, \eta^R, K\right) \tag{1}$$

where B^L and B^U are the *lower* and *upper boundary values* of the interval (i.e., the two pins bounding the interval), η^L and η^R denote the *left* and *right index* values associated to the fuzzy sets the interval is part of, while K is the class ID of the interval. An interval can be associated with multiple classes, in such case its K value is -1.

Figure 3 shows a detailed illustration of the a given $\mu(X_i)$ fuzzy membership function (Fig. 3(a)) for a given attribute i (or more precisely, the first attribute of the example shown in Fig. 2). For each interval, its left and right boundaries (B^L and B^U) are stored (Fig. 3(c)), as well as the index values (η^L and η^R) that link the fuzzy sets to the corresponding FMF in the next layer (which, in this particular example, are the 3 FMFs that can be seen in the second layer in Fig. 2). Note: if an interval is a plateau ($\mu(X_i) = 1$), then its η^R value is -1 (since it is only associated to one fuzzy set).

A class label K is also assigned to each interval to help the training process, and in the last layer the system regards K to gain the output of the system. The values of K in the example have been created from the class labels of the training data: the K_j value belong to an interval j is the same as the class of the sample used to make the leftward pin of the interval. In case there are multiple training samples with the same input attribute value but different class labels, then the K value belonging to their pin is set to -1. The main role of this notation is in the training phase: in the algorithm proposed in Sect. 3.2, it indirectly results in the creation of a new interval each time the training value falls into an interval marked with the K value of -1.

In order to quickly find the interval belonging to the input value X_i, a *self-balancing binary search tree* (BSTs, [6]) is built for each FMF (Fig. 3(b)), In a BST, each intermediate (non-leaf) node has an associated value, which is used to determine which of the

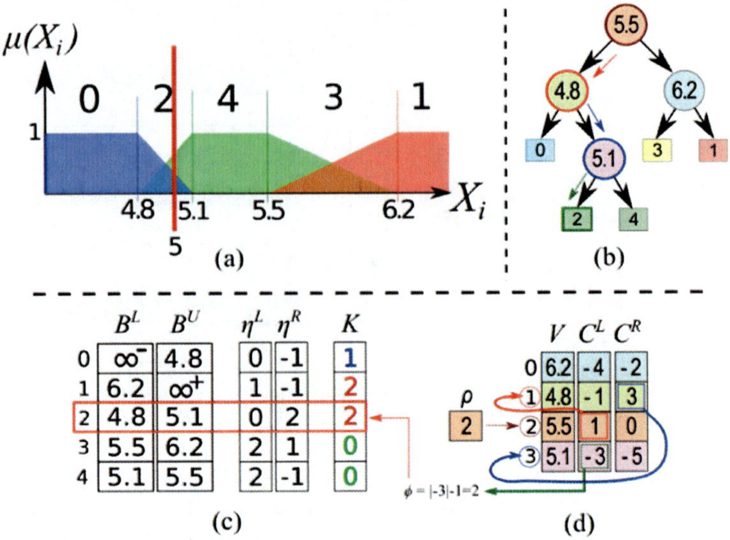

Fig. 3. The parts of a fuzzy membership function in any given layer i: a visual representation of (a) a fuzzy membership function and (b) the corresponding self-balancing fuzzy search tree that is used to quickly determine which interval a given input value X_i falls into. Bellow them are the indexing table representations of (c) the intervals and (d) the search tree. An example for the evaluation can be followed as well: for $X_i = 5$, the search tree determines that it falls into interval #2, by starting at the "root" (stored in ρ) node of the search tree arrays and following the indices till a leaf is reached ($C_3^L = -3$), from which the interval number #2 is gained. The corresponding interval values (B_2^L and B_2^U) are then used to calculate the fuzzy membership function value $\mu(X_i)$, and the index values (η_2^L and η_2^U) provide the index number of the two FMFs that need to be regarded in the next layer in the next step of the evaluation.

two child nodes are to be examined next. The interval IDs (i.e., the index numbers) are stored in the leaves. The tree nodes are also stored in indexing arrays (Fig. 3(d)): their condition value (V), alongside with the indices of their left and right children nodes (C^L and C^R). In the latter ones, nonnegative values indicate inner node IDs, while negative ones are leaves (E.g. "-1" corresponds to interval ID "0", "-2" to ID "1", etc.).

The evaluation algorithm of the indexing table representation of the BST can be followed in Fig. 4, for a given input value Z. Let j denote the ID of the node that is being investigated. The starting value of j is the ID of the root node. In a loop, first the value of node j is checked against Z: if the input value is smaller, then j is set to the ID of the left child of j ($j = C_j^L$), otherwise it is set to the right child ($j = C_j^R$). If the value of j is nonnegative, then the loop continues, otherwise the output value is calculated ($\phi = |j| - 1$).

An example for the usage of BSTs can be seen in Fig. 3. The root ρ is set to node ID 2, so the input value ($Z = 5$) is checked against V_2 ($V_2 = 5.5$). Since Z is smaller, $C_2^L = 1$ will be the new value of j. In the next iteration, the algorithm finds that Z is larger than V_1, and thus, $j = C_1^R = 3$. After that, since Z is smaller than $V_3 = 5.1$ and the left child of node 3 is negative, the evaluation stops and returns the value of 2. From

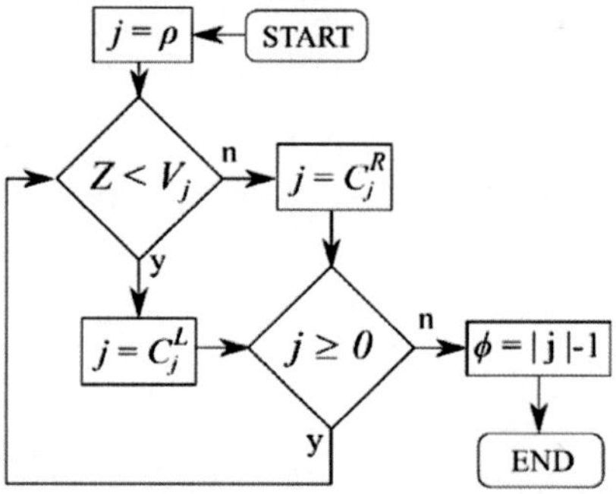

Fig. 4. Evaluation algorithm for the indexing array representation of a binary search tree.

this, it can also be deduced that the index values belonging to the sought interval are $\eta_3^L = 0$ and $\eta_3^R = 2$, and the class value is $K_3 = 2$ (although the latter is only significant in the last layer during the evaluation phase.)

3.2 The Training of SFITs

To train the proposed classifier, an incremental algorithm has been developed where the training data samples are processed one by one, each one either modifying the already existing FMFs in the structure, or creating new ones if necessary.

Let X denote an unprocessed input sample, t the class of the sample and X_i the value belong to its ith attribute. Let S_i^μ be the number of FMFs in layer i (i.e., belonging to the ith attribute), while $S_{i,j}^I$ is the number of intervals in the jth FMF of layer i.

In each layer i ($\forall i \in [0, N-1]$), there are two main tasks: analyze the current structure and create new or modify existing FMFs if necessary, and get the index value (ω) of the FMF that is needed to be regarded in the next layer. This value is 0 when the first layer is examined.

If the FMF referenced by ω does not exist yet ($S_i^\mu \leq \omega$) in the layer, then a new FMF is made with two intervals:

$$I_{i,\omega,0}\left(\infty^-, X_i, S_{i+1}^\mu, -1, t\right) \tag{2}$$

$$I_{i,\omega,1}\left(X_i, \infty^+, S_{i+1}^\mu, -1, t\right)$$

As it can be seen, the initial setting for each FMF consists of two plateaus around the initial value X_i ((∞^-, X_i) and $[X_i, \infty^+)$, respectively). This is changed to triangular or trapezoidal shapes later on, as new values are added to the FMF with class values that are different from that of the initial pin. Because of a new FMF is being created in layer i, another new one will necessarily be made in the next layer ($i + 1$), which will have the ID of S_{i+1}^μ, thus, the left index value of the two intervals in layer i are both set to this value. Remark: since the indexing used in this research starts at 0, any given S_i^μ value not only stores the number of FMFs in layer i, but also the index of the FMF that is yet to be created (because the indices of an array with N elements range from 0 to $N - 1$,

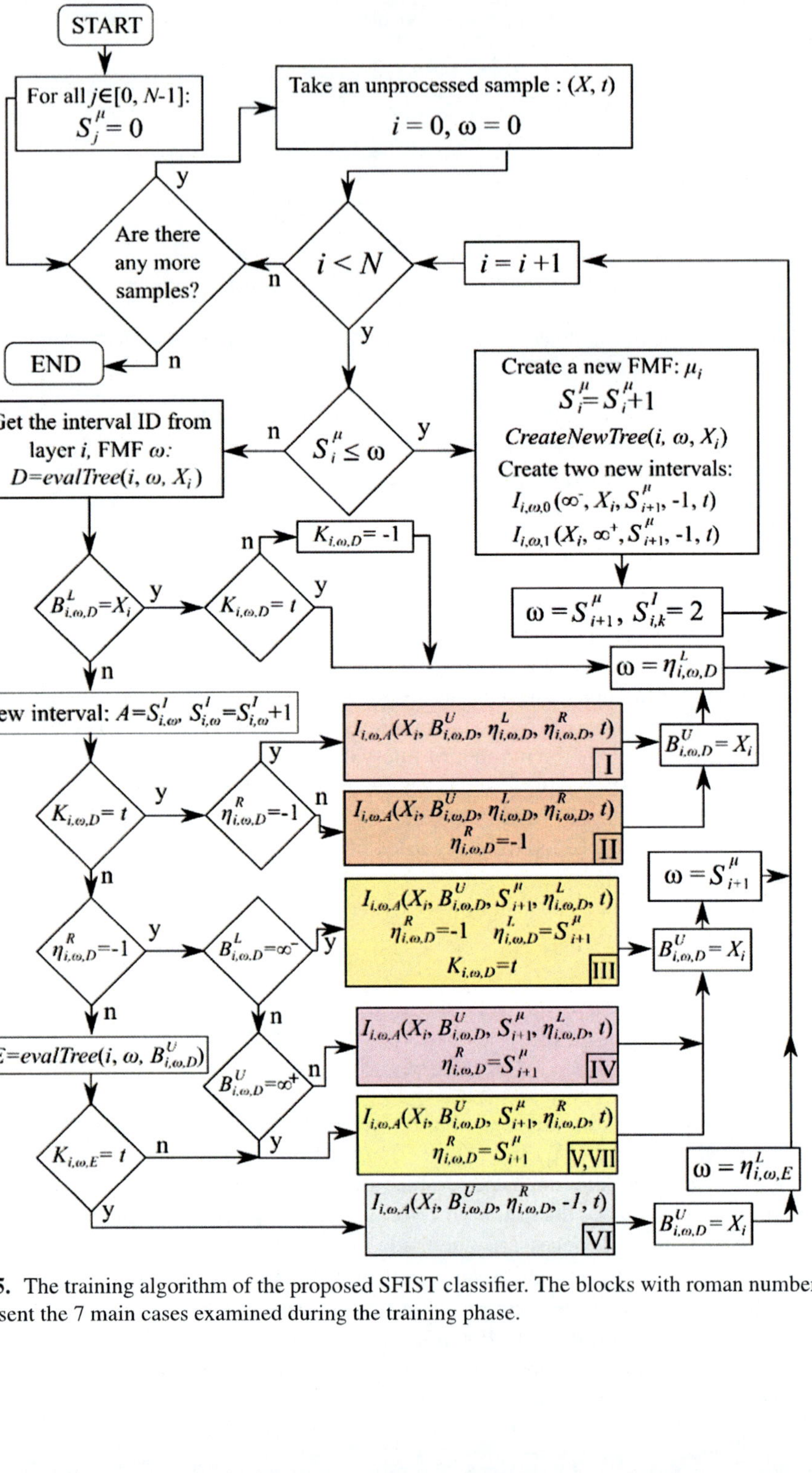

Fig. 5. The training algorithm of the proposed SFIST classifier. The blocks with roman numbers represent the 7 main cases examined during the training phase.

the Nth element does not exist (yet)). This is why the index value S^{μ}_{i+1} is used before it gets incremented in the next layer (during the next step of the training process) (Fig. 5).

A new BST is also created (for attribute i, FMF ω), where a single node is made with value X_i and leaves 0 and 1 (i.e., $V_0 = X_i$, $C^L_0 = -1$, $C^R_0 = -2$, , meaning that the node arrays will only have one element each). The number of FMFs is incremented ($S^{\mu}_i = S^{\mu}_i + 1$), and the next FMF index is set to the number of FMFs in the next layer ($\omega = S^{\mu}_{i+1}$), ensuring that a new FMF will be created in the next step, then the training moves onto the next layer.

If there is already an FMF at ω, then three main details are needed to be examined:

- if the input value is the same as the lower boundary value,
- if the class value of the input sample is the same as that of the interval D that the input value falls into,
- the type of said interval, and its position within the FMF (leftmost, rightmost or somewhere in the middle).

If the input value (X_i) is the same as the value of the lower boundary ($B^L_{i,\omega,D}$) of interval D, then this value has already occurred during course of the training. Furthermore:

- If the class label of the interval is also the same as the label of the training sample ($K_{i,\omega,D} = t$), then there is no need to change anything, proceed to the next layer using the left index of the interval ($\omega = \mu^L_{i,\omega,D}$).
- If the class label is different ($K_{i,\omega,D} \neq t$), then the class label ($K_{i,\omega,D}$) of the interval is changed to -1, before proceeding to the next layer using the left index of the interval ($\omega = \eta^L_{i,\omega,D}$).

If the input value is not the same as the one stored in $B^L_{i,\omega,D}$, then a new interval (with interval ID $A = S^I_{i,\omega}$) is inserted ($S^I_{i,\omega} = S^I_{i,\omega} + 1$). A new interval is created by virtually splitting an existing interval into two: the old interval will retain its index value and most of its parameters, while the new one will always be to the right side of the old one. A new node is also inserted into the BST with value X_i and leaf A ($V_b = X_i$ and $C^R_b = A$, where $b = S^I_{i,j}$).

If the input value (X_i) is *not* the same as the value of the lower boundary ($B^L_{i,\omega,D}$) of interval D, then there are 7 main cases that need to be investigated, since the parameters of the new interval depend on which existing interval is divided into two.

The first two cases deal with the sample having the same class as interval D:

- **Case I:** If the input sample is of the <u>same class</u> as interval D ($K_{i,\omega,D} = t$) and D is a <u>plateau</u> ($\mu^R_{i,\omega,D} = -1$), then interval D is split into two by X_i, and the new interval inherits its indices (Fig. 6(b)):

$$I_{i,\omega,A}\left(X_i, B^U_{i,\omega,D}, \eta^L_{i,\omega,D}, -1, t\right), \tag{3}$$

then the upper boundary value of the old interval D is updated, and the next index value ω is set to its left index:

$$B^U_{i,\omega,D} = X_i, \ \ \omega = \eta^L_{i,\omega,D} \tag{4}$$

- **Case II:** If the input sample is of the same class as interval D ($K_{i,\omega,D} = t$) and D is a transition ($\eta^R_{i,\omega,D} > -1$), then the new interval extends the old one, inheriting most of its parameters (Fig. 6(c)):

$$I_{i,\omega,A}\left(X_i,\ B^U_{i,\omega,D},\ \eta^L_{i,\omega,D},\ \eta^R_{i,\omega,D},\ t\right), \tag{5}$$

then the upper boundary value of the old interval D is updated, its right index value set to -1 (turning the old interval into a plateau), and the next index value ω is set to its left index:

$$\eta^R_{i,\omega,D} = -1,\ \ B^U_{i,\omega,D} = X_i,\ \ \omega = \eta^L_{i,\omega,D} \tag{6}$$

The next 3 cases deal with the sample having a different class label, but falls into a plateau interval, thus, a new fuzzy set is inserted (and consequently, a new FMF will be created in the next layer for all 3):

- **Case III:** If the class of the input sample is different from that of interval D ($K_{i,\omega,D} \neq t$), D is a plateau ($\eta^R_{i,\omega,D} = -1$) and D is the leftmost interval ($B^L_{i,\omega,D} = \infty^-$), then the new interval will be a transition (Fig. 6(d)):

$$I_{i,\omega,A}\left(X_i,\ B^U_{i,\omega,D},\ S^{\mu}_{i+1},\ \eta^L_{i,\omega,D},\ t\right) \tag{7}$$

The old interval D is changed into a plateau, its class is changed to that of the sample and the next index value ω is set to the value of a not yet existing FMF in the next layer:

$$\eta^R_{i,\omega,D} = -1,\ \ B^U_{i,\omega,D} = X_i,\ \ \eta^L_{i,\omega,D} = S^{\mu}_{i+1} \tag{8}$$

$$K_{i,\omega,D} = t,\ \ \omega = S^{\mu}_{i+1} \tag{9}$$

- **Case IV:** If the class of the input sample is different from that of interval D ($K_{i,\omega,D} \neq t$), D is a plateau ($\eta^R_{i,\omega,D} = -1$) and D is the rightmost interval ($B^U_{i,\omega,D} = \infty^+$), then the new interval will be a plateau (Fig. 6(e)):

$$I_{i,\omega,A}\left(X_i,\ B^U_{i,\omega,D},\ S^{\mu}_{i+1},\ \eta^R_{i,\omega,D},\ t\right) \tag{10}$$

The old interval D is changed into a transition, and the next index value ω is set to the value of a not yet existing FMF in the next layer:

$$\eta^R_{i,\omega,D} = S^{\mu}_{i+1},\ \ B^U_{i,\omega,D} = X_i,\ \ \omega = S^{\mu}_{i+1} \tag{11}$$

- **Case V:** If the class of the sample is different from that of interval D ($K_{i,\omega,D} \neq t$), D is a plateau ($\eta^R_{i,\omega,D} = -1$) and D is not either the leftmost or the rightmost interval ($\infty^- < B^L_{i,\omega,D}$ and $B^U_{i,\omega,D} < \infty^+$), then the new interval will be a transition (Fig. 6(f)):

$$I_{i,\omega,A}\left(X_i,\ B^U_{i,\omega,D},\ S^{\mu}_{i+1},\ \eta^L_{i,\omega,D},\ t\right) \tag{12}$$

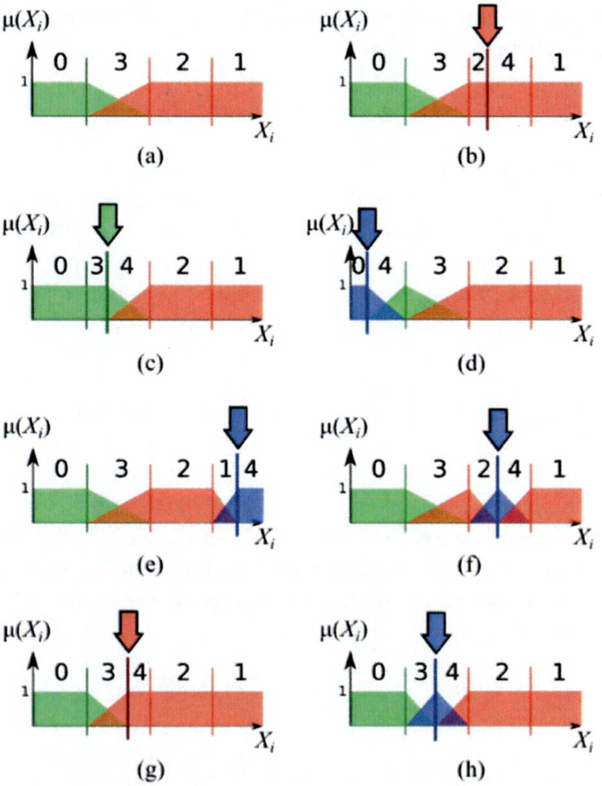

Fig. 6. An illustration for the main cases of inserting a new pin into the FMF: (a) the reference FMF, (b) insertion into same-classed plateau, (c) insertion into same-classed transition, (d) insertion into different-classed leftmost plateau, (e) insertion into different-classed rightmost plateau, (f) insertion into different-classed non-extrema plateau, (g) insertion into different-classed transition (where the interval to the right has the same class), (h) insertion into transition with both neighboring pins having a different class.

The old interval D is also changed into a transition, and the next index value ω is set to the value of a not yet existing FMF in the next layer:

$$\eta^R_{i,\omega,D} = S^{\mu}_{i+1},\ B^U_{i,\omega,D} = X_i,\ \omega = S^{\mu}_{i+1} \tag{13}$$

The last two cases deal with an input sample with a different class label that falls into a transition type interval:

- **Case VI:** If the class of the sample is different from that of interval D ($K_{i,\omega,D} \neq t$), D is a transition ($\eta^R_{i,\omega,D} > -1$) and the class of the sample is the same as the pin to the right ($K_{i,\omega,E} = t$, where interval ID E is gained by evaluating the BST with the upper bound value of interval D), then the new interval is a leftward extension of interval E (Fig. 6(g)):

$$I_{i,\omega,A}\left(X_i, B^U_{i,\omega,D}, \eta^R_{i,\omega,D}, -1, t\right) \tag{14}$$

The lower boundary value of the old interval D is set to the input value and the next index value ω is set to its left index:

$$B^U_{i,\omega,D} = X_i, \ \ \omega = \eta^L_{i,\omega,D} \tag{15}$$

- **Case VII:** If the class of the sample is different from that of interval D ($K_{i,\omega,D} \neq t$), D is a transition ($\eta^R_{i,\omega,D} > -1$) and the class of the sample is different from that of the pin to the right ($K_{i,\omega,E} \neq t$), then the old transition interval is split into two transition intervals, with the new fuzzy set in its center (Fig. 6(h)). A new FMF will be created in the next layer. Its formulas are the same as Eqs. (10) and (11).

3.3 Evaluation

Let us consider a sequence of FMFs in the structure (determined by a series of ω indices) from the first to the last layer as a *path*. In the structure of the proposed classifier, a path can branch at transition type intervals, and either continue or end at plateau type intervals.

In the evaluation phase, the goal is to evaluate all paths, and the one that ends in the last layer with the largest fuzzy membership value will determine the output class label.

In order to calculate the fuzzy membership function value for a given input value, in this paper simple linear transitions are used. Let Z be the input value ($Z \in [A, B]$, where $A < B$), so the FMF value for an increasing slope is calculated as:

$$\mu(Z, A, B) = \frac{Z - A}{B - A}, \tag{16}$$

while the value for a decreasing slope is calculated from Eq. (16):

$$\mu(Z, A, B) = 1 - \frac{Z - A}{B - A} \tag{17}$$

For each transition that the evaluation encounters (i.e., the input value falls into one such interval in a given FMF), the evaluation splits into 2 paths, while each plateau determines if a path is continued (if the closest pin falls within the range of an arbitrary threshold τ) or is closed down (if the input value is too far from both neighboring pins).

This threshold can be chosen simply to be a given percentage of the domain size of the given attributes (calculated from the largest and smallest value from the values of the training data for each attribute i):

$$\tau_i = \theta \cdot \left(\max_{\forall k}(X_{k,i}) - \min_{\forall k}(X_{k,i}) \right) \tag{18}$$

where θ is an arbitrary value ($\theta \in [0, 1]$). For example, for $\theta = 0.1$, the allowed distance from neighboring pins will be 10% of the total domain size of the given attribute.

Let $\beta_{p,i}$ denote the fuzzy membership function value that is calculated in the ith layer for path p. For the first layer, there is always just one FMF ($\omega = 0$). The first step for each layer i ($i \in [0, N - 1]$) is determining which interval the corresponding input attribute value (X_i) falls into, by evaluating the corresponding BST. Let us denote the ID of this interval with D.

If interval D is a plateau, then the fuzzy membership function is calculated for the value for both of the intervals bounds ($B^L_{i,\omega,D}$ and $B^U_{i,\omega,D}$):

$$\beta_{p,i} = \max\Big(1 - \mu\Big(X_i, B^L_{i,\omega,D}, B^L_{i,\omega,D} + \tau_i\Big), \mu\Big(X_i, B^U_{i,\omega,D} - \tau_i, B^U_{i,\omega,D}\Big)\Big) \tag{19}$$

If the resulting $\beta_{p,i}$ value is not 0, then the evaluation resumes in the next layer ($i = i + 1$), using the left index of D ($\omega = \eta^L_{i,\omega,D}$) for determine which FMF to regard.

If interval D is a transition, then the evaluation continues on both ways of the branching path. The 2 FMF values ($\beta^L_{p,i}$ and $\beta^R_{p,i}$, respectively) are calculated for the lower and higher bound values. In order to improve the *flexibility* of the classifier, both branches are always considered, with an arbitrary σ value:

$$\beta^L_{p,i} = \max\Big(\sigma, 1 - \mu\Big(X_i, B^L_{i,\omega,D}, B^U_{i,\omega,D}\Big)\Big) \tag{20}$$

$$\beta^R_{p,i} = \max\Big(\sigma, \mu\Big(X_i, B^L_{i,\omega,D}, B^U_{i,\omega,D}\Big)\Big)$$

where σ is a low, arbitrary value used to grant a certain level of flexibility to the system, in order to widen the search space. In this paper, $\sigma = 0.1$ is used.

After that, the evaluation path defined by the left index value of D ($\omega = \eta^L_{i,\omega,D}$) is resumed with value $\beta^L_{p,i}$, , while the other branch is continued on the right index value ($\omega = \eta^R_{i,\omega,D}$) with value $\beta^R_{p,i}$.

When the processing of a path reaches past the last layer ($i = N - 1$), the total fuzzy value of the path p is calculated:

$$\beta_p = \min_{\forall i}\big(\beta_{p,i}\big) \tag{21}$$

In the simplest solution, the β_p values are collected in a small, temporary indexing table T that has as many elements as many different class labels there are in the problem. Upon reaching the last layer ($i = N - 1$), the element of T indexed by $K_{N-1,\omega,D}$ is set to β_p, if the new value is larger than the old one:

$$T_{K_{N-1,\omega,D}} = \max\big(T_{K_{N-1,\omega,D}}, \beta_p\big) \tag{22}$$

After all paths have been evaluated, the output of the system is the label that reached the highest fuzzy membership function value:

$$y = \Big\{k \mid \max_{\forall k}(T_k)\Big\} \tag{23}$$

4 Performance

4.1 Experimental Results

In order to measure the classification performance of the proposed classifier, it has been trained with different benchmark datasets from the UCI Machine Learning Repository [7], of which two is presented in detail in the following, comparing its results to that of the predecessor SFIT classifier [5]. The available datasets are divided into (non-overlapping) training and testing datasets in given ratios (90:10%, 80:20% and 70:30). Furthermore, the order of the data samples have been changed randomly, so in each iteration a different training and testing datasets have been used. Both classifiers have been trained with these datasets (100 times each), and the average results have been observed.

Remark: the classification accuracy in this paper is calculated from the ratio between the correctly classified testing samples and the total number of training samples.

Table 1 shows the comparison of classification accuracy between the SFIST classifier and its predecessor classifier, using the Wisconsin Breast Cancer dataset [8]. The data consists of $N = 9$ attributes and 2 classes, among a total of 683 data samples. As it can be seen, the new classifier has achieved a higher accuracy (by ~0.17–0.9%).

Table 1. The classification performance of the SFIST classifier on the WBC dataset, compared to that of the SFIT classifier.

Training to Testing set ratio (%)	SFIST	SFIT
70:30	***94.21%***	*94.04%*
80:20	***94.80%***	*93.72%*
90:10	***95.60%***	*94.7%*

Table 2 shows the comparison of classification accuracy using the Seismic Bumps dataset [9], which consists of $N = 15$ attributes (originally 18, but 3 of them are all zero valued, so they have been omitted) and 2 classes, among a total of 2584 data samples. On this dataset too, the new classifier has achieved a higher accuracy rate (by ~1.71–2.07%).

Table 2. The classification performance of the SFIST classifier on the Seismic bumps dataset, compared to that of the SFIT classifier.

Training to Testing set ratio (%)	SFIST	SFIT
70:30	*93.25%*	*91.38%*
80:20	*93.28%*	*91.21%*
90:10	*93.41%*	*91.70%*

Table 3 shows the effect of this using *flexible transitions* ($\sigma = 0.1$) on the Iris dataset. The dataset consists of $N = 3$ attributes and 3 classes among a total of 150 samples. As it can be seen, the usage of the flexible transitions can drastically enhance the classification accuracy of the classifier (by 7.67–9.33%).

Table 3. The effect of the flexible transitions in the SFIST classifier considering the Iris dataset.

Training to Testing set ratio (%)	With flexible transitions	With rigid transitions
70:30	*85.67%*	*77.39%*
80:20	*86.47%*	*78.80%*
90:10	*89.93%*	*80.60%*

4.2 Time Complexity

In comparison to that of its predecessor classifier, the time complexity of the proposed SFIST classifier is higher by a logarithmic factor, since the corresponding BST is needed to be evaluated in each layer (which logarithmically depends on the number of the fuzzy sets (and subsequently, the intervals) in the layer). Thus, the spatial complexity of the SFIST classifier is $O(N \cdot \log_2 M)$, where M is the average number of fuzzy sets per fuzzy membership function in the whole structure.

Remark: Although theoretically there are 2^N paths to process for any given input sample, in practice only the first few attributes cause branching in the path (due to having more fuzzy sets in the FMFs in average), while for the rest of the layers the FMFs consists of only 1 or 2 fuzzy sets, so the overall number of paths per input sample is typically low (e.g. for the WBC dataset, only 1–3 paths are typically processed on average).

4.3 Spatial Complexity

For the proposed SFIST classifier, the two main structural parameters that have to be regarded are the average number of FMFs ($\overline{S^\mu}$), and the average number of intervals ($\overline{S^I}$) among the FMFs in each of the N layers. Thus, the spatial complexity is $O\left(N \cdot \overline{S^\mu} \cdot \overline{S^I}\right)$. As it can be seen, the size of the SFIST structure does not depend on neither the size of the training dataset, nor the size of the domain of the attributes, making it scale much better in both aspects.

To provide a more tangible quantification of the structure sizes, let us regard the elements of B^L, B^U and V as 3 real numbers (8 bytes each) while C^L, C^R, η^L, η^R and K are 5 integer numbers (4 bytes each). The following simplified formula can be used to calculate the approximate size of the structure (in bytes):

$$SIZE_{SFIST} = 3 \cdot 8 \cdot \widehat{S^I} + 5 \cdot 4 \cdot \widehat{S^N} \tag{24}$$

where $\widehat{S^I}$ is the total number of intervals (for which real number arrays B^L, B^U and V are stored) and $\widehat{S^N}$ stands for the total number of nodes (for which integer arrays C^L, C^R, η^L, η^R and K are stored) in the structure.

On the other hand, the size of the predecessor classifier can be calculated from the domain size of the attributes (D_i, $\forall i \in [0, N-1]$), which can be bounded by the number of training samples (P):

$$SIZE_{SFIT} = \sum_{i=0}^{N-1} ((4+8) \cdot \min(D_i, P)) \tag{25}$$

where the fuzzy arrays are real valued (8 bytes per element), while the index arrays are integer valued (4 bytes per element).

A typical SFIST structure trained with the *WBC dataset* ($P = 615$, considering 90:10 training to testing sample ratio), results in 729 intervals and 538 nodes, 29020 bytes (~28.34 KB). The structure of the predecessor classifier (after training it with the WBC data) consists of 38010 valued array elements, taking 456120 bytes (~445.42 KB) memory space (~15 times more).

For the *seismic bumps dataset* ($P = 2326$, using 90:10 training to testing sample ratio), the proposed SFITS classifier structure consists of 8212 intervals and 6329 nodes (~293.04 KB). The structure of the predecessor classifier requires ~95.27 MB, due to the larger training sample set and domain sizes (~333 times larger). This shows that the proposed classifier indeed provides a much more compact structure than its predecessor.

5 Conclusion

In this paper, a new method is proposed for fuzzy inference-based classification. Similarly to its predecessor, the SFIT classifier, the proposed method uses a layered architecture to create a dimensional decomposition of the problem space, but in a more flexible and concise way.

Instead of directly storing the precalculated fuzzy values for known parts of the problem space, the new classifier only stores the fuzzy membership functions and their search trees in the form of indexed arrays. This requires much less memory space, without sacrificing much of the operating speed (by a logarithmic factor of the average number of fuzzy sets in the FMFs).

Experiments on benchmark data sets showed that the proposed classifier provides a better classification accuracy (by 1–2%) compared to that of its predecessor classifier, since path branching and flexible transitions make sure that the evaluation process takes a larger area of the problem space into account.

Although for 2-class problems the classifier provides a good classification performance (~93–95%), though for multiclass problems there is room for improvement (~89.9%). In future work, the system will be further improved to better handle those problems as well.

Furthermore, the application of different shapes for fuzzy sets will be investigated as well, as well as different types of training (e.g. clustering-based) algorithms.

Acknowledgement. Supported by the ÚNKP-23-4-II-OE-59 New National Excellence Program of the Ministry for Culture and Innovation from the source of the National Research, Development and Innovation Fund. This publication is also the result of the Research & Innovation Operational Programme for the Project: "Support of research and development activities of J. Selye University in the field of Digital Slovakia and creative industry", ITMS code: NFP313010T504, co-funded by the European Regional Development Fund.

References

1. Zadeh, L.A.: Fuzzy logic. Computer **21**(4), 83–93 (1988)
2. Guzman-Urbina, A., Ouchi, K., Ohno, H., Fukushima, Y.: FIEMA, a system of fuzzy inference and emission analytics for sustainability-oriented chemical process design. Appl. Soft Comput. **126**, 109295 (2022)
3. Paul, S.K., Chowdhury, P., Ahsan, K., Ali, S.M., Kabir, G.: An advanced decision-making model for evaluating manufacturing plant locations using fuzzy inference system. Expert Syst. Appl. **191**, 116378 (2022)
4. Várkonyi-Kóczy, A.R., Tusor, B., Tóth, J.T.: A multi-attribute classification method to solve the problem of dimensionality. In: Jabłoński, R., Szewczyk, R. (eds.) Recent Global Research and Education: Technological Challenges. AISC, vol. 519, pp. 403–409. Springer, Cham (2017). https://doi.org/10.1007/978-3-319-46490-9_54
5. Várkonyi-Kóczy, A.R., Tusor, B., Tóth, J.T.: Active problem workspace reduction with a fast fuzzy classifier for real-time applications. In: Proceedings of 2016 IEEE International Conference on Systems, Man, and Cybernetics (SMC), Budapest, Hungary, pp. 004423–004428 (2016)
6. Senbel, S.A.: Interesting exercise to demonstrate self-balancing trees. J. Comput. Sci. Coll. **38**(3), 185 (2022)
7. Dua, D., Graff, C.: UCI Machine Learning Repository (2017). http://archive.ics.uci.edu/ml
8. Mangasarian, O.L., Wolberg, W.H.: Cancer diagnosis via linear programming. SIAM News **23**(5), 1–18 (1990)
9. Sikora, M., Wróbel, L.: Application of rule induction algorithms for analysis of data collected by seismic hazard monitoring systems in coal mines. Arch. Min. Sci. **55**, 91–114 (2010)

Optimization of Quartz Sol-gel Glass Cutting Parameters by Elliptical Laser Beams Using Neural Network Simulation and Genetic Algorithm

Yuri Nikityuk(✉), Vladislav Prokhorenko, Alina Semchenko, and Dmitry Kovalenko

Francisk Skorina Gomel State University, 104 Sovetskaya Str., 246019 Gomel, Belarus
nikitjuk@gsu.by

Abstract. This study provides the optimization of double-beam cutting of quartz plates through laser cleaving. Neural network simulation and the authors' version of the modified genetic algorithm were used to determine the optimal processing parameters. Finite element calculations of temperature and thermoelastic stress fields were performed to create the training data array and the array data for testing neural networks. Neural networks and their training algorithms were implemented with the Keras library in Python. The optimal neural network architectures for approximating the maximum values of tensile stresses and temperature during laser cutting of quartz plates were determined. The genetic algorithm was used to find the optimal parameter values for quarts plate laser cutting process.

Keywords: Neural Network Modeling · Laser Cutting · Genetic Algorithm · Parameter Optimization

1 Introduction

Quartz glasses exhibit unique properties. These features are also observed in quartz glasses synthesized via the sol-gel method. This method makes it possible to produce activated and high-purity quartz glasses for laser technology, fiber optics, and optoelectronics [1]. Conventional quartz glass processing methods face significant limitations. Laser cleaving is an effective method for processing glasses, ceramics and crystals, which is based on forming a crack in the material by successive laser heating and exposing the surface under processing to a refrigerant. Several significant advantages of this process can be identified, including a high level of separation accuracy, a high processing speed, the absence of debris, and the increased strength of the resulting products. At the same time, using double-beam schemes increases the efficiency of this technology [2–7]. Currently, artificial neural networks are successfully used to simulate laser processing of materials, and genetic algorithms are used to optimize the corresponding technological parameters [8, 9].

Y. Ono and J. Kondoh (Eds.): Inter-Academia 2023, LNNS 939, pp. 310–315, 2024.
https://doi.org/10.1007/978-3-031-54450-7_34

Artificial neural networks possess the capacity to serve as universal approximation models, hence offering the potential of producing favorable outcomes in the simulation of complex technical systems. Genetic algorithms can be classified as a particular type of evolutionary approaches, which are based on the principles of collective learning within a population and employ the simulation of natural selection. Genetic algorithms provide the search for better solutions by inheriting and enhancing useful properties of a set of objects in the process of simulating their evolution.

This study employed numerical finite element simulations to develop neural network models for the double-beam laser cleaving of sol-gel glasses with subsequent determination of effective processing modes using a genetic algorithm.

2 Finite Element Analysis

Calculation of temperatures and thermoelastic stresses in sol-gel glass during double-beam laser cleaving (see Fig. 1) was performed using the finite element method using APDL programming language (Ansys parametric design language).

Figure 1 illustrates the correspondence between different positions and their respective characteristics [10]. The same experiment scheme is applied in this study for another class of material (sol-gel glasses). Position 1 represents an elliptical laser beam with a wavelength of 10.6 μm, position 2 stands for a circular laser beam with a wavelength of 1.06 μm, position 3 denotes a refrigerant, positions 4 and 5 depict cross-sections of laser beams with wavelengths of 10.6 μm and 1.06 μm, respectively. Lastly, position 6 signifies a zone of refrigerant impact.

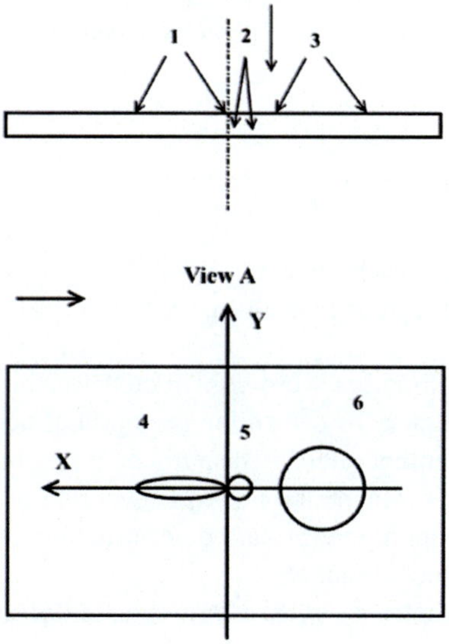

Fig. 1. Schematic of mutual arrangement of laser beams and refrigerant impact zones.

For eight factors (P1-P8): P1 represents the processing speed V; P2 is the quartz plate thickness H; P3 and P4 refer to the major semi-axis A and minor semi-axis B of an elliptical beam with emission wavelength λ = 10.6 μm; P5 is the radius of the emission spot R of a beam with emission wavelength λ = 1, 06 μm; P6 denotes the power P of the laser with emission wavelength λ = 10.6 μm; P7 is the power P_0 of the laser with emission wavelength λ = 1.06 μm; P8 stands for the distance h between the impact zones of the laser with emission wavelength λ = 1.06 μm and the refrigerant. Calculations were performed for 100 combinations of input parameters. The following output parameters were selected for study: maximum temperature in the laser treatment zone T and maximum tensile stresses σ_{yy} in the treatment zone.

Figure 2 illustrates the typical distributions of temperature and thermoelastic stress fields during the implementation of finite element modeling of double-beam laser cleaving on quartz glasses.

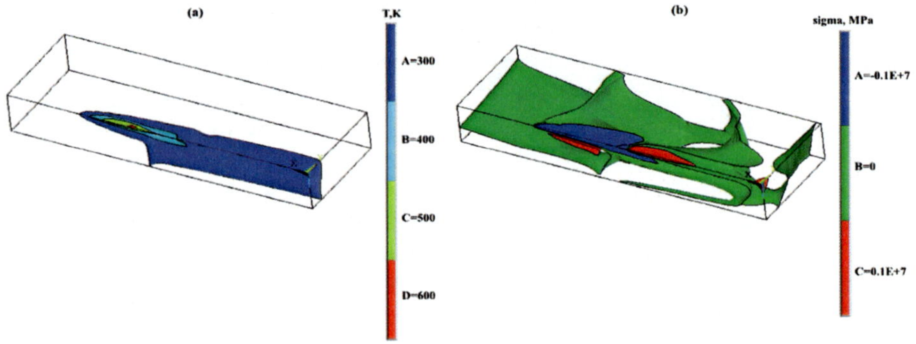

Fig. 2. Computed distributions of temperatures (a) and thermoelastic stresses (b) during the implementation of double-beam laser cleaving on quartz glasses

3 Neural Network Modeling

The process of double-beam laser cleaving of sol-gel glasses was simulated using artificial neural networks in accordance with the algorithm used in [10] for silicate glass cleaving.

Neural network approximators were constructed for the parameters σ_{yy} and T in Python using the Keras library. To determine the optimal neural network architecture, three-layer networks were enumerated by the number of neurons in hidden layers ranging from 8 to 96 with a step of 4. The neural networks were trained by minimizing the MSE loss function with early stopping. MSE and determination coefficient R2 metrics were used to evaluate the constructed models.

Figure 3 displays the graphs of typical metric changes throughout the training process of the candidate models.

Figure 4 depicts the distribution of mean-square error (MSE) values for neural network models when testing the models on normalised data for the problem of determining the values of maximum tensile stress (a) and maximum temperature (b). The number of neurons in the first and second hidden layers of the artificial neural network are marked

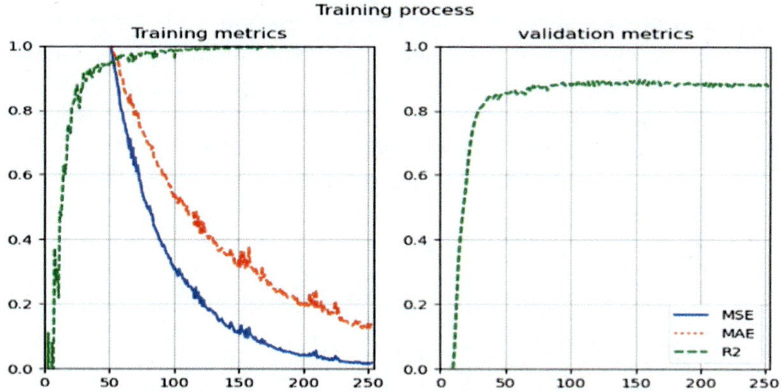

Fig. 3. Training process of neural networks with normalized values of metrics

on the x (horizontal) and y (vertical) axes, respectively. The intensity of color coding represents the extent of error: the error increases from light to dark.

During the numerical experiment, it was found that the artificial neural network with [8-24-60-1] architecture provides the most accurate results when approximating σ_{yy}, and the artificial neural network with [8-38-54-1] architecture shows the best results when approximating temperature T.

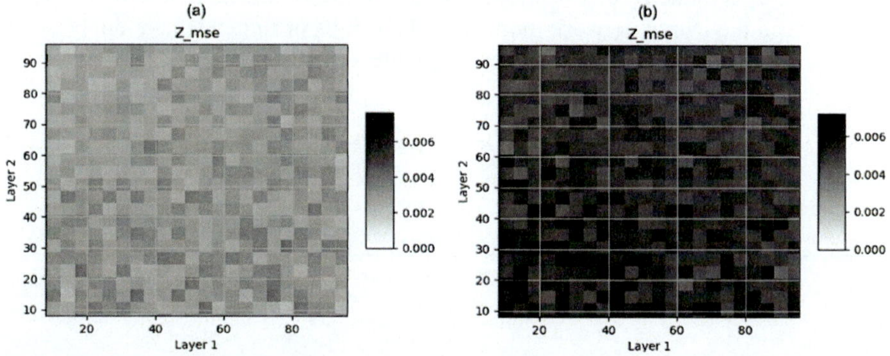

Fig. 4. Map of MSE metric values during cross-validation of three-layer neural network architectures for σ_{yy} (a) and T (b) approximation.

4 Genetic Algorithm Optimization

The authors' version of the modified genetic algorithm implemented in Python [11] was used to solve the multicriteria problem of finding the input parameters to maximize the tensile stress σ_{yy} under the condition of maximizing the velocity V and limiting the temperature value $T < 1473$ K.

The approach employed for creating further generations of the population was derived from the genome crossing technique outlined in reference [11]. Additionally, genome mutation was achieved by introducing random changes in factors. The objective function included the values of the σ_{yy} and T estimates predicted by neural network approximators and velocity V. Furthermore, penalties were included for exceeding permissible ranges of normalized factors and maximum temperature values:

$$L(A, B, V, P) = -(\alpha_1\sigma_{yy} + \alpha_2 V) + \beta_1 E_1 + \beta_2 E_2 + \beta_3 E_3 + \beta_4 E_4 + \beta_5 E_5 + \beta_6 E_6 + \beta_7 E_7 + \beta_8 E_8 + \beta_9 E_9,$$

$$E_1 = \{1, V \notin [0.1]0, V \in [0.1], E_2 = \{1, H \notin [0.1]0, H \in [0.1], E_3 = \{1, A \notin [0.1]0, A \in [0.1], E_4 = \{1, B \notin [0.1]0, B \in [0.1],$$

$$E_5 = \{1, R \notin [0.1]0, R \in [0.1], E_6 = \{1, P \notin [0.1]0, P \in [0.1], E_7 = \{1, P_0 \notin [0.1]0, P_0 \in [0.1], E_8 = \{1, h \notin [0.1]0, h \in [0.1],$$

$$E_9 = \{1, T \geq 1473K\ 0, T < 1473K,$$

$$\alpha_1 = \alpha_2 = 0.5,\ \beta_i = 1.0,\ i = 1, \ldots, 9 \quad (1)$$

Table 1 displays the results of multicriteria optimization. The required correspondence between the neural network model and the results of finite element analysis was established, validating their reciprocal correspondence and accuracy.

Table 1. Optimization results

V m/s	H m	A m	B m	R m	P W	P0 W	h M	T K	σ_{yy} MPa
0.03	*0.002*	*0.002*	*0.001*	*0.001*	*29*	*73*	*0.001*	*1468*	*7.03*

5 Conclusion

This study presents a multicriteria optimization of double-beam cutting of quartz plates via laser cleaving, using neural network simulation. Effective neural network architectures have been established for estimating the maximum of tensile stresses and the maximum of temperature in the laser radiation and refrigerant exposure area when performing double-beam cleaving of quartz plates. The authors' version of the modified genetic algorithm was used to determine the optimal parameters for laser processing.

References

1. Shalupaev, S.V., Semchenko, A.V., Nikitjuk, Y.V.: Silica gel glasses after laser radiation. Mater. Sci. **21**(4), 495–501 (2003)
2. Nisar, S., Li, L., Sheikh, M.: Laser glass cutting techniques - a review. J. Laser Appl. **25**(4), 042010-1–042010-11 (2013)
3. Serdyukov, A.N., Shalupaev, S.V., Nikitjuk, Y.: Features of controlled laser thermal cleavage of crystalline silicon. Crystallography **55**(6), 933–937 (2010)
4. Saman, A.M., Furumoto, T.: Evaluation of separating process for different materials by thermal stress cleaving technique. Int. J. Eng. Technol. **7**(4.27), 158–161 (2018)
5. Shalupaev, S.V., Shershnev, E.B., Nikityuk, Y., Sereda, A.A.: Two-beam laser thermal cleavage of brittle nonmetallic materials. J. Opt. Technol. **73**(5), 356–359 (2006)
6. Liu, J., Lu, J., Ni, X., Dai, G., Zhang, L., Chen, Y.: Numerical study on thermal stress cutting of silicon wafer using two-point pulsed laser. Opt. Appl. **XLI**(1), 247–255 (2011)
7. Junke, J., Xinbing, W.: Cutting glass substrates with dual-laser beams. Opt. Lasers Eng. **47**, 860–864 (2009)
8. Bakhtiyari, A.N., Wang, Z., Wang, L., Zheng, H.: A review on applications of artificial intelligence in modeling and optimization of laser beam machining. Opt. Laser Technol. **135**, 1–18 (2021)
9. Nikityuk, Y.V., Serdyukov, A.N., Aushev, I.Y.: Optimization of two-beam laser cleavage of silicate glass. J. Opt. Technol. **89**, 15–24 (2016)
10. Nikitjuk, Y., Serdyukov, A.N., Aushev, I.Y.: Determination of the parameters of two-beam laser splitting of silicate glasses using regression and neural network models. J. Belarusian State Univ. Phys. **1**, 35–43 (2022)
11. Nikitjuk, Y.V., Prokhorenko, V.A., Kulyba, A.I.: Multi-criteria optimization of quartz glass laser cutting parameters using neural network simulation and genetic algorithm. Probl. Phys. Math. Tech. **3**, 26–31 (2023)

Execution Efficiency of a Sequential, Multi-Threaded, and a Parallel Search for a Structured Data in a Symbol Table Implemented in a Binary Search Tree, an Array, and in a Linked List

Igor Košt'ál(✉)

Faculty of Economic Informatics, University of Economics in Bratislava, Dolnozemská cesta 1, 852 35 Bratislava, Slovakia
igor.kostal@euba.sk

Abstract. A symbol table is an abstract mechanism used in many kinds of applications, including search applications, to store key-value pairs (information), where we can later search for and retrieve the value by specifying a key. For an effective symbol table implementation, it is important to choose an appropriate data structure. We have created a C# .NET application that allows to search for values, structured data of persons, according to given keys, e.g., first names and surnames, or phone numbers, in a symbol table (a phone book) that is implemented in a binary search tree, an array, and in a linked list sequentially, on multiple threads and in parallel using its instance methods, while the application measures the execution times of particular searches. By comparing these execution times, we were finding which of these data structures allows more efficient searching in a sequential search, multi-threaded, or in a parallel search.

Keywords: Sequential Search · Multi-threaded Search · Parallel Search · Symbol Table · Binary Search Tree · Array · Linked List

1 Introduction

Many applications use a symbol table for storing information. If we want to create an effective symbol table implementation, then we have to choose an appropriate data structure for it. We were interested in which of data structures a binary search tree (BST), an array, and a linked list is more suitable for a sequential, multi-threaded, and parallel search for, e.g., a data set of persons or phone numbers, by the C# .NET application search methods in the symbol table that contains a phone book. We have created a C# .NET application that allows to search for such data sets according to given keys, e.g., first names and surnames, or phone numbers, in these three implementations of a symbol table sequentially, on multiple threads and in parallel using its instance methods. This application also measures the execution times of all searches and writes them into a disc log file. Using these execution times, we were finding which of these data structures

Y. Ono and J. Kondoh (Eds.): Inter-Academia 2023, LNNS 939, pp. 316–332, 2024.
https://doi.org/10.1007/978-3-031-54450-7_35

allows more efficient searching in a sequential search, multi-threaded, or in a parallel search. The results and an evaluation of this experiment are presented in the paper.

In the relevant sources, the Web of Science, and the Scopus, we did not find studies or papers dealing with our topic *Execution Efficiency of a Sequential, Multi-Threaded, and a Parallel Search for a Structured Data in a Symbol Table Implemented in a Binary Search Tree, an Array, and in a Linked List* or similar to our topic. More authors deal with parts of our topic only, for example, with a BST implementation [1–3], an array implementation [1–4], and a linked-list implementation [1–3] of a symbol table.

The research described in this paper is a continuation of our research from [5], in which we have dealt with similar topic *Execution Efficiency of a Sequential Search, Multi-Threaded, and a Parallel Search for a Data in a Symbol Table Implemented in an Array, and in a Linked List*. However, in [5] we did not work with an implementation of a symbol table in a binary search tree and our C# application from [5] worked with simple key-value pairs, URL - IP address pairs, only in two implementations of a symbol table, in an array and linked-list implementation. Now our C# application works with structured data stored in three implementations of a symbol table, in a binary search tree, an array and in a linked list. A symbol table contains structured data of persons of a phone book.

2 A Symbol Table, Its Use in Applications, and Its Implementations

A symbol table is a data structure of items with keys that supports two basic operations: insert a new item, a key-value pair, and return an item (a value) with a given key [1]. The fact that the symbol table is often used in different applications is confirmed by the many kinds of applications in which it is implemented. These are e.g., applications that work with a phone book, a dictionary, accounts, an e-book, DNS records [3] etc. From the given overview of the use of the symbol table its wide use in various kinds of applications is evident, therefore it makes sense to consider its use in the development of such or similar kinds of applications.

A symbol table can be implemented using a binary search tree, an array, and a linked list. A *binary search tree* (BST) is a binary tree where each node has a comparable key (and an associated value) and satisfies the restriction that the key in any node is larger than the keys in all nodes in that node's left subtree and smaller than the keys in all nodes in that node's right subtree [2].

For a symbol table implementation, we also use an unordered dynamic array of objects and an unordered linked list of objects, which are dynamic structures. Especially a linked list is a true dynamic data structure that is maximal memory efficient. We can even create an empty list that contains no data elements (nodes), only informational data, in our case the *head* and *tail* references and the *Count* instance variable. We can add new data elements (nodes) to the linked list at its beginning, end, or any place in it during the execution of the application. We can also delete any data element (a node) from the list, it can be the first, last or any of its element (a node). That is, the list can dynamically grow or shrink during the application execution, still maintaining maximum memory efficiency. A dynamic array cannot work like that. However, there are also disadvantages of working with a list, e.g., we cannot access its data element (a node) as easily as using an index to a dynamic array element. [5] In the list, we have to get to the given data element (a node) by going through all previous data elements (nodes) [4].

3 The Task Parallel Library, Kinds of a Parallelism, the *Parallel.Invoke* Method

We have created a C# application, which uses the *System.Threading.Tasks.Parallel* library class, from the Task Parallel Library in its parallelized source code.

This Task Parallel Library (TPL) is a set of public types and APIs (Application Programming Interfaces) in the *System.Threading* and *System.Threading.Tasks* namespaces in the NET Framework. The purpose of the TPL is to make developers more productive by simplifying the process of adding parallelism and concurrency to applications [6].

In parallel programming in the .NET Framework, we recognize two kinds of a parallelism - data and task parallelism.

Data parallelism refers to scenarios in which the same operation is performed concurrently (that is, in parallel) on elements in a source collection or array [6]. We did not use a data parallelism in a parallelized code of our C# application.

The Task Parallel Library (TPL), as its name implies, is based on the concept of a task, which represents an asynchronous operation. The term task parallelism refers to one or more independent tasks running concurrently. Tasks provide two primary benefits [6]:

- More efficient and more scalable use of system resources. Behind the scenes, tasks are queued to the *ThreadPool*, which has been enhanced with algorithms that determine and adjust to the number of threads and that provide load balancing to maximize throughput. This makes tasks relatively lightweight, and we can create many of them to enable fine-grained parallelism.
- More programmatic control than is possible with a thread or work item. Tasks and the framework built around them provide a rich set of APIs that support waiting, cancellation, continuations, robust exception handling, detailed status, custom scheduling, and more.

Task parallelism with the *Parallel.Invoke* method is used in a parallelized code of our C# application. The ***Parallel.Invoke*** method provides a convenient way to run any number of arbitrary statements concurrently. We just pass in an *Action* delegate for each item of work. The easiest way to create these delegates is to use lambda expressions. The lambda expression can either call a named method or provide the code inline [6].

Parallel.Invoke is a static method with two overloaded versions. Here is the signature of the version of *Parallel.Invoke* from [6] that is used in a parallelized code of methods of our C# application.

Syntax (in C#): *public static void **Invoke** (params Action[] actions);*

Parameters: *actions* - an array of *Action* delegates to execute.

4 A C# Application, Its Methods

Our C# application was created in the C# language and built in the Microsoft Visual Studio 2019 development environment. The C# application implements a symbol table (a phone book) in a BST of objects, in an unordered array of objects, and in an unordered linked list of objects, and allows, except of other things, to search data records of persons according to the *first name* and *surname* key, and the *phone number* key in these three implementations of a symbol table sequentially, on multiple threads, and in parallel, while this application measures the execution time of each search in each implementation of the symbol table. The source code of this application contains seven classes, two structures, and five delegates (Fig. 6). The following classes are defined in this source code:

- ***Node*** - objects of this class represent the nodes of a binary search tree with an implemented symbol table. They contain key-value pairs: a first name and a surname - personal data in the *keyFirstSurName* and *personalInfo* instance variables, the *left* and *right* references to the left and right subtree, and the *size* instance variable with the number of nodes in a subtree. The *Set_keyValuePair* member method initializes the *keyFirstSurName* and *personalInfo* instance variables of the *Node* class object, and the *Get_keyFirstSurName* and *Get_personalInfo* member methods get the values of the *keyFirstSurName* and *personalInfo* instance variables of such object.
- ***BST*** - an object of this class containing the *root* reference to a root node of a binary search tree, represents a binary search tree, in which nodes a symbol table is implemented (Fig. 2). The nodes of a binary search tree are objects of the *Node* class. The *BST* class provides member methods for inserting the specified key-value pair into the symbol table, finding the smallest and largest keys in the symbol table, searching for nodes according to various kinds of keys, finding the number of key-value pairs in the symbol table, for removing the specified key and its associated value from the symbol table and so on. For example, the *BST* class provides the *SearchALLNodeKeyVPhoneNInordr* method for searching for all key-value pairs in a symbol table (nodes in a binary search tree) according to the *phone number* key and the *SearchNodeKeyValInorder_str* member method for searching for a key-value pair in this symbol table according to the *first name* and *surname* key (the *keyFirstName* and *keySurName* parameters of the method) (Fig. 1).

```
public string SearchNodeKeyValInorder_str(Node s, string keyFirstName, string keySurName) {
   if (s == null)  return "";
   int compare = Compare(s, keyFirstName, keySurName);
   if (compare > 0)  return SearchNodeKeyValInorder_str(s.left, keyFirstName, keySurName);
   else if (compare < 0)
      return SearchNodeKeyValInorder_str(s.right, keyFirstName, keySurName);
   else
      return GetNodeData_str(s);  }
```

Fig. 1. The source code of the *SearchNodeKeyValInorder_str* member method.

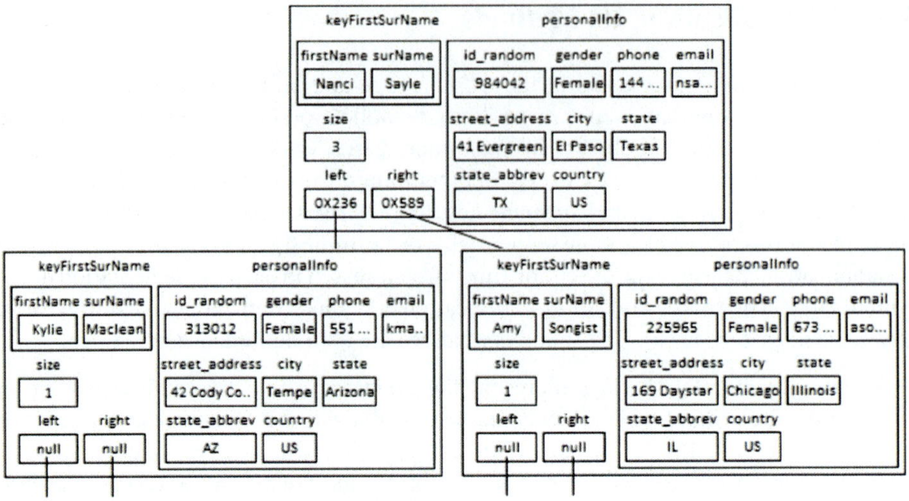

Fig. 2. The small part of a symbol table (a phone book) implemented in a BST (with 3 nodes - the *Node* class objects) that contains structured personal data of persons, key-value pairs: the *keyFirstSurName* key - the *personalInfo* value.

The *SearchNodeKeyValInorder_str* method works recursively: if the tree is empty, it terminates the search as unsuccessful. If the search key is equal to the key in the node, it terminates the search as successful (by returning the value associated with the key). If the search key is smaller than the key in the node, it searches (recursively) in the left subtree. If the search key is greater than the key in the node, it searches (recursively) in the right subtree [3].

The *SearchALLNodeKeyVPhoneNInordr* and *SearchNodeKeyValInorder_str* methods have searched for data records of persons, key-value pairs: first names and surnames - personal data of persons according to 5 given keys a ***phone number*** and 5 given keys a ***first name*** and a ***surname*** in a symbol table (a phone book) sequentially (Fig. 3), on multiple threads (Fig. 5), and in parallel in an experiment.

The **asynchronous execution** of these methods (also others in our C# application) on multiple threads is realized using the *BeginInvoke* and *EndInvoke* instance methods of the objects of the delegate types that we had to declare, and which signature must match the signature of the method to which the delegate, the object of the delegate type, will point to [5].

The common language runtime automatically defines *BeginInvoke* and *EndInvoke* methods for these delegate types, with the appropriate signatures. The *BeginInvoke* method initiates the asynchronous call to a target method on a secondary thread. It has the same parameters as the method that we want to execute asynchronously, plus two additional optional parameters. *BeginInvoke* returns immediately and does not wait for the asynchronous call to complete. *BeginInvoke* returns an *IAsyncResult*, which can be used to monitor the progress of the asynchronous call. The *EndInvoke* method retrieves the results of the asynchronous call. It can be called any time after *BeginInvoke*. If the asynchronous call has not completed, *EndInvoke* blocks the calling thread until it completes. The parameters of *EndInvoke* can include the out and ref parameters of the method that we want to execute asynchronously, plus the *IasyncResult* returned by *BeginInvoke* [6].

The **parallel execution** of 5 the *bst.SearchALLNodeKeyVPhoneNInordr* instance methods and 5 the *bst.SearchNodeKeyValInorder_str* instance methods (Fig. 4) is realized using the *Parallel.Invoke* static method that uses lambda expressions.

```
BST bst = new BST();  // Creates the 'bst' object of the 'BST' class.      . . .
string[] results = new string[10];
         . . .
stopwatch.Reset(); // Stops time interval measurement and resets the elapsed time to zero.
stopwatch.Start(); // Starts, or resumes, measuring elapsed time for an interval.
results[0] = bst.SearchNodeKeyValInorder_str(bst.Get_root(), splitInput[0], splitInput[1]);
results[1] = bst.SearchNodeKeyValInorder_str(bst.Get_root(), splitInput[2], splitInput[3]);
results[2] = bst.SearchNodeKeyValInorder_str(bst.Get_root(), splitInput[4], splitInput[5]);
results[3] = bst.SearchNodeKeyValInorder_str(bst.Get_root(), splitInput[6], splitInput[7]);
results[4] = bst.SearchNodeKeyValInorder_str(bst.Get_root(), splitInput[8], splitInput[9]);
stopwatch.Stop(); // Stops measuring elapsed time for an interval.
TimeSpan ts = stopwatch.Elapsed;// Gets the total elapsed time measured by the current instance.
string strIntervalStopWatch = ts.ToString();
double msIntervalStopWatch = ts.TotalMilliseconds;
```

Fig. 3. The section of source code of the *Main* method with the sequential searching for 5 nodes in a BST by the *SearchNodeKeyValInorder_str* instance method.

```
string[] results = new string[n];
stopwatch.Reset();
stopwatch.Start();
// Performs five tasks in parallel on the source array
Parallel.Invoke(
 () => {results[0] = bst.SearchNodeKeyValInorder_str(bst.Get_root(), splitInput[0], splitInput[1]);},
 () => {results[1] = bst.SearchNodeKeyValInorder_str(bst.Get_root(), splitInput[2], splitInput[3]);},
 () => {results[2] = bst.SearchNodeKeyValInorder_str(bst.Get_root(), splitInput[4], splitInput[5]);},
 () => {results[3] = bst.SearchNodeKeyValInorder_str(bst.Get_root(), splitInput[6], splitInput[7]);},
 () => {results[4] = bst.SearchNodeKeyValInorder_str(bst.Get_root(), splitInput[8], splitInput[9]);}
); // closes the 'Parallel.Invoke' method
stopwatch.Stop();
TimeSpan ts = stopwatch.Elapsed;
string strIntervalStopWatch = ts.ToString();
double msIntervalStopWatch = ts.TotalMilliseconds;
```

Fig. 4. The section of source code of the *Main* method with the parallel searching for 5 nodes in a BST by the *SearchNodeKeyValInorder_str* instance method.

- ***NodeSTarray*** - objects of this class are elements of a dynamic array of objects, containing key-value pairs, a first name and a surname - personal data in the *keyFirstSurName* and *personalInfo* instance variables. These elements represent the items of an implemented symbol table. The *refSTNodesArray* reference points to the array of such objects. The *SetKeyValuePairSTarr* member method initializes the *keyFirstSurName* and *personalInfo* instance variables of the *NodeSTarray* class object, and the

```
DlgSrchNodKVlnrdStr[] refDlgSearchNodKVlnrdSt;
refDlgSearchNodKVlnrdSt = new DlgSrchNodKVlnrdStr[n];
// Creates 'kx' objects of the 'DlgSrchNodKVlnrdStr' delegate type.
// These objects refer to 'SearchNodeKeyValInorder_str' instance method.
for (int kx = 0; kx < n; kx++)
refDlgSearchNodKVlnrdSt[kx] = new DlgSrchNodKVlnrdStr(bst.SearchNodeKeyValInorder_str);

stopwatch.Reset();
stopwatch.Start();
for (int z = 0, d = 0; z < n; z++, d += 2) {
   Monitor.Enter(obj); // Acquires an exclusive lock on the 'obj' object.
   try { // The 'BeginInvoke' method initiates the asynchronous call.
      refAsyncResults[z] = refDlgSearchNodKVlnrdSt[z].BeginInvoke(bst.Get_root(),
                                          splitInput[d], splitInput[d + 1], null, null);
   }
   finally { Monitor.Exit(obj); } // Releases an exclusive lock on the 'obj' object.
}
string[] returned_values = new string[n];
for (int c = 0; c < n; c++) // The 'EndInvoke' method retrieves the results of the asynchronous call.
   returned_values[c] = refDlgSearchNodKVlnrdSt[c].EndInvoke(refAsyncResults[c]);
stopwatch.Stop();
TimeSpan ts2MT1 = stopwatch.Elapsed;
string strIntervalStopWatchMT1 = ts2MT1.ToString();
double msIntervalStopWatchMT1 = ts2MT1.TotalMilliseconds;
```

Fig. 5. The section of source code of the *Main* method with the multi-threaded searching for 5 nodes in a BST by the *SearchNodeKeyValInorder_str* instance method.

Get_keyFirstSurName and *Get_personalInfo* member methods get the values of the *keyFirstSurName* and *personalInfo* instance variables of such object.

- ***STarray*** - an object of this class, containing the *refSTNodesArray* reference to an array of references to objects of the *NodeSTarray* class (elements of an array) and their number in the *n* instance variable, represents an unordered array, in which elements a symbol table is implemented. The *STarray* class provides member methods for inserting the specified key-value pair into the symbol table, searching for nodes according to various kinds of keys, finding the number of key-value pairs in the symbol table, for removing the specified key and its associated value from the symbol table and so on. For example, the *STarray* class provides the *SearchALLNodesKeyPhoneNumb* method for searching for all key-value pairs in a symbol table (elements in an unordered array) according to the phone number key and the *SearchPersonalInfoOfKey* member method for searching for a key-value pair in this symbol table according to the *first name* and *surname* key. The *SearchALLNodesKeyPhoneNumb* and *SearchPersonalInfoOfKey* methods have searched for data records of persons, key-value pairs: first names and surnames - personal data of persons according to 5 given keys a ***phone number*** and 5 given keys a ***first name*** and a ***surname*** in a symbol table (a phone book) sequentially, on multiple threads, and in parallel in an experiment.

- ***NodeLLUns*** - objects of this class that contain key-value pairs, a first name and a surname - personal data in the *keyFirstSurName* and *personalInfo* instance variables, and the *Next* reference to the next data element (a node), represent the data elements (nodes) of an unordered linked list with an implemented symbol table. The *Set_keyValueNodeLLUnsort* member method initializes the *keyFirstSurName* and *personalInfo* instance variables of the *NodeLLUns* class object, and the *GetKeyFirstSurNNodeLLUns* and *GetPersonalInfoNodeLLUns* member methods get the values of the *keyFirstSurName* and *personalInfo* instance variables of such object.
- ***LinkedListSTUnsort*** - an object of this class containing the head reference to the first, the tail reference to the last data element (a node) of the linked list and their number in the Count instance variable, represents an unordered link list, in which data elements (nodes) a symbol table is implemented. The data elements (nodes) of an unordered linked list are objects of the *NodeLLUns* class. The *LinkedListSTUnsort* class provides member methods for inserting the specified key-value pair into the symbol table, searching for nodes according to various kinds of keys, finding the number of key-value pairs in the symbol table, for removing the specified key and its associated value from the symbol table and so on. For example, the *LinkedListSTUnsort* class provides the *SearchALLNodesKey PhoneNumb* method for searching for all key-value pairs in a symbol table (nodes in an unordered link list) according to the phone number key and the *SearchNodeKeyValue_node* member method for searching for a key-value pair in this symbol table according to the first name and surname key. The *SearchALLNodesKeyPhoneNumb* and *SearchNodeKeyValue_node* methods have searched for data records of persons, key-value pairs: first names and surnames - personal data of persons according to 5 given keys a ***phone number*** and 5 given keys a ***first name*** and a ***surname*** in a symbol table (a phone book) sequentially, on multiple threads, and in parallel in an experiment.

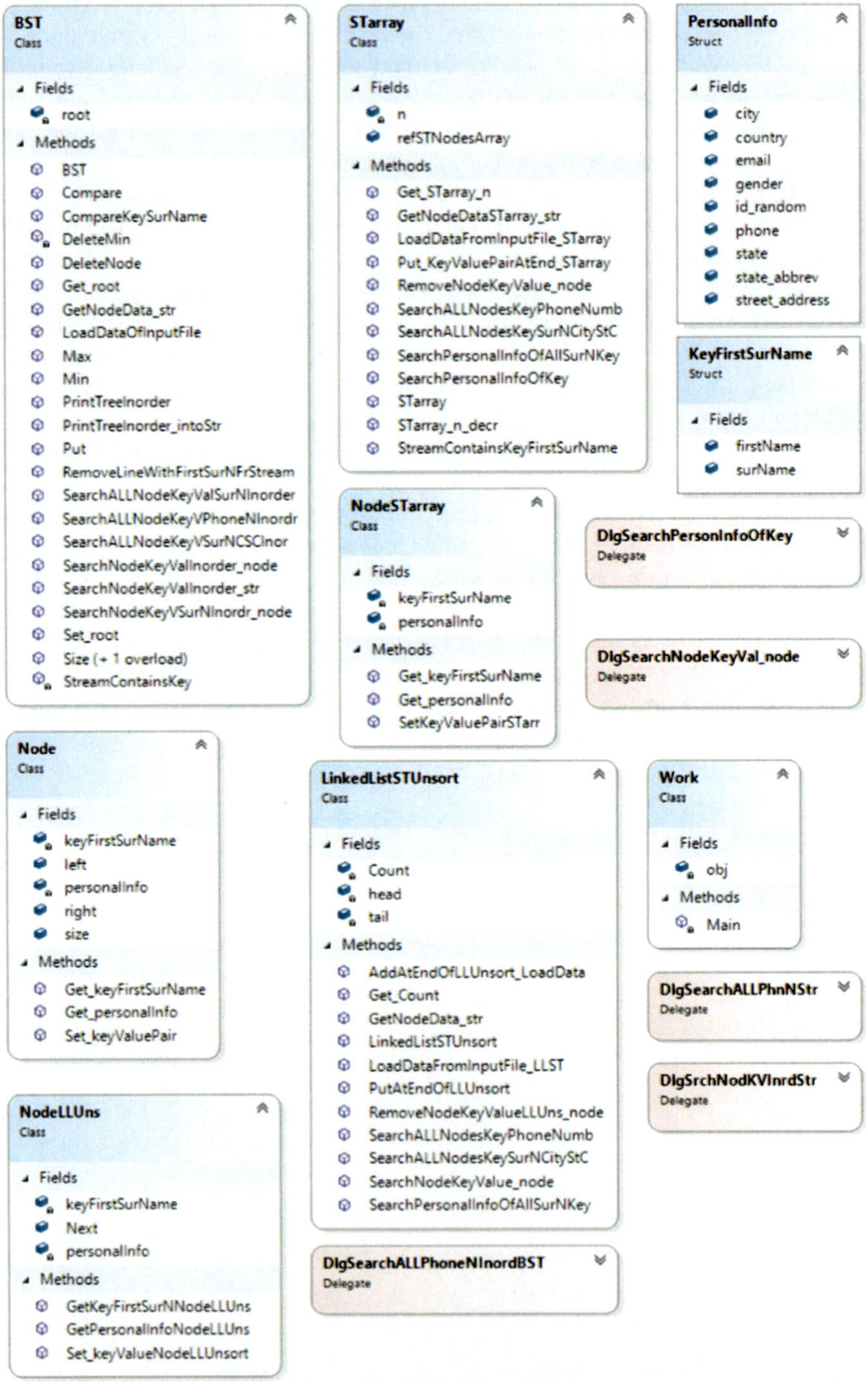

Fig. 6. The class diagram of the C# application.

As we mentioned above, our C# application can perform search operations in three implementations of a symbol table (a phone book), in a BST of objects, in an unordered array of objects, and in an unordered linked list of objects, and simultaneously it is able to measure the execution times of particular search operations. By a comparison of these execution times, we were finding which of these three data structures allows more efficient searching in a sequential search, multi-threaded, or in a parallel search in an experiment.

5 Experiment and Its Results

We assume that an implementation of a symbol table in an unordered linked list is more appropriate for sequential, multi-threaded, and for a parallel search for the same set of structured data in a symbol table (a phone book) by the search methods of a C# application than its implementation in a BST, and an unordered array.

To verify this hypothesis, we performed an experiment using our C# application that has searched for the sets of values associated with the sets of given keys inserted by a user in a BST, in an unordered array, and in an unordered linked list, sequentially, on multiple threads, and in parallel. This application has also measured the execution times of all sequential, all multi-threaded, and all parallel searches that were executed by its search methods. The C# application wrote all execution times together with results of all searches into the *LogFile.txt* disc log file. Using these execution times, we want to confirm or refuse our hypothesis.

We have carried out two kinds of searches using our C# application in BST of objects, in an unordered array of objects, and in an unordered linked list of objects for the purpose of this experiment. In each symbol table implementation (a phone book), we have searched for data records of persons according to:

- the following identical set of 5 keys: *first name* and *surname*: Stephen King, Stan Smith, Gill Andersson, Andrew Cowley, Paul Smith
- and the following identical set of 5 keys: *phone number*: +1 (915) 144-4242, +61 (999) 717-6787, +1 (770) 693-8938, +61 (135) 388-3603, +1 (602) 324-6011

The data record of each person, which creates a BST node, an element of an unordered array and a linked list node, contains: the *first name* and *surname* key and the assigned value, personal data of the person: ID, a gender, a phone, an e-mail, an address, a city, a state, a state abbreviation, and a country.

During the mentioned searches, the same sets of data records of 100, 200, 300, 400, 500, 600, 700, 800, 900 and 1000 persons stored in all 3 implementations of the symbol table (a phone book), i.e., in the BST of objects, in the unordered array of objects and in unordered linked list of objects were searched.

Searches for data records of persons according to the same set of 5 keys *first name* and *surname* were executed using the following instance methods of the C# application:

- a sequential, multi-threaded, and parallel search in a BST - *bst.SearchNodeKeyValInorder_str*
- a sequential, multi-threaded, and parallel search in an unordered array - *sTarray.SearchPersonalInfoOfKey*

- a sequential, multi-threaded, and parallel search in an unordered linked list - *linkedListSTUnsort.SearchNodeKeyValue_node*

Searches for data records of persons according to the same set of 5 keys *phone number* were executed using the following instance methods of the C# application:

- a sequential, multi-threaded, and parallel search in a BST - *bst.SearchALLNodeKeyVPhoneNInordr*
- a sequential, multi-threaded, and parallel search in an unordered array - *sTarray.SearchALLNodesKeyPhoneNumb*
- a sequential, multi-threaded, and parallel search in an unordered linked list - *linkedListSTUnsort.SearchALLNodesKeyPhoneNumb*

The execution time of each of these searches was measured by the *Main* method, and along with the results of the searches, the method wrote them into the console and into the *LogFile.txt* disk file.

The C# application was running on the computer with the following basic hardware configuration: Intel Core i5-8250U Processor (6MB Cache, 1.60 GHz (Processor Base Frequency), 3.40 GHz (Max Turbo Frequency)), 4 GT/s (Bus Speed), 4 Cores, 8 Threads, RAM: 8 GB. The Microsoft Windows 10 Home, 64-bit operating system and the Microsoft .NET Framework 4 were installed on this computer.

A part of one of the outputs of our C# application stored in the *LogFile.txt* disc log file after a sequential, multi-threaded, and parallel search for the set of data records of persons according to the same set of 5 keys *first name* and *surname*: *Stephen King, Stan Smith, Gill Andersson, Andrew Cowley, Paul Smith*, in all 3 implementations of the symbol table (a phone book), i.e., in a BST with 1000 nodes, in an unordered array of objects with 1000 elements, and in an unordered linked list with 1000 nodes, by the search methods of this application is shown in Fig. 7. The C# application wrote the similar output into the system console. The C# application wrote the results of other searches into the same *LogFile.txt* file, too.

The execution times of all sequential, all multi-threaded, and all parallel searches data records of persons according to the set of 5 given keys a *first name* and a *surname* and the set of 5 given keys a *phone number* that were executed by the C# application search methods during the experiment are shown in Fig. 8 and Fig. 9.

Brief Results Analysis. Several facts result from a comparison of the execution times of sequential, multi-threaded, and parallel searches for the sets of data records of persons according to the same sets of 5 given keys a *first name* and a *surname* and 5 given keys a *phone number* inserted by a user in a BST of objects, in an unordered array of objects, and in an unordered linked list of objects by the C# application search methods.

Findings for searching for data records of persons according to the same set of 5 given keys a ***first name*** and a ***surname*** in three implementations of a symbol table (a phone book):

- The shortest execution times in the *sequential* search are achieved by the search in the BST of objects and the longest execution times are achieved by the search in an unordered array of objects. However, the differences between the shortest and longest execution times are very small, less than 0.7 ms.

```
********** PARALLEL searching according to first names and surnames **********
   Persons with the 'Stephen King, Stan Smith, Gill Andersson, Andrew Cowley, Paul Smith'
first names and surnames found / NOT found in PARALLEL in a symbol table:
   The key-value pairs found (4) / NOT found (1) in a symbol table implemented in a BST (1000):
         The key-value pair with the 'Stephen King' key found:
      Stephen King, (Male, 984053), +1 (915) 144-4242, sking0@miitbeian.gov.cn,
      52 Evergreen Court, El Paso, Texas, TX, United States

         The key-value pair with the 'Stan Smith' key found:
      Stan Smith, (Male, 117426), +1 (770) 693-8161, ssmith90@google.jp,
      1459 Mallory Junction, Atlanta, Georgia, Georgia, United States

         The key-value pair with the 'Gill Andersson' key found:
      Gill Andersson, (Male, 577303), +1 (516) 626-8006, ganderssone@wisc.edu,
      305 3rd Point, Port Washington, New York, NY, United States

         A person with the first name and surname 'Andrew Cowley' NOT found!

         The key-value pair with the 'Paul Smith' key found:
      Paul Smith, (Male, 117653), +1 (770) 693-8909, psmith9@google.jp,
      80 Mallory Junction, Atlanta, Georgia, Georgia, United States
               [Execution time:  00:00:00.0053621, 5,3621 ms]
   The key-value pairs found (4) / NOT found (1) in a symbol table implemented in an unordered
ARRAY of objects (1000):
         The key-value pair with the 'Stephen King' key found:
      Stephen King, (Male, 984053), +1 (915) 144-4242, sking0@miitbeian.gov.cn,
      52 Evergreen Court, El Paso, Texas, TX, United States

         The key-value pair with the 'Stan Smith' key found:
      Stan Smith, (Male, 117426), +1 (770) 693-8161, ssmith90@google.jp,
      1459 Mallory Junction, Atlanta, Georgia, Georgia, United States

         The key-value pair with the 'Gill Andersson' key found:
      Gill Andersson, (Male, 577303), +1 (516) 626-8006, ganderssone@wisc.edu,
      305 3rd Point, Port Washington, New York, NY, United States

         A person with the first name and surname 'Andrew Cowley' NOT found!

         The key-value pair with the 'Paul Smith' key found:
      Paul Smith, (Male, 117653), +1 (770) 693-8909, psmith9@google.jp,
      80 Mallory Junction, Atlanta, Georgia, Georgia, United States
               [Execution time:  00:00:00.0003535, 0,3535 ms]
```

Fig. 7. The part of the output of the C# application stored in the *LogFile.txt* disc log file that shows results of the parallel search in a BST and an array implementation of a symbol table with 1000 items.

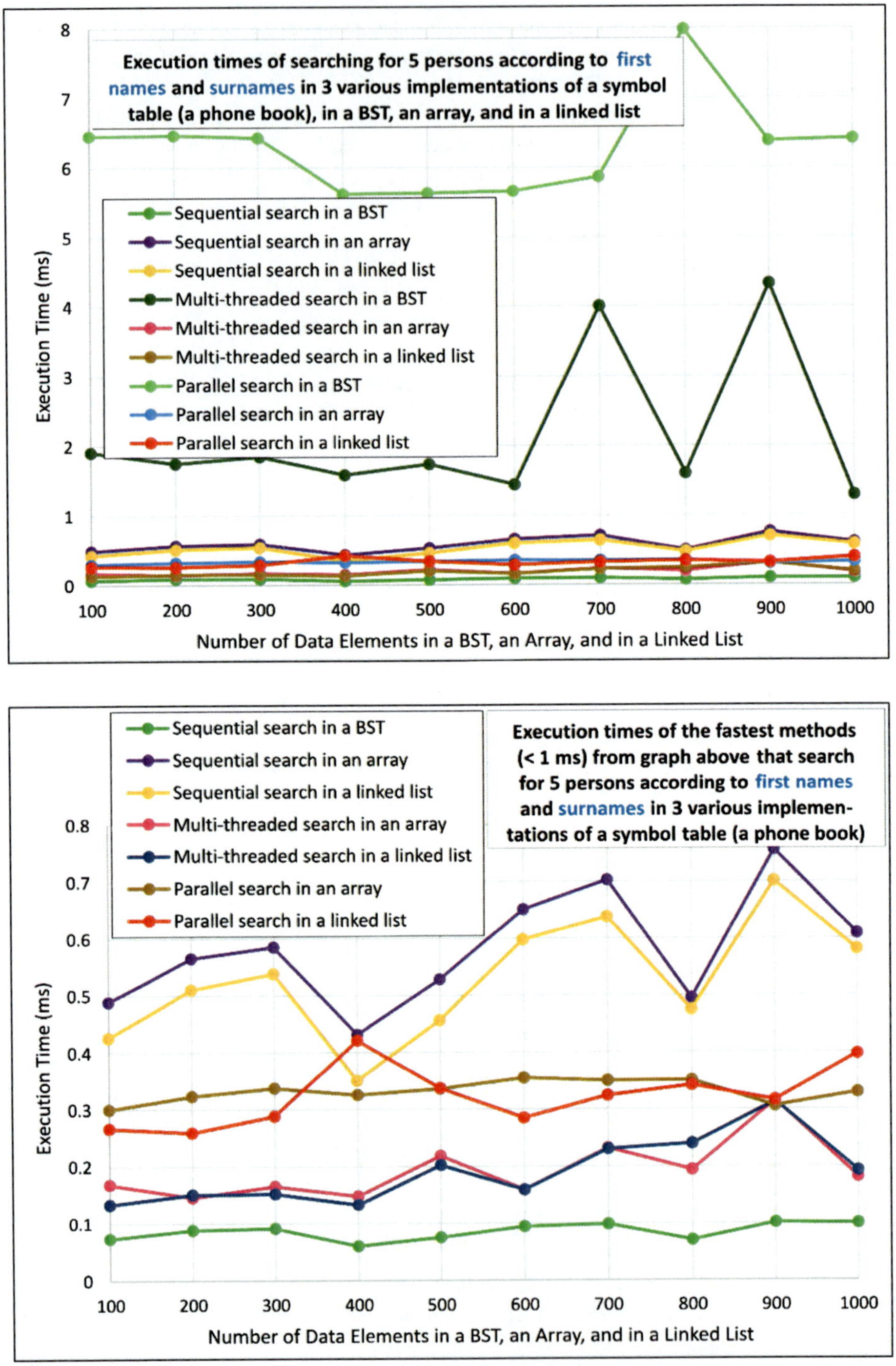

Fig. 8. Execution times of searching for 5 persons according to first names and surnames in 3 various implementations of a symbol table (a phone book), in a BST, an array, and in a link. list.

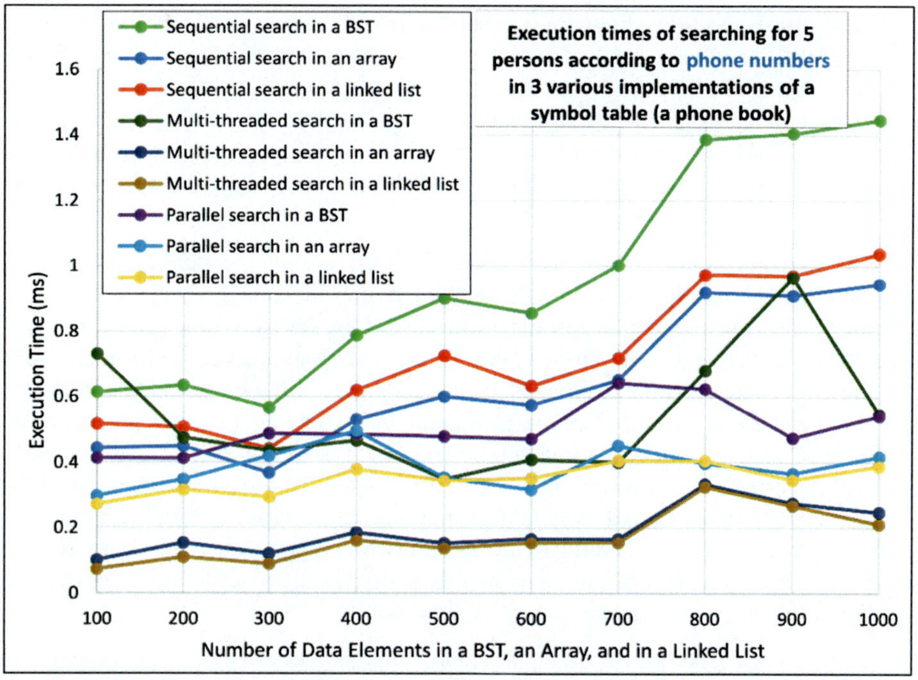

Fig. 9. Execution times of searching for 5 persons according to phone numbers in 3 various implementations of a symbol table (a phone book), in a BST, an array, and in a linked list.

- The shortest execution times in *multi-threaded* search are achieved by the search in an unordered linked list of objects and the longest execution times are achieved by the search in a BST of objects.
- The shortest execution times in *parallel* search are achieved by the search in an unordered linked list of objects and the longest execution times are achieved by the search in a BST of objects.
- The shortest execution time of all searches according to the same set of 5 given keys a *first name* and a *surname* is achieved by *sequential* search in BST of objects, and the longest execution time of all searches is achieved by *parallel* search in a BST of objects.

Findings for searching for data records of persons according to the same set of 5 given keys a ***phone number*** in three implementations of a symbol table (a phone book):

- The shortest execution times in the *sequential* search are achieved by the search in an unordered array of objects and the longest execution times are achieved by the search in a BST of objects.
- The shortest execution times in *multi-threaded* search are achieved by the search in an unordered linked list of objects and the longest execution times are achieved by the search in a BST of objects.

- The shortest execution times in *parallel* search are achieved by the search in an unordered linked list of objects and the longest execution times are achieved by the search in a BST of objects.
- The shortest execution time of all searches according to the same set of 5 given keys a *phone number* is achieved by a *multi-threaded* search in an unordered linked list of objects, and the longest execution time of all searches is achieved by a *sequential* search in a BST of objects.

The execution times of the search itself in all three implementations of a symbol table (a phone book) are affected by the composition of the key.

If it consists of two components, that is our case of the key a *first name* and a *surname*, then it is advantageous to search in a BST during a sequential search. This search was also the most execution efficient of all searches according to the same set of 5 given keys a *first name* and a *surname* in three implementations of a symbol table (a phone book).

For a search according to a one-component key, that is our case of a key a *phone number*, a multi-threaded search in an unordered linked list of objects seems to be the most execution efficient and a sequential search in a BST seems to be the least efficient.

For a parallel search according to a one- and two-component key we do not recommend using a search in a BST, in this case it is more advantageous to use a search in an unordered linked list of objects.

A *parallel search* according to a one- and two-component key in a BST implementation of a symbol table achieves the longest execution times. This is caused by the construction of a BST (Fig. 2), which is more complex than the construction of an unordered array and unordered linked list, and the rule that is used in the construction of a BST: The key in a node is larger than the key of every node in its left subtree and smaller than the key of every node in its right subtree [6]. When creating an unordered array and unordered linked list, no rule is used, new data elements (or nodes) are inserted into an array and a list sequentially in the order as they are read from an input data stream. Therefore, a search operation in a BST is more complex than search operations in an unordered array and unordered linked list, which may be the reason for longer execution times of a parallel search according to one- and two-component keys in BST compared to a parallel search in an unordered array and unordered linked list. If we compare these search operations in these three implementations of a symbol table, it is obvious, which of them is the most complex.

Search in a BST: We suppose that we want to search for a node with a given key in a BST. A recursive algorithm follows [3]:

- If the tree is empty, we terminate the search as unsuccessful.
- If the search key is equal to the key in the node, we terminate the search as successful (by returning the value associated with the key).
- If the search key is smaller than the key in the node, we search (recursively) in the left subtree.
- If the search key is greater than the key in the node, we search (recursively) in the right subtree.

Search in an unordered array/an unordered linked-list implementation of a symbol table: we scan through an array/a list, using some comparative method to compare the search key with the key in each element/node in an array/a list. If we find the match, we return the associated value; if not, we return null [7].

From the above results that by a task parallelization of the source code we do not always have to achieve the shortest execution times for each data structure.

Sequential searches in different implementations of a symbol table have the following properties [1]: Sequential search in an array and linked-list implementations of symbol table of *N* unordered items uses *N/2* comparisons for search hits (on the average) and *N* comparisons for search misses (always).

Sequential search in a BST implementation of symbol table of *N* ordered items uses *N* comparisons (the worst case) or *lg N* comparisons (on the average) for search hits and for search misses.

In a sequential search in a BST according to a two-component key, our search method probably performed *lg N* comparisons. It was faster than other search methods that searched in an array and linked-list implementation of the symbol table.

In a sequential search in a BST according to a one-component key, our search method probably performed *N* comparisons (the worst case). This method achieved the longest execution times of all searches according to an one-component key.

The number of comparisons in a BST depends on the location of the search node in it. If the search node is located close to the beginning (to the root) of the BST, then *lg N* comparisons are performed (on the average). This case occurred during a sequential search according to a two-component key (a *first name* and a *surname*) in a BST by our search method. If the search node is close to the end of a BST, then *N* comparisons are performed (the worst case). This case occurred during a sequential search according to a one-component key (a *phone number*) in a BST by our search method.

6 Conclusion

From the results of the comparison of the execution times of sequential, multi-threaded, and parallel searches for the sets of data records of persons according to the same sets of 5 given two-component keys and 5 given one-component keys in three implementations of a symbol table (a phone book), i.e., in a BST of objects, in an unordered array of objects, and in an unordered linked list of objects by the C# application search methods, it is obvious that for a two-component key, it is advantageous to search in a BST in a sequential search.

This search was also the most execution efficient of all searches according to the same set of 5 given two-component keys in three implementations of a symbol table (a phone book). For multi-threaded and parallel searches according to a two-component key, it is advantageous to use the search in an unordered linked list of objects.

From the results of the comparison of these execution times it is also obvious that for a one-component key it is advantageous to search in an unordered array of objects in a sequential search, while in a multi-threaded and parallel search it is advantageous to search in an unordered linked list of objects. The shortest execution time of all searches according to the same set of 5 given one-component keys is achieved by the multi-threaded search in an unordered linked list of objects.

From the results of our experiment, it seems to be the most optimal choice for sequential, multi-threaded and parallel searching for the sets of data records of persons according to the sets of given one- and two-component keys using an unordered linked-list implementation of a symbol table (a phone book).

References

1. Sedgewick, R.: Algorithms in C, Third Edition, Parts 1–4, Fundamentals, Data Structures, Sorting, Searching. Addison-Wesley Publishing Company, Inc., Boston, San Francisco, New York (1998)
2. Sedgewick, R., Wayne. K.: Algorithms, Fourth Edition, p. 02116. Pearson Education, Inc., Boston, MA (2011)
3. Sedgewick, R., Wayne. K.: Symbol Tables. https://introcs.cs.princeton.edu/java/44st/. Accessed 22 Aug 2023
4. Drozdek, A.: Data Structures and Algorithms in C++, 4th edn. Cengage Learning, Boston (2013)
5. Košt'ál, I.: Execution efficiency of a sequential search, multi-threaded, and a parallel search for a data in a symbol table implemented in an array, and in a linked list. In: TIEES 2022: Trends and Innovations in E-Business, Education and Security: Tenth International Scientific Web-conference of Scientists and PhD. Students or Candidates, pp. 65–82. The Slovak Society for Economic Informatics, Bratislava, Slovakia (2022)
6. Microsoft Corp.: Microsoft documentation. https://learn.microsoft.com. Accessed 22 Aug 2023
7. Sedgewick, R., Wayne. K.: Elementary Symbol Tables. https://algs4.cs.princeton.edu/31elementary/. Accessed 18 Aug 2023

Author Index

Y. Ono and J. Kondoh (Eds.): Inter-Academia 2023, LNNS 939, pp. 333–334, 2024.
https://doi.org/10.1007/978-3-031-54450-7